# FIRE AND EMERGENCY LAW CASEBOOK

*Thomas D. Schneid*

## Online Services

**Delmar Online**
For the latest information on Delmar Publishers new series of Fire, Rescue and Emergency Response products, point your browser to:

**http://www.firesci.com**

## Online Services

**Delmar Online**
To access a wide variety of Delmar products and services on the World Wide Web, point your browser to:

**http://www.delmar.com**
or email: info@delmar.com

**thomson.com**
To access International Thomson Publishing's home site for information on more than 34 publishers and 20,000 products, point your browser to:

**http://www.thomson.com**
or email: findit@kiosk.thomson.com

A service of I(T)P®

# FIRE AND EMERGENCY LAW CASEBOOK

*Thomas D. Schneid*

## Delmar Publishers

I(T)P an International Thomson Publishing Company

Albany • Bonn • Boston • Cincinnati • Detroit • London • Madrid
Melbourne • Mexico City • New York • Pacific Grove • Paris • San Francisco
Singapore • Tokyo • Toronto • Washington

## NOTICE TO THE READER

Cover photo courtesy of Jody R. Warner

**Delmar Staff**

Publisher: Robert D. Lynch
Acquisitions Editor: Mark Huth
Developmental Editor: Jeanne Mesick

Project Editor: Thomas Smith
Production Coordinator: Toni Bolognino
Art and Design Coordinator: Michael Prinzo

COPYRIGHT © 1997
By Delmar Publishers
an International Thomson Publishing Company

The ITP logo is a trademark under license

Printed in the United States of America

For more information, contact:

Delmar Publishers
3 Columbia Circle, Box 15015
Albany, New York 12212-5015

International Thomson Publishing Europe
Berkshire House 168-173
High Holborn
London, WC1V7AA
England

Thomas Nelson Australia
102 Dodds Street
South Melbourne, 3205
Victoria, Australia

Nelson Canada
1120 Birchmount Road
Scarborough, Ontario
Canada M1K 5G4

International Thomson Editores
Campos Eliseos 385, Piso 7
Col Polanco
11560 Mexico D F Mexico

International Thomson Publishing Gmbh
Königswinterer Strasse 418
53227 Bonn
Germany

International Thomson Publishing Asia
221 Henderson Road #05-10
Henderson Building
Singapore 0315

International Thomson Publishing - Japan
Hirakawacho Kyowa Building, 3F
2-2-1 Hirakawacho
Chiyoda-ku, 102 Tokyo
Japan

2 3 4 5 6 7 8 9 10 XXX 02 01 00 99 98 97

Library of Congress Cataloging-in-Publication Data

Schneid, Thomas D.
    Fire and emergency law casebook / Thomas D. Schneid.
       p.   cm.
    Includes index.
    ISBN 0-8273-7342-2
    1. Fire departments—Law and legislation—United States—Cases.  2. Fire
Fighters—Legal status, laws, etc.—United States—Cases.  3. Assistance in
emergencies—United States—Cases.  I. Title.
KF3976.A7S36    1996
344.73'0537—dc20
(347.304537)                                                          96-30155
                                                                        CIP

# Contents

# Table of Cases

# Acknowledgments

The author and Delmar Publishers would like to thank the following reviewers for the comments and suggestions they offered during the development of this project. Our gratitude is extended to:

Tom Harmer
Titusville Fire and Emergency Services
Titusville, FL

John Pangborn
Jersey City Fire Department
Jersey City, NJ

Mary Marchione
Montgomery County Government
Rockville, MD

James Madden
Lake Superior State University
Sault Ste. Marie, MI

Jay Franey
Aims Community College
Greeley, CO

Russell Chandler
J.S. Reynolds Community College
Richmond, VA

Richard Arwood
Memphis Fire Department
Memphis, TN

John Fred Shreve
Mason Volunteer Fire Company
Loveland, OH

David Fultz
Louisiana State University
Eunice, LA

## Special Thanks

To my wife, Janie, and my girls, Shelby and Madison, for keeping me focused and allowing me the time to complete this text.

To Christy and Sheila, for all of their efforts in typing and assembling this text. Without their help, I would never have completed this text on time.

## Cover Photo

1991 Photo—"The Flange"

A mixture of gasoline and diesel fuel is gravity fed through a piping system to the "flange" seen at the center of the fire ball. The flowing mixture is then lit, creating the fire seen here. Two five-person crews then attack the fire with four (4) inch-and-three-quarter hand lines. This training exercise was sponsored by the Fire and Rescue Training Division of Kentucky Tech, the states vocational education program.

# Preface

The *Fire and Emergency Law Casebook* was developed to enable the reader to acquire a basic knowledge of the law in each of the specified subject areas that directly or indirectly affect fire services by providing a basic knowledge of the methodology through which to locate, read, and comprehend the various statutes, regulations and cases that are the framework of the law. As such, it was not intended by the author to be all-inclusive, and by no means covers all the potential areas of the law which may impact a fire and emergency service organization.

The topics and cases selected by the author are consistent with the ultimate educational purpose of this text. Although this text encompasses various legal topics and cases, the law changes virtually on a daily basis. The author advises the reader to research any specific legal questions and to acquire the advice of legal counsel to ensure the status of the particular law.

It is the author's hope that the *Fire and Emergency Law Casebook* will be used in conjunction with classroom lecture and discussions. This text was designed to provide the reader with the tools to locate, read, and brief the cases provided for each chapter. Objectives are given at the beginning of each chapter to guide the reader to the key areas of the chapter. After reviewing the objectives, the reader should review the general information in the specific chapter to ensure a firm grasp on the basic information. The reader should then proceed to review and brief the cases. Specific questions are provided at the end of each chapter to assist the reader and facilitate complete understanding.

The author sincerely hopes that this text is not only informative and educational for readers in the fire and emergency service and others, but is also an enjoyable experience for each reader. The law is a fascinating area of study and the knowledge base acquired by the reader in these selected areas of the law can be utilized to prepare you and your fire or emergency service organization to take a proactive approach to preventing or minimizing potential legal liabilities.

# Foreword

Fire. Emergency. Law. Those three words in today's context may cause some alarm for most of us who provide essential emergency services. Not too long ago, the notion that the good Samaritan was a protection from lawsuits was common. Emergency personnel were often thought of as above reproach, as "untouchable" in the legal arena. To bring any sort of "action" against the fire and emergency services was unconscionable. But current trends and norms of society no longer allow the luxury of those "protections" that were common in the past.

When our public calls for service, it expects and deserves professional responses and that we will endeavor for the best possible outcome. The public deserves no less than the best we can offer, and we must be professional in our dealings with the customers, the public.

This is the age of responsibility and accountability, not only to ourselves, but, more importantly, to our "customers." We must always bear in mind that *we* are responsible for our actions and that we *are* accountable for what we do every day. Litigation is everywhere. I recall my wizened criminal law professor's admonishment, "It really doesn't matter whether you are right or wrong, I will defend you till you run out of money, and then . . . ." Those words are certainly pertinent today, as you may very well have experienced in your own situation. It is very important to heed that advice, and remember that we are *always* in the public spotlight each and every day.

Law libraries are full of case law of all types and outcomes. In this volume, Tom Schneid provides a compendium on a myriad of fire and emergency case law that are the subject of daily lawsuits. He has packed this text with numerous cases that cover some often-missed areas of concern in public service. They include cases involving the ADA, OSHA, and the NLRB, as well as cases involving hiring, wrongful termination, negligence, criminal codes, and civil liability. He has chosen representative cases that serve as an overview and that could have some very real parallels in your own situation. Remember, a case that seems sensational may well have stemmed from a simple event that has gone awry, when the "other" party has invoked legal action.

You are to be commended for taking that first bold step toward your own protection by reading this *Fire and Emergency Law Casebook* and utilizing the lessons learned from its pages. The example cases included in this text emphasize the need for perseverance, common sense, and dedication toward doing the job right, from the outset. We must also subscribe to the age-old proviso of "do no harm," and must be as professional and properly trained to do the job that our public deserves, and demands.

The rules and regulations that govern our daily lives are almost too numerous to mention, but they are there. As a member of the fire and emergency services community, it is up to you to learn to document what you are doing to protect your co-workers or employees, and the customers you serve.

Your key to future survival is education. Numerous example cases in this text can impart valuable lessons. The key to using this text to its fullest is to read the chapter, review the cases, and try to place yourself in that scenario. Then try to extricate yourself from that same situation. You can help yourself by studying the cases and the eventual outcomes.

Any job worth doing is worth doing right. As provided by the sample cases in this book, take the time to study what went wrong, so that we may all benefit from others. Practice, plan, and prepare for your future in this most rewarding service of fire and emergency response, but remember we must always be there for the "customer." After all, isn't that why we work in "public service"?

To serve and protect!

*Michael J. Fagel*
Aurora, Illinois

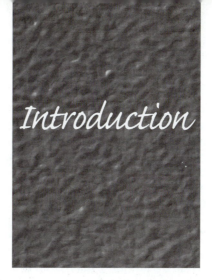

*Introduction*

# Analyzing Case Law and Briefing a Case

**case law**
the accumulation of court decisions that, in essence, shapes and develops new law

**Case law** is the accumulation of court decisions that, in essence, shapes and develops new law. Throughout this text, you will find numerous cases that identify and typify a particular point of law and that exemplify the court's decision-making process by showing how decisions were derived in the particular case. As a basic rule of thumb, when analyzing or briefing a case, you should first read through the case in detail. Then, the second reading, you should identify the basic issues, facts, and holding of the decision as well as any dissenting opinion. On the third reading, you should take the notes that will form the actual brief. It is essential that you take good notes and brief your case extremely well for ease of reference later in your studies.

## FINDING THE CASE

**reporter**
the published volumes of decisions by a court or group of courts

Finding the case in the library is often one of the most difficult parts of your total analysis. As can be seen from the various cases provided in this text, other cases are referenced and numerous "cites" are provided throughout this text. All reported decisions of cases in the U.S. judicial system are listed according to the publication in which the case appears (called a **reporter**), the volume in which the case is located, and the initial page number, for example, 36 S.E. 2d 924. If you were searching for this case, you would go to the South Eastern Reporter

(S.E.), second edition (2d), and look for volume number 36 and find the case at page 924. In the federal judicial system, the cases tend to be published by the region of the country and in most state judicial systems the case will be found in a state reporter. Not all cases are reported and published. Generally, trial court decisions are not published because these decisions do not serve as mandatory precedent for other courts to follow. Usually only decisions of federal and state appellate courts are published.

Statutes, regulations, and standards tend to be within other publications such as the Code of Federal Regulations. This system is set up utilizing the same publication, volume, and page number as with the judicial court decisions. However, the series numbers may reflect the particular regulation or standard. For example, 29 C.F.R. 1910.120 is the Occupational Safety and Health Administration's Standard with regard to hazardous waste operation and emergency response. If you were searching for this particular standard, you would go to the Code of Federal Regulations, which is signified by the C.F.R. designation, volume 29 and section 1910.120.

Some law offices and law libraries also provide an electronic database through which to locate cases. The two major databases are WESTLAW and LEXIS. Each of these databases normally provide training and publications to assist you in locating the particular case. Generally these databases provide a basic menu of the various areas where the law is located and numerous sub-databases to guide your search. For example, if you were searching for a federal decision, you would enter the federal database and then narrow your search to the particular area of law that you are seeking. If you know the case name, it can usually be used to pull up the case. If you are searching for a particular issue of the law, these databases can provide the case cites for your review and evaluation. It is highly recommended that you acquire the particular training or assistance from the librarian at the law library prior to conducting any search on WESTLAW or LEXIS.

It is also highly recommended that you become familiar with the particular library that you will be using during this course. Take a few minutes out of your busy schedule and walk through the library to locate the different publications available and note the location of each of these publications. Thumb through a few of the publications and practice locating particular cases so that you can find cases in a timely manner in the future.

## BRIEFING THE CASE

The basic purpose for briefing a case is to help you understand the particular legal issues of the case and to refresh your memory on the significant portions of the case later. There are various methods of briefing a case and the following format is an example of one of the methods. Your instructor may suggest a different format or you may devise your own system to analyze these cases. No matter what method you adopt, be sure to read the case thoroughly at least once to get a general idea of what the case is about prior to beginning to take notes for your brief.

The basic framework that we recommend for you to brief a case includes the following:

- Issues
- Facts
- Holding (Decision)
- Dissent
- Your Opinion
- Underlying Policy Reason

## Issues

Identify the basic issue or issues before the court. In order to find the basic issue or issues involved, you must identify the rule of law that governs the dispute and decide how it applies to the particular facts of the case. In most circumstances, you will be writing the issue for your case brief as a question that combines the rule of law with the material facts of the case. For example, does the arson statute in the state of Kentucky apply to a minor child?

## Facts

**litigation**
a lawsuit or legal action

The facts of the particular case describe the events between the conflicting parties that led to this particular **litigation** and tell how the case came before the court that is now deciding it. Often included in the facts are the relevance to the issue the court must decide and the basic reasons for its decision. When you first read through the case, you will not know which facts are relevant until you know what the issue or issues are in the particular case. Thus, it is vitally important that you read through the case at least once prior to beginning to summarize the facts of the particular case.

In addition to the specific facts of the situation, it is important to see what court decisions have come prior to the case you are currently reviewing. Often, the decisions that are published are appellate decisions and thus a district court or circuit court has decided the matter previously and now the matter is on appeal. If the particular facts of the situation in an appellate case are not provided in detail, you may want to research and review the district or circuit court decisions in order to acquire the particular facts in your case.

**motions**
documents submitted to a court or judge for the purpose of acquiring a ruling or order directing some act to be done in favor of the person filing the motion

In this section, you should also include the relevant background for this case. You should identify who the plaintiff and the defendant are, the basis of the plaintiff's suit, and the relief the plaintiff is seeking. You may also want to include the procedural history of the case such as Motions to Dismiss and other **motions** that are relevant to the case. In an appeals case, the decisions of the lower courts, the grounds for those decisions, and the parties who appealed should also be noted.

Within this facts section, you should be as brief as possible. However, all pertinent points should be noted. Although this is a judgment call, most statements of facts in a brief should not be more than two or three paragraphs. Given the fact that you would have read the case at least three times while briefing the case, the facts provided in your brief of the case should be the major points used to refresh your memory.

## Holding

The holding is the court's decision on the question that was actually presented before it by the parties. The court may make a number of legal statements but if they do not relate to the question actually before it, this information is considered **dicta**. The holding can normally be identified by the statement the court has decided or what the majority decision is. In essence, a holding, provides the answer to the question you were asking in your issue statement. If there is more than one issue involved in the case, there may be more than one holding in any given case.

**dicta**
opinion of a judge in the minority in the decision; not as binding as a legal precedent

## Dissent

Often with U.S. Supreme Court cases and appellate cases, the majority decision is the decision of the court. However, the minority position is also often provided in order that the opinions of the minority can also be voiced in the written opinion. Although the dissenting opinion is not law and has no bearing on the case, the dissent provides another point of view on the particular issue of the case and also may be referred to in some later case.

## Opinion

After you have reviewed the case at least three times and have analyzed the court decision and briefed your case, you should have a good idea whether you agree or disagree with the court's opinion. In this section, you should provide your personal opinion as to whether you agree or disagree with the court and the reasons why you agree or disagree with the decision.

## Policy

In many cases, there is an underlying social policy or particular social goal that the court wishes to further. When a court explicitly refers to those policies in a particular case, this information should be included in your brief since it will provide you with a better understanding of the court's decision. For example, in the historic case of *Brown v. Board of Education*, the decision of the court was formed through an underlying social policy to eliminate discrimination in our school system. This underlying social policy is often very important in appellate and Supreme Court decisions. Attached is an example brief for your review and evaluation. It is highly recommended that you "test" your skills by briefing several of the cases within this text or other cases prior to your initial briefing of a case for a class. In addition to the foregoing information, the following helpful hints may assist you in briefing the case:

- Try to confine your brief to a maximum of one page. If your brief is over two pages, you have probably provided too much information. Remember, a brief is to refresh your memory at the time you need to recall this information for class or other purposes.

- The case that is printed in the textbook may have been edited and shortened by the textbook author. Normally, a full court opinion may be as short as a couple of pages or can be several hundred. If you find that you are having difficulty understanding the case because information has been deleted from the case in this book, you may want to go to the library and read the full text opinion.

- During your first few attempts at briefing a case, it is often difficult to extract the important elements and issues of a case. Keep in mind that not all judges are expert writers so the opinions may often be confusing or difficult to understand. Additionally, you should realize that not all courts follow the same format in writing opinions, thus some decisions may be more difficult to understand than others. Judges sometimes go off on a tangent and discuss other rules and points of law that are not essential to the determination of the particular case. It is your job to be able to filter through this information to identify the particular issues and laws that are applicable to the case.

- You may often run across unfamiliar latin or "legal" terms. Since you need to have a clear understanding of the terminology used in the particular case, you will have to look up the term in a legal dictionary. It is often a good idea to have a *Black's Law Dictionary*, *Ballantine Law Dictionary*, *Gilmer's Law Dictionary*, or other law dictionary available while you are reading and briefing the case. Standard dictionaries often do not provide these terms or the explanation provided may be incomplete. Use of a law dictionary is essential when reading these cases.

- When reading the cases provided in this text, you may want to look at the particular chapter and section headings of the textbook in which the case appears. If you are having difficulty identifying the particular issue of the case, the issue is normally related to the topic discussed in the chapter or section heading. The cases in this text have been inserted to illustrate the subject matter being discussed in each of the chapters.

- Remember, the issue or issues in the particular case should always be stated in the form of a question. You should never begin your issue with the words "whether or not" because this will not form an interrogatory question. Also, the terminology, "should plaintiff win" or "is there a contract" are not correct forms of stating the particular issue.

- In determining the particular rule of law, ask yourself, "If I had to tell someone who knew nothing about this case what this case is about or what it stands for in one sentence, what would I say?" Often, the rule of law can be determined by taking the issue and putting it into the form of a declaratory statement and adding a few words. For example, in the case of *Miranda v. Arizona*, 384 U.S. 436 (1966), the issue and rule may be as follows:

  *Issue:* When a person is taken into police custody, or otherwise deprived of his freedom of action in a significant way, must his constitutional rights to remain silent and to have an attorney present be explained to him prior to questioning?

  *Rule:* When a person is taken into police custody, the following warnings must be given prior to questioning:

  **1.** That he has the right to remain silent.

  **2.** That any statement he makes may be used against him as evidence.

  **3.** That he has the right to have an attorney present.

  **4.** If he cannot afford an attorney, one will be appointed for him.

- Last, do not use other people's briefs. Without having read a particular case and analyzed the court decision yourself, use of another individual's brief of a case is essentially worthless. A brief is simply the codification of your thoughts and work that you will refer to in the future in order to refresh your memory.

- One last item of importance. There is no substitute for the careful reading and analysis of the applicable cases. Appropriate time and effort should be provided to this important analysis in order to achieve the desired results.

## Example Case Brief

Case Name:    *Estate of Humpty Dumpty v. Wall Company*, xx F.2d xx (1995)

Issue:    Is the defendant liable for its product when the plaintiff assumes the risk of sitting on the wall?

Facts:    On January 1, 1994, Humpty Dumpty sat on the wall designed and built by the Wall Company. While sitting on the top of this wall of ten feet, Humpty Dumpty fell causing numerous injuries ultimately leading to his death on January 2, 1994. The plaintiff asserts that the wall was defectively designed, lacked appropriate warnings of the hazards, and asserts the egg shell rule. Experts for the plaintiff testified that the wall was not properly marked and the design was improper for the setting. The defendant argues that Humpty Dumpty assumed the risk of sitting on the wall given his job as a professional wall sitter and also asserts that the medical attention provided by the King's men was the superceding cause of the death of Humpty Dumpty. Medical experts testified that Humpty Dumpty possessed a rare genetic disease in his shell which made him susceptible to injuries as a result of blunt trauma.

Holding:    For the Estate of Humpty.

Opinion:    I agree with the court in this case. Although Humpty Dumtpy was a professional wall sitter, the Wall Company was responsible for the defective design of the wall. However, the Wall Company should not be responsible for the entire amount of damages because of Humpty's predisposed physical condition and the negligent care provided by the King's men.

Policy:    Although no specific public policy was mentioned in the case, the implied policy was that of sending a message to product manufacturers to ensure that their products are safe before placing them on the market.

# Chapter 1

# Basic Principles of Law

No man is above the law and no man is below it; nor do we ask any man's permission when we ask him to obey it.

*Theodore Roosevelt*

## Learning Objectives

- To acquire a general understanding regarding the formation of laws.
- To acquire a basic understanding of our judicial system.
- To acquire an understanding of the basic differences in laws.

In the United States, we possess a myriad of laws on various subjects on the federal, state, and local levels. All of our laws are based directly or indirectly on the U.S. Constitution. The old axiom is that ignorance of the law is no defense; thus knowledge of the law is still applicable today.

Firefighters and emergency medical services (EMS) personnel may encounter such federal laws as the Environmental Protection Agency (EPA) regulations regarding environmental contamination. State laws usually govern the rights, responsibilities, and prohibitions closely related to fire and emergency services. State laws also normally govern all criminal activities in the state, such as arson, as well as providing specific laws governing such areas as workers' compensation coverage and labor laws. Local laws impact the fire and emergency services in a more specific manner by providing such laws as building restrictions and zoning requirements.

Laws are normally developed using the balance of power concept established by the U.S. Constitution. On a federal level, most laws are developed by the legislative branch (Congress) with the approval of the executive branch (the president). On a state level, laws are similarly developed by the legislative branch with the governor's approval. Local laws are usually developed by city councils or similar bodies with the mayor or city manager's approval. Each level has established a judicial branch to adjudicate and resolve any conflicts that may arise between parties affected by the laws or regulations.

Our judicial system comprises numerous levels of courts encompassing the federal and state systems (see federal court system diagram). **Jurisdiction** and specific rules determines which court has authority to hear a particular case and render an enforceable decision. The federal judicial system includes the U.S. Supreme Court, twelve circuit courts, numerous district courts, and various specialty courts, such as tax courts. Additionally, the U.S. Court of Appeals for the federal circuit has exclusive jurisdiction over all appeals from the U.S. Court of Claims and the Court of International Trade.

State court systems are usually structured similarly to the federal system. There is normally a final court of appeals followed by several appellate courts and numerous circuit or district courts. The names for these courts can differ from those in the federal system (for example, the New York Supreme Court is a trail level court).

On the federal and state levels, it is not unusual for the same courts to hear a case, whether the issue involves a civil or criminal matter. The differences, however, between a civil matter and a criminal matter are significant. According to *Black's Law Dictionary*, **civil law** is "that body of law which every particular nation, commonwealth, or city has established particularly for itself; more properly called 'municipal' law, to distinguish it from 'law of nature' and from international law. Laws concerned with civil or private laws and remedies, as contrasted with criminal law."[1] **Criminal law** as defined is "The substantive

---

[1]*Black's Law Dictionary*, West Publishing Company (1983).

**jurisdiction**

the perimeters of a particular court's power or range to hear a case

**civil law**

the body of law that every particular nation, commonwealth, or city has established particular to itself

**criminal law**

law established to prevent harm to society that declares what conduct is criminal and prescribes that punishment be imposed for such conduct

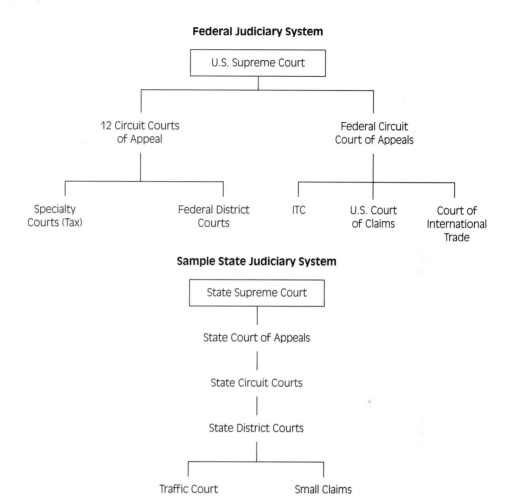

**Federal Judiciary System**

U.S. Supreme Court

12 Circuit Courts of Appeal

Federal Circuit Court of Appeals

Specialty Courts (Tax)

Federal District Courts

ITC

U.S. Court of Claims

Court of International Trade

**Sample State Judiciary System**

State Supreme Court

State Court of Appeals

State Circuit Courts

State District Courts

Traffic Court

Small Claims

*Diagram of the federal court system.*

criminal law is that law which for the purpose of preventing harm to society, (a) declares what conduct is criminal, (b) prescribes that punishment to be imposed for such conduct. It includes the definition of specific offenses and general principals of liability. Substantive criminal laws are commonly codified into criminal or penal codes. . . "[2] In essence, civil law can be thought of as the private individual's pursuit of redress, usually monetary damages, whereas criminal law deals with harms against society, and the redress can be the removal of personal freedoms, such as imprisonment.

The rules and procedures used by the courts for a civil action are substantially different from those in a criminal action. First, the parties in a civil action

[2] *Id.*

**plaintiff**
the party bringing the legal action

**defendant**
the party being sued

are usually different from those in a criminal action. In a civil action, the **plaintiff** (the party bringing the action) is usually a private citizen, private corporation, or other party outside of the government, and the **defendant** (the party being sued) is usually a private party, private corporation, or the government. In a criminal action, the party bringing the action is virtually always the government (U.S. attorney, state attorney, local prosecutor, etc.) and the defendant is the person charged with the crime. Second, the burden of proof is substantially greater in a criminal case than in a civil case. Third, the rules or procedures are significantly different (i.e., civil procedure rules and criminal procedure rules). And, last, the damages sought usually differ.

A civil action is normally brought by one party against another party for the declaration, enforcement, or protection of a prescribed right or redress or prevention of a wrong. In a civil action, the party bringing the action is known as the *plaintiff* and the party being sued is known as the *defendant.* The right to sue is considered a fundamental right of all individuals and entities (i.e., fire service corporations), and the suit can proceed provided there is some material, but yet unresolved, question regarding a right or injury. In most courts, significant deference is provided to parties to bring suit because the courts do not want to forestall a party's opportunity to be heard and to have the claim adjudicated.

**law of tort**
violation of duty in a private or civil wrong

The **law of tort** can be categorized by either (1) the nature of the conduct of the party or (2) the nature of the harm to the injured party. Intentional harms to the person include assault, battery, false imprisonment, intentional infliction of emotional distress, negligence, and liability without fault, or strict liability. Intentional harms may include injury to the person, damage to tangible property, or harms to intangible personal interests.

Under the civil action category of intentional harms to the person, the actions of assault, battery, false imprisonment, and intentional infliction of distress have been utilized in cases relating to the fire service. The intentional torts to property include trespass to land, trespass to chattel (personal property), and conversion.

The defenses available for the foregoing intentional torts fall within the categories of privilege and immunity. Privilege can include self defense, defense of others, defense of property, recapture of property, merchant's privilege, and discipline of children. Immunity includes interspousal, parent and child, charitable, and governmental immunities.

Many cases have been brought against fire service organizations under the tort of negligence. The components of a negligence action include:

1. Duty. Did the fire service organization owe a duty to the plaintiff and, if so, what standard of care did the fire service organization owe to the plaintiff under the circumstances?

2. Breach. Did the fire service organization or individual firefighter, through their conduct, violate that duty of care?

3. Harm and causation. Did the fire service organization's conduct factually or proximately bring about the actual harm to the plaintiff?

4. Damages. What was the harm?

The defenses to a negligence action usually fall within the categories of substantive defenses (contributory negligence or assumption of the risk) or procedural defenses (immunity from prosecution).

Fire service organizations should be aware that the foregoing are but a few of the potential theories in which a civil action can be brought by or against a fire service organization. Other potential actions may lie in the areas of product liability, strict liability, and harm to economic interests such as deceit, negligent misrepresentation, interference with contractual relationships, defamation, malicious prosecution, and invasion of privacy.

Criminal actions are usually brought against an individual or other entity who has harmed the interests of our society. Criminal laws can be divided into five basic categories: (1) homicide; (2) nonhomicide crimes against a person; (3) crimes against property interests; (4) crimes against habitation; and (5) inchoate crimes (see Chapter 3).

---

(This case has been edited for the purposes of this text.)

**PEOPLE** of the State of Michigan, Plaintiff-Appellee,

v.

Loren **TYLER** and Robert Tompkins, Defendants-Appellants.

250 N.W.2d 467 (1977)
Supreme Court of Michigan

LEVIN, Justice.

Loren Tyler and Robert Tompkins were convicted of conspiracy to burn real property.[1] Tyler was also convicted of burning real property[2] and burning insured property with intent to defraud.[3]

The fire occurred in premises leased by Tyler where he conducted a retail furniture business. Tompkins was a business associate of Tyler.

Physical evidence, taken by police and fire officials without a search warrant from the premises after the fire was extinguished, was admitted at the trial over objection.

The question is whether the authorities may enter fire-damaged premises without a warrant after the fire is extinguished for the purpose of investigation and, if discovered, collection of evidence of arson.

The Court of Appeals held that the provisions of the Fourth Amendment and the corresponding provisions of this state's constitution,[4] prohibiting "unreasonable searches and seizures," do not "apply to the investigation of burned premises to determine whether the fire was the result of arson where some evidence of arson is found during the process of extinguishing the fire." The court reasoned that "the investigation of a fire to determine if arson has been committed does not place a person under

---

[1]M.C.L.A. § 750.157a; M.S.A. § 28.354(1).

[2]M.C.L.A. § 750.73; M.S.A. § 28.268

[3]M.C.L.A. § 750.75; M.S.A. § 28.270.

[4]Const. 1963, art. 1, § 11.

criminal investigation. It places the cause of the fire under investigation."[5]

We reverse and remand for a new trial.

The fire broke out shortly before midnight on January 21, 1970. The fire department arrived soon thereafter. Fire Chief See discovered and seized two plastic containers, one partially filled with a flammable liquid, before the firefighters left. Defendants do not challenge the admissibility of that evidence.

Chief See conferred with Detective Webb of the police department at the scene shortly before the firefighters left. Webb's efforts to take pictures of the interior of the building were unsuccessful.

By 4 a.m. the fire was extinguished. The premises were thereafter left unattended until 8 a.m., when See returned with an assistant fire chief and together they briefly surveyed the interior of the building.

The officials again returned to the scene of the fire between 9 and 9:30 a.m. and discovered a thin linear burn in the carpet of the room. The burn circled the room, went through a door and continued down a stairway to an exit. Pieces of carpet and wood containing the burn marks were removed, and at the trial admitted over objection.

Four days later, on January 26, a sergeant of the Michigan State Police, Arson Section, took photographs of the interior of the building which were lost in the mail. The sergeant returned with Tyler three days later, on January 29, but no evidence was then obtained. The sergeant returned again without Tyler three weeks later, on February 16, and took more pictures and removed part of a fuse found in the building and several pieces of glass, which were admitted in evidence over objection.

The Court of Appeals found that "consent for the numerous searches was never obtained from defendant Tyler."[6] While Tyler did accompany the State Police sergeant when he visited the premises on January 29, and may not have objected to that inspection, none of the evidence admitted over objection was obtained at that time.

[1] The people contend that Tompkins who, in contrast with Tyler, did not have a leasehold interest

---

[5]*People v. Tyler*, 50 Mich.App. 414, 418, 419, 213 N.W.2d 221, 224 (1973).

[6]*Id.*, p. 418, 213 N.W.2d p. 223.

in the burned premises, has no standing to raise the search and seizure issue.[7] The prosecutor did not raise this issue in the court of Appeals,[8] and, therefore, it will not be considered.[9]

**I**

[2] The primacy of the warrant requirement is well established.[10] "[E]xcept in certain careful defined classes of cases, a search of private property without proper consent is 'unreasonable' unless it has been authorized by a valid search warrant." *Camara v. Municipal Court of City & County of San Francisco*, 387 U.S. 523, 528–529, 87 S.Ct. 1727, 1731, 18 L.Ed.2d 930 (1967).

The proscription of "unreasonable searches and seizures" and the warrant requirement

"must be read in light of 'the history that have rise to the words'—a history of 'abuses so deeply felt by the Colonies as to be one of the potent causes of the Revolution. . .' [*United States v. Rabinowitz*], 339 U.S. [56], 69, 70 S.Ct. 436, [94 L.Ed. 653 [1950]. The Amendment was

---

[7]*See Jones v. United States*, 362 U.S. 257, 80 S.Ct. 725, 4 L.Ed. 2d 697 (1960).

The standing issue is discussed in Allen, *The Wolf Case: Search and Seizure, Federalism and the Civil Liberties*, 45 Ill.L.Rev. 1 (1950); Kamisar, *Illegal Searches or Seizures and Contemporaneous Incriminating Statements: A Dialogue on a Neglected Area of Criminal Procedure*, 1961 U. of Ill.L Forum 78; Traynor, *Mapp v. Ohio At Large in the Fifty States,* 1962 Duke L.J.319.

The Supreme Court of California, in *People v. Martin*, 45 Cal.2d 755, 290 P.2d 855, 857 (1955), allowed a co-defendant to raise the search and seizure issue. This rule was reaffirmed in *Kaplan v. Superior Court of Orange County*, 6 Cal.3d 150, 98 Cal.Rptr. 649, 650, 491 P.2d 1 (1971). *Cf. United States v. Valencia*, 541 F.2d 618 (CA 6, 1976). The American Law Institute, in its proposed Model Code or Pre-Arraignment Procedure, has accepted the view that a motion to suppress things seized may be made by a codefendant who may himself be without standing. ALI Model Code of Pre-Arraignment Procedure (Official Draft No. 1, 1972), § 290.-1(5).

[8]Defendants' joint belief in the Court of Appeals discusses the question of Tyler's standing, but does not refer to Tompkin's standing. In the People's brief, only Tyler's standing was challenged.

[9]*See Long v. Pettinato*, 394 Mich. 343, 349, 230 N.W.2d 550 (1975).

[10]*Vale v. Louisiana*, 399 U.S. 30, 35, 90 S.Ct. 1969, 26 L.Ed.2d 409 (1970); *Coolidge v. New Hampshire*, 403 U.S. 443, 454–455, 91 S.Ct. 2022, 29 L.Ed.2d 564 (1971); *Jones v. United States*, 357 U.S. 493, 497–498, 78 S.Ct. 1253, 2 L.Ed.2d 1514 (1958); *People v. White*, 392 Mich. 404, 410, 221 N.W.2d 357 (1974).

in large part a reaction to the general warrants and warrantless searches that had so alienated the colonists and had helped speed the movement for independence. In the scheme of the Amendment, therefore, the requirement that 'no Warrants shall issue, but upon probable cause,' plays a crucial part. As the Court put it in *McDonald v. United States*, 335 U.S. 451, 69 S.Ct. 191, 93 L.Ed. 153 [(1948)]:

" 'We are not dealing with formalities. The presence of a search warrant serves a high function. Absent some grave emergency, the Fourth Amendment has interposed a magistrate between the citizen and the police. This was done not to shield criminals nor to make the home a safe haven for illegal activities. It was done so that an objective mind might weigh the need to invade that privacy in order to enforce the law. The right of privacy was deemed too precious to entrust to the discretion of those whose job is the detection of crime and the arrest of criminals. . . . And so the Constitution requires a magistrate to pass of the desires of the police before they violate the privacy of the home. We cannot be true to that constitutional requirement and excuse the absence of a search warrant without showing those who seek exemption from the constitutional mandate that the exigencies of the situation made that course imperative.' *Id.*, at 455–456,69 S.Ct. at 193." *Chimel v. California*, 395 U.S. 752, 761, 89 S.Ct. 2034, 2039, 23 L.Ed.2d 685 (1969).[11]

In the development of the probable cause and warrant requirements, the United States Supreme Court has recognized three kinds of searches.

The first is the regulatory search, in which inspection is a "crucial part of the regulatory scheme." "[I]f inspection is to be effective and serve as a credible deterrent, unannounced, even frequent, inspections are essential. In this context, the prerequisite of a warrant could easily frustrate inspection; and if the necessary flexibility as to time, scope, and frequency is to be preserved, the pro-

tections afforded by a warrant would be negligible." *United States v. Biswell*, 406 U.S. 311, 315, 316, 92 S.Ct. 1593, 1596, 32 L.Ed.2d 87 (1972).[12] The Court concluded that "effective regulation of licensed firearm businesses is impractical without resort to a broad inspection power."[13]

[3] Unannounced prophylactic inspections by fire department officials of theaters, department stores and other places where large crowds gather may be necessary to assure that unblocked exits and adequate fire extinguishers are maintained. In light of the public nature of the premises and the relative unintrusiveness of the inspection, a warrant may not be required.

A second kind of search is the so-called administrative search. Where an investigation is to determine the cause of a fire (*e.g.*, faulty wiring, malfunctioning furnace, natural gas leak) and to prevent such fires from occurring and recurring, the need for a warrant and the standard of probable cause are governed by *Camara, supra*.

In *Camara*, the United States Supreme Court held that a warrant was required for a routine annual inspection for housing code violations, rejecting the argument that "the individual and his private property are fully protected by the Fourth Amendment only when the individual is suspected of criminal behavior."

"[W]e cannot agree that the Fourth Amendment interests at stake in these [administrative] inspection cases are merely 'peripheral' It is surely anomalous to say that the individual and his private property are fully protected by the Fourth Amendment only when the individual is suspected of criminal behavior. For instance, even the most law-abiding citizen has a very tangible interest in limiting the circumstances under which the sanctity of his home may be

---

[11]*See also Boyd v. United States*, 116 U.S. 616, 625–627, 6 S.Ct. 524, 29 L.Ed. 746, 749–750 (1886).

[12]Regulatory searches at or near international borders are accorded similar treatment. *See United States v. Martinez-Fuerte*, 428 U.S. 543, 96 S.Ct. 3074, 49 L.Ed.2d 1116 (1976), upholding a warrantless border search at a permanent checkpoint even though there was no reason to believe the vehicle contained an illegal alien.

[13]Greenberg, *The Balance of Interests Theory and the Fourth Amendment: A Selective Analysis of Supreme Court Action Since Camara and See*, 61 Cal.L.Rev. 1011, 1021 (1973).

broken by official authority, for the possibility of criminal entry under the guise of official sanction is a serious threat to personal and family security. . . . [I]nspections of the kind we are here considering do in fact jeopardize 'self-protection' interests of the property owner. Like most regulatory laws, fire, health, and housing codes are enforced by criminal processes. In some cities, discovery of a violation by the inspector leads to a criminal complaint. Even in cities where discovery of a violation produces only an administrative compliance order, refusal to comply is a criminal offense, and the fact of compliance is verified by a second inspection, again without a warrant. Finally, as this case demonstrates, refusal to permit an inspection is itself a crime, punishable by fine or even jail sentence." *Camara*, *supra*, 387 U.S. p.p. 530–531, 87 S.Ct. p. 1731.

In the companion case of *See v. City of Seattle*, 387 U.S. 541, 87 S.Ct. 1737, 18 L.Ed.2d 943 (1967), the Court applied the *Camara* analysis in holding that a warrant was required for an administrative search of business premises.

The Court reaffirmed *Camara's* requirement of a warrant for an administrative search in *G.M. Leasing Corp. v. United States*, 428 U.S. 543, 97 S.Ct. 619, 50 L.Ed.2d 530 (1977), where it held a warrantless search by Internal Revenue Service agents, pursuant to a civil statute, unconstitutional.[14]

In *Camara*, the Court also held that probable cause for an administrative search would be determined by a less rigorous standard than in criminal investigations:

" '[P]robable cause' to issue a warrant to inspect must exist if reasonable legislative or administrative standards for conducting an area inspection are satisfied with respect of a particular dwelling. Such standards, which will vary with the municipal program being enforced, may be based upon the passage of time, the nature of the building (*e.g.*, a multi-family apartment house), or the condition of the entire area, *but they will not necessarily depend upon specific knowledge of the condition of the particular dwelling.*" *Camara*, *supra*, 387 U.S. p. 538, 87 S.Ct. p. 1736 (emphasis supplied).

[4] A warrant for an administrative search could thus be issued pursuant to "reasonable legislative or administrative standards" designed to permit fire officials to conduct investigations of the causes of fires to prevent their occurrence or recurrence.

[5] It is apparent, however, that probable cause (although measured by a reduced standard) and, absent exigent or other special circumstances (prophylactic regulatory search, plain view, consent), a warrant is required even if the investigation of a fire is to determine "the cause of the fire under investigation" and does not "place a person under criminal investigation." The Court of Appeals erred in reasoning to contrary conclusion.

[6] The third kind of search is the criminal investigation, where officials seek evidence to be used against persons in a criminal prosecution. Such investigations require warrants based on probable cause to believe that evidence of a crime will be found.

[7] The rationale for distinguishing, on the basis of the nature of the search, between regulatory, administrative and criminal searches, is the relative intrusiveness of the search. As stated in *Camara*, "a routine inspection of the physical condition of private property is a less hostile intrusion than the typical policeman's search for fruits and instrumentalities of crime." *Camara*, *supra*, p. 530, 87 S.Ct. p. 1731. The differing probable cause requirements reflect the greater need to protect against the more extensive and more intrusive criminal investigative search.[15]

While it may be no easy task under some cir-

---

[14]*See, also, Barlows, Inc. v. Usery*, 424 F.Supp. 437 (D.Idaho 1976) (3 judge court), holding a provision in the Occupational Safety and Health Act, 29 U.S.C. § 657, authorizing the Secretary of Labor "upon presenting appropriate credentials" "to enter without delay and at reasonable items" any business establishment covered by the act, to be unconstitutional.

[15]"[A] lesser quantum of evidence is constitutionally required for [administrative] inspections because the search involved is less of an intrusion on personal privacy and dignity than that which generally occurs in the course of criminal investigation. This is a real and meaningful distinction. The concern of the inspector is directed

cumstances to distinguish as a factual matter between an administrative inspection and a criminal investigation, in the instant case the Court is not faced with that task. Having lawfully discovered the plastic containers of flammable liquid and other evidence of arson before the fire was extinguished, Fire Chief See focused his attention on assembling proof of arson and began a criminal investigation.[16] At that point there was probable cause for issuance of a criminal investigation search warrant.[17]

[8,9] As the facts of this case illustrate, requir-

ing a warrant would not burden unduly officials whose duty it is to extinguish fires and investigate their causes. Evidence acquired while firefighters are lawfully on the premises putting out the fire is admissible under the plainview doctrine.[18] The owner of the property will often consent to a search. In other cases, as here, there will be probable cause to justify the issuance of a search warrant as part of a criminal investigation. If there is insufficient evidence to justify issuance of a warrant as part of a criminal investigation but there is a sufficient basis for an investigation within "reasonable legislative or administrative standards" to determine the cause of the fire, a warrant might issue on that basis.

[10] If there are exigent circumstances, such as reason to believe that the destruction of evidence is imminent or that a further entry of the premises is necessary to prevent the recurrence of the fire, no warrant is required and evidence discovered is admissible.[19]

The exigent circumstances exception does not, however, justify a search after the emergency no longer obtains, and the justification for the excep-

---

toward such facilities as the plumbing, heating, ventilation, gas, and electrical systems, and toward the possible accumulation of garbage and debris. These matters may be looked into in a much shorter period of time than it often takes to search for evidence of crime, and certainly no rummaging through the private papers and effects of the householder is required. Nothing is seized. A police search for evidence brings with it 'damage to reputation resulting from an overt manifestation of official suspicion of crime.' A routing inspection that is part of a periodic or area inspection plan does not single out any one person as the object of official suspicion. The search in a criminal investigation is made by armed officers, whose presence may lead to violence, and is perceived by the public as more offensive than that of the inspector. Police searches are conducted at all times of the day and night, while routine inspections are conducted during regular business hours. By their very nature and purpose, police searches usually must be conducted by surprise. In contrast, some inspections programs involve advance notice that the inspector will call on a certain date, and an inspector on his rounds will sometimes agree to return at a more convenient time if the house holder so requests. This permits the owner or occupant to remove or conceal anything that might be embarrassing to him." *LaFave, Administrative Searches and the Fourth Amendment: The Camara and See Cases,* 1967 Sup.Ct.Rev. 1, 18–20.

[16]The assistant fire chief testified concerning the searches on the morning following the fire:

"Q: So your purpose in being in the building was solely and exclusively for investigation and gathering of evidence?

"A: Yes, Sir.

"Q: You were not there for any purpose of fire fighting?

"A: No, Sir, I was not."

[17]The Supreme Court declared in *Katz v. United States,* 389 U.S. 347, 88 S.Ct. 507, 19 L.Ed.2d 576 (1967), that it

"has never sustained a search upon the sole ground that officers reasonably expected to find evidence of a particular crime and voluntarily confined their activities to the least intrusive means consistent with that end. Searches conducted without warrants have been held unlawful 'notwithstanding facts unquestionably showing probable cause,' *Agnello v. United States,* 269 U.S. 20, 33, 46 S.Ct. 4, 6, 70 L.Ed. 145, for the Constitution requires 'that the deliberate, impartial judgment of

a judicial officer . . . be interposed between the citizen and the police . . . ' *Wong Sun v. United States,* 371 U.S. 471, 481–482, 83 S.Ct. 407, 414, 9 L.Ed.2d 441. 'Over and again this Court has emphasized that the mandate of the [Fourth] Amendment requires adherence to judicial process,' *United States v. Jeffers,* 342 U.S. 48, 51, 72 S.Ct. 93, 95, 96 L.Ed. 59, and that searches conducted outside the judicial process, without prior approval by judge or magistrate, are *per se* unreasonable under Fourth Amendment—subject only to a few specifically established and well-delineated exceptions."

[18]*Cardwell v. Lewis,* 417 U.S. 583, 94 S.Ct. 2464, 41 L.Ed.2d 325 (1974); *Coolidge v. New Hampshire,* 403 U.S. 443, 91 S.Ct. 2022, 29 L.Ed.2d 564 (1971).

[19]*See, e.g., United States v. Gargotto,* 510 F.2d 409 (CA 6, 1974). The court declared that where the fire had been recently extinguished and the firefighters were still working in the area where the records were found there were exigent circumstances justifying a warrantless search.

*Similarly see Romero v. Superior Court of Los Angeles County,* 266 Cal.App.2d 714, 72 Cal.Rptr. 430 (1968) (search for hidden fires after explosions was reasonable); *United States v. Green,* 474 F.2d 1385, 1389 (CA 5, 1973) (upholding warrantless search for cause of fire after fire was extinguished, but before fire officials had determined fire's source and were uncertain whether it would recur).

tion had ceased to exist. We cannot accept the bald assertions of other courts to the contrary.[20]

[11] In the instant case there was not exigent circumstances justifying the searches made hours, days or weeks after the fire was extinguished. As expressed in *G.M. Leasing Corp., supra,*—U.S. p. —, 97 S.Ct. p. 631: "[t]he agents' own actions . . . in their delay for two days following their first entry, and for more than one day following the observation of materials being moved from the office, before they made the entry during which they seized the records, is sufficient to support the District Court's implicit finding that there were no exigent circumstances in this case."

## II

The people argue that the evidence was taken by the fire and police officials pursuant to their statutory duties under M.C.L.A. § 29.6; M.S.A. § 4,559(6), providing:

> "The director or any officer is authorized to investigate and inquire into the cause or origin of a fire occurring in this state resulting in loss of life or damage to property, and for that purpose may enter, without restraint or liability for trespass, any building or premises and inspect the same and the contents and occupancies thereof."

The argument is that because the Legislature has specifically authorized such searches, they are constitutionally sound.

In *Department of Natural Resources v. Seaman*, 396 Mich. 299, 315, 240 N.W.2d 206, 213 (1976), we considered a statute authorizing conservation officers to conduct a search on probable cause without a warrant.[21] We said: "[A] search without a warrant in order to be reasonable under the Fourth Amendment requires more than probable cause—exigent circumstances must also be present. *People v.*

*White*, 392 Mich. 404, 221 N.W.2d 357 (1974)." We held that the statute must be construed to preserve its constitutionality, and read it as permitting a warrantless search only where there are exigent circumstances.[22]

[12] We agree with an earlier panel of the Court of Appeals[23] which noted that rather than explicitly authorizing warrantless searches the statute before us is silent on the question. To preserve the constitution of the statute a warrant requirement should be read into it. The Supreme Court of Indiana, considering a similar statutory provision, reached the same conclusion. *State v. Buxton*, 238 Ind. 93, 148 N.E.2d 547 (1958). Similarly see *G.M. Leasing Corp. v. United States, supra.*

Although post-fire searches made solely for the administrative purpose of determining the cause and source of determining the cause and source of the fire may properly occur under reasonable guidelines and with *Camara's* reduced standard of probable cause, where the investigation turns to the collection of criminal evidence constitutional requirements cannot be subordinated to a sweeping grant of statutory authority.[24]

Even if the search is administrative, under *Camara* a warrant is required. To justify a warrantless search by labeling it "administrative" (*Bennett v.*

---

[20]*Bennett v. Commonwealth*, 212 Va. 863 S.E.2d 215 (1972) (warrantless search occurring day after the fire is one that "the law has traditionally upheld in emergency situations"). *People v. Kulick*, 57 Mich.App. 126, 225 N.W.2d 709 (1974) (search the day after fire was only a continuation of search valid at time of fire).

[21]M.C.L.A. § 300.12; M.S.A. § 13.1222.

[22]"Although M.C.L.A. § 300.12; M.S.A., § 13.1222 does not mention the requirement of exigent circumstances to justify warrantless searches, it is clear that this provision was enacted to cover those situations in which it is not feasible to obtain a warrant. We are 'duty bound' under the Michigan Constitution to construe a statute in such a way as to uphold its validity. *People v. Bricker*, 389 Mich. 524, 528, 208 N.W.2d 172 (1973). Consequently, we construe M.C.L.A. § 300.12; M.S.A. § 13.1222 as, *inter alia*, requiring the presence of exigent circumstances before conservation officers are authorized to conduct warrantless searches under the act." *Department of Natural Resources v. Seaman*, 396 Mich. 299, 315, 240 N.W.2d 206, 213 (1976).

[23]*People v. Dajnowicz*, 43 Mich.App. 465, 204 N.W.2d 281 (1972).

[24]We are unpersuaded by decisions of courts in other jurisdictions upholding warrantless post fire searches on the ground that the searches were made pursuant to statutory authority. *Stone v. Commonwealth*, 418 S.W.2d 646 652 (Ky. 1967), *cert. den.* 390 U.S. 1010, 88 S.Ct 1259, 20 L.Ed.2d 161 (1968); *State v. Rees*, 258 Iowa 813, 139 N.W.2d 406 (1966) (relying on *Frank v. Maryland*, 359 U.S. 360, 79 S.Ct. 804, 3 L.Ed.2d 877 [1959], which authorized civil inspections without warrants; *Frank* was overruled by *Camara, supra*, in 1967).

*Commonwealth*, 212 Va. 863, 188 S.E.2d 215 (1972) is therefore erroneous.

While the Legislature is not free to generally except searches and seizures for evidence of arson from the constitutional limitation, it may, as outlined in *Camara*, prescribe, in a narrowly drawn statute, reasonable standards to facilitate the issuance of administrative search warrants to determine the causes of fires to prevent their occurrence or recurrence.

### III

[13] The people contend alternatively that there is no reasonable expectation of privacy in property that has been burned. We can accept the premise of that contention, "the fourth amendment protects people, not places [and that] [w]hat a person knowingly exposes to the public, even in his own home or office, is not a subject of Fourth Amendment protection," *Katz v. United States*, 389 U.S. 347, 88 S.Ct. 507, 19 L.Ed.2d 576 (1967), without accepting the conclusion.

[14] Simply because a person's home or place of business has been burned does not mean that he has no expectation of privacy regarding whatever of his possessions remain. A fire is not an invitation to any or all to enter to satisfy their curiosity or for any other purpose. It does not open the property to public scrutiny of governmental officials.

Personal papers, family heirlooms and other objects may survive the fire; the owner has a justifiable interest in protecting such property. Although the premises may be uninhabitable, personal possessions may remain undestroyed; unless uninhabitability becomes tantamount to actual abandonment, there may still be justifiable expectation of privacy.[25]

[15] Some courts have upheld warrantless post-fire searches on the theory that there is no reasonable expectation of privacy in the burned premises[26] or that the premises have become uninhabitable.[27] We agree that a warrantless entry would not invade a constitutionally protected interest in privacy if the owner or occupant of the burned premises abandons the property[28] or if the premises are so completely destroyed that there no longer are recognizable objects of personal property. The record does not factually support a conclusion that Tyler had abandoned the fire-damaged premises.

### V

[16] Finally, the People contend that warrantless post-fire searches can be justified by the public interest in preventing fires, saving lives, protecting property and preventing insurance fraud.

There is public interest in preventing and solving all crimes against persons or property. Constitutionally infirm searches and seizures may not be justified by the generalized, undifferentiated interest in protecting the public safety.

The United States Supreme Court, in distinguishing between regulatory, administrative and criminal investigative searches, has provided a framework for achieving a workable balance between the need for investigation of the causes of fires and protection of the individual's right of privacy.

[17] If there are exigent circumstances, or the evidence is in plain view, no warrant is required. Nor is a warrant required for a prophylactic regulatory inspection of public places.

[18] If there has been a fire, the blaze extin-

---

[25]"It seems that the primary defect with 'habitability' as a standard for determining fourth amendment protection where arson investigations are involved is that it should not be the only factor to receive consideration. Other factors which seem to bear on an expectation of privacy are: the amount of personal property remaining on the premises, the extent of destruction caused by the fire, the possibility of danger to the community from the fire and affirmative acts on the premises, the extent of destruction caused by the fire, the possibility of danger to the community from the fire and affirmative acts on the part of the individual demonstrating the existence, or lack thereof, of any expectation of privacy. It is also suggested that courts consider these other factors along with 'habitability' in determining whether fire-damaged premises are entitled to fourth amendment protection." Note, *Arson Investigations and the Fourth Amendment*, 30 Wash & Lee L.Rev. 133, 145, fn. 86 (1973).

[26]*State v. Murdock*, 160 Mont. 95, 500 P.2d 387, 391 (1972).

[27]*State v. Vader*, 114 N.J. Super. 260, 276 A.2d 151 (1971).

[28]*State v. Felger*, 19 Or.App. 39, 526 P.2d 611, 615 (1974): "We need not decide whether a warrant was necessary for these subsequent inspections of the premises by fire and police officials because . . . defendant had already abandoned the tenancy."

guished and the firefighters have left the premises, a warrant is required to reenter and search the premises, unless there is consent or the premises have been abandoned.

[19, 20] The standard of probable cause will vary depending on the nature of the post-fire search. If the authorities are seeking evidence to be used in a criminal prosecution, the usual standard (probable cause to believe that evidence of a crime will be found) will apply. Where the cause is undetermined, and the purpose of the investigation is to determine the cause and to prevent such fires from occurring and recurring, an administrative search may be conducted pursuant to a warrant issued in accordance with reasonable legislative or administrative standards or, absent their promulgation, judicially prescribed standards; if evidence of wrongdoing is discovered, it may, of course, be used to establish probable cause for the issuance of a criminal investigative search warrant or in prosecution.

The warrant requirement protects individual privacy from unrestrained exercise of governmental power. Enforcement of the principle that warrantless searches are an exception, justified only by special circumstances, and that warrants may issue only on a determination by a neutral magistrate of probable cause, albeit in this context a varying standard of probable cause, can be harmonized with effective law enforcement.

We hold that the warrantless searches were unconstitutional and that the evidence obtained was therefore inadmissible.

REVERSED and REMANDED for a new trial.

KAVANAGH, C.J., and WILLIAMS and FITZGERALD, JJ., concur.

COLEMAN, J., concur in result.

RYAN and MOODY, JJ., not participating.

# Review Questions

1. What is the highest federal court in the United States?

2. What is the highest court in your state?

3. Outline the hierarchy of the federal court system.

4. Outline the hierarchy of your state court system.

5. What are the names of the U.S. Supreme Court justices?

6. What are the names of the justices of the highest court in your state?

7. Where would you find the criminal code for your state?

8. Where would you find the statutes or laws governing civil actions in your state?

*Chapter*

# 2

# Civil Actions against Fire and Emergency Service Organizations

### *Learning Objectives*

- To gain an understanding of the potential civil liabilities against fire and emergency service organizations.
- To gain an understanding of civil liability.
- To gain an understanding of negligence in a fire and emergency setting.
- To gain an understanding of the concept of immunity.

A definite distinction must be made between criminal and civil liability in the area of a fire officer's accountability. In virtually all situations, the fire or emergency service officer is responsible for his or her own criminal acts (see Chapter 3). In the area of civil liability, the lines are often blurred, depending on the facts of the situation.

*Respondeat Superior*
concept by which the master is responsible for the actions of the servant; the employer for the employees

As a general rule for civil liability, fire and emergency service officers are usually provided the same immunity as the fire service organization. Under the concept of ***Respondeat Superior***, so long as the fire or emergency service officer is functioning within the scope of his or her employment, the fire or emergency service organization usually is the named responsible party. However, if a fire service officer is outside the scope of employment, or if he or she purposefully or willfully endangers another through their actions or omissions, personal civil and potentially criminal liability may result. Under many federal and state statutes, civil and criminal liability may result where the officer willfully disregards a duty as proscribed by law (for example, Occupational Safety and Health Administration (OSHA), EPA, and other regulations).

Additionally, civil actions seek monetary damages and the "deep pocket" in most civil actions is the fire or emergency service organization or municipality rather than the individual fire officer. However, when a fire or emergency service officer is beyond the scope of duty, and injury or harm occurs, potential personal liability may attach. The question of whether a fire or emergency service organization is liable tends to fall within the following categories:

1. Was the activity in which the harm occurred a proprietary function in which no judgment was required?
2. Was the activity in which the harm occurred a governmental function?
3. Was the activity in which the harm occurred within the category of judicial, quasi judicial, legislative, or quasi legislative function?
4. Was the activity in which the harm occurred outside of any or all of the foregoing functions?

**immunity**
a special privilege providing freedom or exemption from penalty, burden, or duty

These categorizations are necessary in evaluating potential civil liability because of the potential protections afforded through **immunity**. However, as previously noted, the concept of immunity is fast deteriorating because of the number of exceptions being provided by the courts.

The second category of questions that should be asked in evaluating potential civil liability relate to the facts of the situation that resulted in the harm:

1. Was a duty created by statute or other law that required fire or emergency service organizations to respond? Was a special duty created through actions? Was that duty breached?
2. What was the nature of the duty violated (proprietary, governmental, ministerial)?
3. Was the negligent individual a firefighter, an EMT, an agent, or person otherwise employed by the fire or emergency service organization?
4. Was the firefighter's act beyond the scope of his or her responsibility or authority?

5. If the firefighter was acting outside the scope of authority, did the fire service organization or municipality subsequently ratify the act?

6. Were the fire or emergency service organization, its firefighters, EMTs, agents, or employees negligent?

7. Was the injured employee a firefighter or employee of the fire or emergency service organization or municipality, that is, do workers' compensation laws apply?

8. Did the injured individual assume the risk or extend the harm ?

As with individual liability, the major area of potential liability for fire or emergency service organizations is under the theory of tort liability (i.e., negligence). Most fire and emergency service organizations possess a duty created by statute, law, or the creation of a special relationship; when that duty is breached and harm occurs, the potential of civil liability is present. Activities such as fire suppression, tactics and strategies, or 911 system failures are only a few of the activities performed by fire or emergency service organizations that are highly susceptible to potential liability when the activity malfunctions and harm occurs.

In the past, fire and emergency service organizations were often placed within the category of a governmental agency and thus, under the concept of "the king can do no wrong," fire service organizations were rarely, if ever, found liable under a tort liability theory. Under the general rule, fire and emergency service organizations owe a duty to the community as a whole and thus were not designed to protect an individual's interest. The duty owed by the fire department or emergency service organization was considered to be limited in nature, and restricted to protecting the community in total. In the concept of immunity, fire and emergency service organizations, as a governmental undertaking, were virtually immune to any type of tort liability for losses of individuals. Through the years, this well-established principle of law has begun to erode. This age old concept that a fire service organization owed a duty to all but to no one in particular has received extensive criticism and is slowly deteriorating. Through the years, courts have established many exceptions to this general rule of nonliability. For example, many courts have established exceptions to the rule concerning construction and maintenance of fire department facilities, repair of fire alarm systems, and operation of fire service vehicles. The basic premise in which fire and emergency service organizations can determine the extent of their immunity from tort liability usually focuses on the exercise of professional judgment or lack thereof. In general, the court decisions regarding fire and emergency service organization liability or immunity tend to fall into the basic concept that when fire service organization activity is within the area of proprietary functions (i.e., no professional experience is required), no immunity exists for the fire or emergency service organization. Thus, the fire and emergency service organization and the municipality and the other governmental agency may possibly be liable under the same tort liability theory as a private corporation.

(This case has been edited for the purposes of this text.)

Thomas **AYRES** and Helen **AYRES**

v.

**Indian Heights Volunteer Fire Department, Inc.**
Court of Appeals of Indiana
Second District
Sept. 9, 1985
482 N.E.2d 732. (1985)

SULLIVAN, Judge.

On January 20, 1983, at or about 10:30 a.m., the Plaintiffs, Thomas and Helen Ayres, had a fire in their enclosed Ford truck in the driveway of their residence at 5206 Algonquin Trail, Kokomo, Indiana. The Indian Heights Volunteer Fire Department, Defendants, was called by a neighbor of the Plaintiffs and upon arrival at the scene told a neighbor who was extinguishing the fire with his hand extinguisher to get out of the way; whereupon the firemen sprayed a large fire extinguisher into the rear of the truck with such force that it blew the burning materials out of the truck and against the fiberglass door of the Plaintiffs' garage causing it to burn.

The firemen had a large fire hose, but were unable to get it to work until after setting the garage door afire; then, when they got the hose working, they ignored the request of Plaintiff, Helen Ayres to enter the service entrance and spray from the inside so as to keep the fire from entering the garage where Plaintiffs had stored valuable merchandise; instead, they sprayed from the outside, blowing the fire from the burning door into the garage and totally destroying the garage and its contents.

On September 7, 1983, the Plaintiffs filed a complaint against the volunteer fire department for alleged negligence in fighting the fire in which they lost their garage and its contents. They also brought action against the township trustee, Billy D. Myers, because of his statutory duty to furnish the owners of real estate within his jurisdiction with reasonable and safe fire protection.

On September 20, 1983, the Fire Department filed its answer to the complaint denying the general allegations and also filed a Trial Rule 12(B)(6) motion to dismiss based upon the doctrine of governmental immunity alleging that its actions were discretionary functions for which the Indiana Tort Claims Act provides immunity. In his answer three days later, the Trustee admitted his statutory duty to provide fire protection but denied that he had breached his duty. Trustee also accompanied his answer with a motion to dismiss under Trial Rule 12(B)(6).

On October 12, 1983, the Plaintiffs filed interrogatories to each defendant inquiring as to the existence and for the production of any written contract for fire protection. On November 16 and 17, 1993, the Defendants filed answers to these interrogatories admitting the existence of such a contract and producing copies of it. On December 15, 1983, the trial court heard arguments on the motion to dismiss and on January 17, 1984, the court entered its judgment.

The Court issued a summary judgment for each of the Defendants pursuant to Trial Rule 56 and found that there was no material issues of fact.

Plaintiffs appealed the decision of the trial court. They contend that the trial court improperly converted the defendant's motions to dismiss for failure to state a claim under Trial Rule 12(B)(6) into motions for summary judgment under Trial Rule 56. Plaintiffs also alleged that the specific acts of negligence on the part of the Fire Department were ministerial acts and therefore the Indiana Tort Claims Act does not provide immunity. They contend that the Trustee is liable for the negligent acts of the Fire Department because the selection of the Fire Department by the Trustee was made pursuant to a statute, I.C. 36-8-13-2, and, therefore, was a ministerial act for which the Indiana Tort Claims Act does not provide immunity.

The Court of Appeals of Indiana held that the trial court incorrectly applied the law in granting the Fire Department's motion for summary judgment but that it did not constitute reversible error. The Court also held that because the volunteer fire department proved to be an independent contractor, it was not immune from liability for its alleged negligence. The Court held that the Trustee was immune from liability for alleged negligence of the fire department.

Summary judgment for Trustee is AFFIRMED; summary judgment for the Fire Department is REVERSED.

BUCHANAN, C.J., and SHIELDS, J., concur.

*Decision Explanation:*
*The court of appeals agreed with the district court in granting summary judgment to the trustee, however* *the decision to grant summary judgment to the volunteer fire department was reversed and sent back to district court for trial.*

(This case has been edited for the purposes of this text.)

Bonita K. **GORDON**, Glynn A. Gordon, and William Garner,
SUING Individually and as Next Friends of Jesse W. Garner,
Crystal Renee Johnson, Lois Johnson Garner, and Lydia Russell, Appellants,
v.
**CITY OF HENDERSON**, Tennessee and Henderson **FIRE DEPARTMENT**,
Appellees
Supreme Court of Tennessee, at Jackson
Feb. 27, 1989
766 S.W.2d 784

COOPER, Judge.

On November 27, 1984, four residents of Henderson, Tennessee died from smoke inhalation and asphyxiation, when fire destroyed the home in which they were residing. The Plaintiffs, which were relatives of the deceased, filed a complaint alleging that the deaths were proximately caused by the negligence of the City of Henderson Fire Department.

The Plaintiffs alleged that when the Fire Department was called about the fire, the firemen were absent from their regular duty station and that the Henderson Fire Department had to locate the firemen. The response time of the firemen, after they were located, was at least fifteen minutes, when the proper response time, considering the location of

the fire and the fire house should have been less than five minutes. When the firemen arrived at the fire, some had the smell of liquor on their breath and were unable to respond to the fire as trained professional firemen and they incorrectly placed their equipment in operation.

The Plaintiffs allege that a timely response of the firemen and professional manner would have prevented the deaths of the Plaintiff's relatives. The Defendants claimed that they were unable to respond promptly due to the fact that their fire truck driver was out sick and they did not have a back-up driver.

Plaintiffs' complaint was dismissed by the trial court upon defendants' motion to dismiss for failure to state a claim upon which relief could be granted. The Plaintiffs appealed and the Court of Appeals affirmed the action of the trial court and concluded that the activities of the defendants that were the bases of the complaint were "discretionary functions", for which the defendants were immune from suit. The Plaintiffs appealed again.

The Plaintiffs appealed on the latter section of the Tennessee Governmental Tort Liability Act which allows the Act to remove immunity "for injury caused by negligent act or omission of employees," with numerous exceptions.

The Supreme Court of Tennessee found that the Tuscumbia Fire Department had a duty to respond immediately to the call that the {Plaintiffs'} house was on fire and that, even though their fire truck driver was sick, they acted unskillfully by not having a back-up driver who could have immediately taken the place of the sick driver. The Fire Department lacked proficiency.

The judgment of the Court of Appeals sustain-

ing the Rule 12 motion and dismissing the action is reversed. The cause is remanded to the trial court for further proceedings.

HARBISON, C.J., and FONES, DROWOTA and O'BRIEN, JJ., concur.

*Decision Explanation:*
*The Supreme Court reversed the decision of the dis-* *trict court and appellate court and sent the case back to the district court for trial.*

## Review Questions

1. What is the basic issue in the *Ayres* case?
2. Who is the plaintiff in the *Ayres* case? The defendant?
3. What type of action did the plaintiff bring in the *Ayres* case?
4. What was the decision of the Court of Appeals of Indiana in the *Ayres* case?
5. What was the issue in the *Gordon* case?
6. What was the decision in the *Gordon* case?
7. Does your state possess any statutes or other laws creating a duty for your fire service organization?
8. What is your scope of authority as a firefighter in your state?
9. If a firefighter is injured on the job in your state, is he or she covered under workers' compensation?
10. Has your fire service organization evaluated the potential risks and liabilities?
11. Is your fire service organization aware of the potential liabilities in its operation? In its structures? In its activities?

# 3

# Criminal Actions against Firefighters

Crime, like virtue, has its degrees.

*Racine*

## Learning Objectives

- To gain an understanding of criminal law.
- To understand that individual fire fighters and EMS personnel are responsible for their own criminal acts.

**burden of proof**

the necessity or duty to prove the facts positively in a dispute on the issue raised between the parties in a cause

Criminal laws can be either federal or state. In most circumstances, the location of the criminal act or the type of act is the deciding factor as to whether the crime is prosecuted by state or federal authorities. For example, if a murder were committed on a city street, the murder would usually be prosecuted by the state authorities. If the murder were committed on federal property, the murder charge would be prosecuted by federal authorities. Certain crimes involving interstate actions (such as mail fraud) usually are prosecuted by federal authorities.[1] Each particular crime has elements that must be proved by the federal or state prosecutor; the **burden of proof** is on the prosecutor. Trial by jury is usually permitted in all criminal cases and the penalty for the crime is usually proscribed by law.[2]

Fire service organizations and individual firefighters should be aware that they are not immune from committing a criminal act or being victimized by a criminal act. Criminal actions can be brought against individual firefighters, the fire service organizations, or other entity that has harmed the interests of society. Criminal laws are divided into five basic categories: (1) homicide; (2) nonhomicide crimes against a person; (3) crimes against property interests; (4) crimes against habitation; and (5) inchoate crimes. A firefighter can be susceptible to the lure of criminal acts because of his or her specialized training and expertise and the access that employment within the fire service can provide to an individual inclined to such criminal activities.

Homicide is the death of one human being caused by another. In most states, homicide is categorized under various levels including murder and degrees thereof and, voluntary and involuntary manslaughter.

Murder is usually defined as a criminal homicide committed with malice aforethought. Malice aforethought is a technical term that encompasses the intent to kill (i.e., expressed malice or willful and wanton disregard, in the commission of a crime, and other technical definitions). For first-degree murder, premeditation and deliberation by the person committing the murder or attempting commission of certain dangerous felonies (felony-murder rule) are required. First-degree murder can also include such situations as torture or extreme cruelty prior to murder, use of poison in the murder, and lying in wait prior to the murder. Second-degree murder is usually any murder that does not meet the definitions set forth for first-degree murder, that is, without premeditation and deliberation.

Voluntary manslaughter is intentional homicide under extenuating circumstances that mitigate but do not justify or excuse the killing. Usually the element of malice aforethought, which is required for murder, is not present. The most common types of voluntary manslaughter involve the intentional killing while in the heat of passion (absent an adequate cooling off period) caused by adequate provocation (i.e., loss of control).

---

[1] On the federal level, the U.S. Justice Department is usually the prosecuting authority.
[2] See, for example, the Federal Sentencing Guidelines, 18 U.S.C. § 3577.

Involuntary manslaughter is an unintentional homicide without malice that is neither justified nor excused. Involuntary homicide can include criminal negligence and a killing during the commission of a misdemeanor (misdemeanor-manslaughter rule).

The common-law[3] felony of robbery is larceny from the person through violence or intimidation. Extortion is the corrupt demanding or receiving by a public official of a fee for services that should have been performed gratuitously. Kidnapping is an aggravated form of false imprisonment involving any transportation or secrecy. Rape is unlawful sexual intercourse with a female[4] without her consent. Rape can be by force or through fraud or inducements. Statutory rape usually involves unlawful intercourse with a female below the age of consent (usually age sixteen) and consent is usually not a defense.

Crimes against property include larceny, embezzlement, obtaining property by false pretenses, and receiving stolen goods. The basic difference between larceny and embezzlement is that larceny occurs if a firefighter takes a tool from the firehouse and keeps the tool, whereas embezzlement occurs if the fire chief gives the firefighter $50 to buy a tool for the fire service, the firefighter buys the tool from the dealer and then keeps it.

Of particular importance to fire service personnel is the category of crimes against habitation. These crimes include burglary and arson. At **common law**, burglary is the breaking and entering of the dwelling of another in the nighttime with the specific intent to commit a felony therein. Most states have broadened this category to eliminate the breaking and the nighttime requirement and added such other areas as cars and all buildings.

Arson is a common-law crime that has been extensively expanded by state legislatures. At common law, arson was defined as the malicious burning of the dwelling of another. Common-law arson required that the individual possessed the intent to burn the dwelling, actually set the fire, and the building was required to be another person's home. Most modern statutes have expanded the scope to include all buildings or other personal property, such as automobiles, and provided for separate laws when an individual burns his or her own property for insurance proceeds.

Inchoate crimes include attempts to commit a crime, solicitation, and conspiracy to commit a crime. Under these laws, the individual may not be required to actually have completed the crime and must withdraw within a specified time period in order to not accept responsibility. These crimes are often included with other crimes or used to compound the offense.

**common law**

court-created law as compared to statutory law created by the government's legislative or executive branch.

---

[3]Common law is defined as "as distinguished from the law created by the enactment of legislatures, the common law comprises the body of those principals and rules of action, related to the government and security of persons and property, which derives their authority solely from usage and custom of immemorial antiquity, or from the judgments and decrees of the courts recognizing, affirming, and enforcing such usages and customs; and, in this sense, particularly the ancient unwritten laws of England." (Black's Law Dictionary (5th Edition.))
[4]Rape has evolved via statute to include male/male and female/female cases.

(This case has been edited for the purposes of this text.)

**UNITED STATES** of America, Plaintiff-Appellant,
v.
John Leonard **ORR**, Defendant-Appellee.
United States Court of Appeals, Ninth Circuit.
Argued and Submitted Aug. 21, 1992.
Decided Oct. 14, 1992
944 F.2D 593

WILLIAM A. NORRIS, REINHARDT and TROTT, Circuit Judges.

This case involves an interlocutory appeal from the district court's order to exclude evidence of a manuscript authored by John Orr.

Orr, the Defendant-Appellee, was the chief arson investigator in the Glendale Fire Department who was charged with five counts of arson and three counts of attempted arson. His manuscript was about a Los Angeles firefighter who sets serial arson fires. The arsonist's modus operandi in the manuscript was strikingly similar to the modus operandi of the actual arsonist.

The Defendant filed a motion in limine to exclude the manuscript and related letters to Orr's literary agent as hearsay. The district court granted the motion to exclude the manuscript and any references to it, not because it was hearsay, but because the court believed the manuscript would have too great a prejudicial effect on the jury when examined under the Federal Rules of Evidence 403.

The district court found the manuscript to be prejudicial because "the points of origins and methods of starting the arson fires described in the manuscript do not appear to . . . be so unique as to justify the conclusion that . . . defendant must have had first-hand knowledge of the facts and circumstances underlying the arson fires charges against him."

The United States Court of Appeals reviewed the order and held that the district court abused its discretion in excluding the manuscript. The Court stated that the manuscript and letters were highly probative of modus operandi and thus the identity of the arsonist. The key similarities the court found were:

1. Both were firefighters.
2. Both were non-smokers.
3. Both used a delay incendiary device designed to fully ignite the fire approximately ten to fifteen minutes after the device was in place.
4. One draft of the manuscript described a match attached to a cigarette and placed inside a paper bag—similar to the actual facts: matches attached to a cigarette and placed inside yellow lined paper.
5. Both started fires in retail stores located in Los Angeles during business hours and both placed the incendiary device in combustible materials located in the store.
6. Both started fires in the drapery section of a Los Angeles fabric store.
7. Both started fires in a display of styrofoam products.
8. Both started fires in hardware stores.
9. Both started fires at several retail stores in close proximity to one another within a short time span on the same day.
10. Both started fires in the same locations while both the character and actual arsonist were traveling to or from arson investigator's conferences in Fresno.

The Court of Appeals concluded that the evidence of the case had no collateral aspects capable of generating prejudice against the defendant. The evidence was directly relevant to the issue of the identity of the arsonist responsible for the fires at issue. The district court was mistaken in viewing the evidence as "prejudicial." Therefore, the Court of Appeals held that the evidence . . . was properly admissible because it was so highly relevant to proof of modus operandi and identity.

REVERSED.

REINHARDT, Circuit Judge, dissenting.

*Decision Explanation:*
*In this limited appeal, the court of appeals found that the evidence was admissible and thus permitted to be used against the defendant.*

(This case has been edited for the purposes of this text.)

### McCOY
v.

### The STATE
Supreme Court of Georgia
Feb. 5, 1993
425 S.E. 2D 646 (Ga.1993)

BENHAM, Justice.

Appellant was indicted for felony murder and arson in the first degree. His first trial resulted in a conviction for arson in the first degree and a mistrial on the felony murder charge due to the jury's inability to reach a verdict. A second trial on the felony murder count resulted in a verdict of guilty. The trial court merged the arson conviction into the felony murder conviction and sentenced only for the felony murder. Appellant asserts on appeal that the trial court erred in refusing to give certain requested jury instructions and that the evidence did not support the verdict.

The evidence at trial authorized the jury to find that the appellant and his co-indictee left a party and walked to an abandoned house. After exploring the house, and having noticed the presence of a well closed by a wooden cover, appellant borrowed his companion's lighter and deliberately set the house afire. Two volunteer firemen who responded to the fire were directed to take a hose to the back of the house to prevent the fire from spreading to other property. In the darkness and the dense smoke from the fire, one of the firemen fell into the well, the cover of which had been burned in the fire. The well was filled with smoke and ashes and the fireman, unable to obtain sufficient oxygen, died of acute carbon monoxide poisoning associated with smoke inhalation and oxygen depletion.

It being clear from the evidence that appellant deliberately set the house afire, that the victim came to the scene as a direct result of appellant having set the fire, that the protective cover over the well was burned away by the fire appellant set, and that the victim died as a result of breathing the concentrated smoke from the fire which appellant set, we hold that the evidence was sufficient to authorize a rational trier of fact to find appellant guilty beyond a reasonable doubt of felony murder with arson in the first degree as the underlying felony. Appellant's reliance on *State v. Crane*, 279 S.E. 2d 695 (1981), and *Hill v. State*, 295 S.E. 2d 518 (1982), is unwarranted. The felony murder statute was inapplicable in those cases because the deaths were not caused by the defendant but by the victim and a police officer, respectively, whereas the death in this case was directly attributable to appellant's felonious conduct in setting the fire.

In support of his effort to show that the victim's death was not due to his criminal conduct in burning the house, but to the negligence of the landowner in leaving an abandoned well unfilled, appellant requested a charge on the duty to fill in abandoned wells. The requested charge, however, was not a complete and accurate statement of the law in that it did not specify whose responsibility it was to fill in abandoned wells; indeed, it could be inferred from the requested charge and the evidence that appellant, having discovered the well, was under a duty to report it. Since the requested charge was not accurate and was not adjusted to the evidence, there was no error in refusing to give it.

Appellant requested charges on the offenses of involuntary manslaughter, reckless conduct, and criminal trespass, arguing that his conduct could have been found to constitute one of the latter two misdemeanors, authorizing the jury to find him guilty of involuntary manslaughter rather than felony murder. However, the uncontradicted evidence in this case showed the completion of the greater offense of arson in the first degree, rendering it unnecessary that the trial court charge on the lesser offenses.

JUDGMENT AFFIRMED.

*Decision Explanation:*
*The Supreme Court of Georgia found that the district court did not err in permitting the evidence and affirmed the conviction.*

## Review Questions

1. Who is the plaintiff in the *U.S. v. Orr* case? Who is the defendant?

2. What is the issue in the *Orr* case?

3. What is the decision of the U.S. Court of Appeals for the 9th Circuit in the *Orr* case?

4. What is the "felony murder" statute referred to in *McCoy v. State*?

5. What is the arson statute in your state? Are there degrees of arson?

6. What is a misdemeanor? What is a felony in your state? Are there degrees of misdemeanors and felonies in your state?

7. Where is the criminal code in your state located?

8. Do firefighters possess arrest powers in your state?

9. Are firefighters exempt from criminal actions while in the course of their work?

# Chapter 4

# Negligence Actions

Lawyers spend a great deal of time shovelling smoke.

*Oliver Wendell Holmes, Jr.*

## *Learning Objectives*

- To understand the elements of a negligence action.
- To identify areas of potential liability in your fire and emergency service organization.
- To learn methods for avoiding potential liability for negligence.

Numerous actions have been brought against fire and emergency service organizations under the theory of negligence. In these cases, someone was injured and that person argues that the fire or emergency service organization did something wrong or failed to do something it was required to do, thereby directly or indirectly causing injury. Thus the injured person or entity is entitled to recover damages from the fire or emergency service organization. Suppose your house had a minor fire in the basement and you called the fire department. The fire department personnel failed to respond because they didn't feel like taking the vehicle out in the rain. Your house burned to the ground as a result of the fire department not responding to assist you. Was the fire department negligent? Did the fire department have a duty to respond?

Basically a negligence action must prove:

1. Duty. Did the fire service organization owe a duty to the plaintiff and, if so, what standard of care did the fire service organization owe to the plaintiff under the circumstances?

2. Breach. Did the fire service organization or individual firefighter, through their conduct, violate that duty of care?

3. Harm and causation. Did the fire service organization's conduct factually or proximately bring about the actual harm to the plaintiff?

4. Damages. What was the harm?

Under the first component of duty, the general rule is that the law does not impose the duty to act in any way upon individuals. A duty can be created by law, by the relationship between the parties, or by one party creating the harm to another. In the case of most fire and emergency service organizations, the duty to act has been created by statute or law.

The degree of knowledge and professional qualifications are weighed in the category of duty. Children below the age of seven are usually presumed incapable of negligence. Between the ages of seven and fourteen, some courts apply the rebuttable presumption that the child is incapable of negligence. Children performing adult activities, such as operating a snowmobile, airplane, or boat, are usually held to an adult standard. An individual's mental deficiencies are usually not considered. However, physical infirmities are considered in determining the component of duty. Professionals, such as physicians, are held to a professional standard of care.

Fire and emergency service organizations should be aware that virtually all states have established the special rule for infant and children trespassers known as the **attractive nuisance doctrine.**[1] This doctrine provides that the possessor of land is subject to liability for physical harm to children trespassing thereon caused by an artificial condition on the land if: (1) the possessor knows or has reason to know that children are likely to trespass in the location of the danger

**attractive nuisance**
an attractive object or feature—created by someone either on his or her own premises, a public place, or another's premises—that may be reasonably observed as being a source of danger thereby requiring that precautions be taken

[1]Restatement (Second) of Tort § 339.

and (2) the professor knows or has reason to know that the condition will involve an unreasonable risk of death or serious bodily harm to such children and the children, because of their youth, do not discover the condition or realize the risk involved and (3) the burden of eliminating the danger is slight as compared with the risk to children involved and (4) the possessor fails to exercise reasonable care to eliminate the danger or otherwise to protect the children. Given the bells, alarms, sirens, activity, and other items that tend to attract small children to fire stations, this is an especially important area of concern for fire and emergency service organizations.

The second component necessary for a negligence action is a **breach** of the duty of care. The burden of proving that the defendant violated the duty owed is on the plaintiff. The three basic methods of proving a breach of the duty are:

1. *Res Ipsa Loquitur.* "The thing speaks for itself" (circumstantial evidence);

2. Violation of a statute; or

3. Direct evidence of negligence.

In most cases involving fire and emergency service organizations, the breach component is easily achieved by the plaintiff because the duty is created by law, statute, or ordinance. This is often know as *negligence per se.* A breach can be by commission or omission.

The third component in a negligence action is proof of actual or proximate causation and harm. Plaintiffs must prove that they have suffered a harm to themselves or their property. There is usually no recovery for pure economic harm, such as lost profits or wages.

The two basic tests used for this determination are the "but for" test and the "substantial factors" tests. In most cases, the but for test is utilized (this is also known as the *sine qua non* rule). Under this test, the court determines where the harm to the plaintiff would have occurred "but for" the negligent acts of the defendant. The substantial factors rule is used in special cases where, under the but for test, the defendant would escape liability because of other contributing causes.

Where two or more defendants commit negligent acts but it is impossible to show which injuries were caused by which defendant, most courts, under the theory of joint and several liability, hold all of the wrongdoers liable for all of the plaintiff's injuries.[2]

Of importance to fire fighters and EMS personnel is the use of the **rescue doctrine** in determining the foreseeability of the risk. Under this doctrine, a defendant who negligently places a victim in a perilous position owes a duty to the person coming to the aid of the victim. Thus, if the rescuer is injured in effecting a rescue, the defendant will be liable to the rescuer as long as the rescuer exercises reasonable care.[3]

**breach**
the breaking or violating of a law, right, obligation, engagement, or duty either by commission or omission

**rescue doctrine**
an invitation to rescue, with danger present, providing for liability in the case of injuries sustained by the rescuer

---

[2]See *Maddux v. Donaldson*, 363 Mich. 425, 108 N.W.2d 33 (1961).
[3]See *Wagner v. International Railway*, 232 N.Y. 2176, 133 N.E. 437 (1921).

**compensatory damages**
payment awarded to an injured party for the restoration of his or her position prior to injury

**punitive damages**
payment awarded to an injured party over and above the compensatory damages awarded

**assumption of risk doctrine**
a concept whereby one may not redeem for an injury resulting from voluntary entering into a situation known to be dangerous

The last component in a negligence action is damages. Damages usually consist of **compensatory damages** (medical costs, lost wages, future losses, pain and suffering) and, in some states, **punitive damages** (damages provided to deter future actions or to set an example). For damages to be recovered for mental distress brought about by the negligent conduct of the defendant, most states require that the plaintiff prove physical harm.[4]

The defenses that may be available to fire and emergency service organizations to combat a negligence action include the substantive defenses of contributory negligence and assumption of the risk. The procedural defenses may include immunity among other defenses.

Most states have adopted some form of comparative negligence rather than the contributory negligence theory. Under the comparative negligence theory, the plaintiff is permitted to recover his or her damages reduced by the percentage of negligence attributable to the plaintiff. Under a pure contributory negligence theory, if the plaintiff is somewhat negligent, this negligence could bar recovery of damages.

Closely aligned with the theory of contributory negligence is the defense of assumption of the risk (also known as *volent non fit injuria*). The defense of assumption of the risk requires proof that the plaintiff knowingly entered into, or stayed in, a position of danger. In many comparative negligence jurisdictions, the assumption of the risk defense has been abolished as a separate defense and merged into the comparative negligence analysis.

For firefighters and EMS personnel, the **assumption of risk doctrine** generally bars recovery by firefighter or EMS personnel who have accepted an unreasonable risk and have been injured as a result of the voluntary acceptance of such risk. The basis for this doctrine is that the fire and emergency service organization should not be liable for injuries that the firefighter or other related personnel knowingly and voluntarily accepted as part of the job. Risks that the employer possesses no greater knowledge of and which are not unusual in nature should not serve as the basis for the fire service organization's liability. For example, when an individual joins a fire service organization, the individual assumes the risk of working in burning buildings as an inherent part of the job responsibilities. If the individual is injured while fighting a fire inside of a building, the individual will be able to acquire coverage under most workers' compensation laws but could be barred under the doctrine of assumption of the risk for other actions sounding in tort.

Assumption of the risk clauses are often used in contracts for employment. In most circumstances, the fire and emergency service organization or employer cannot be sued where the fire fighter was placed in a position of danger due to the fire service organization's negligence. The fire and emergency service organization or employer must be free of negligence in order for the assumption of the risk defense normally to be applicable.[5]

[4]See *Bosley v. Andrews*, 393 Pa. 161, 142 A.2d 263 (1958).
[5]See *Jackson v. City of Kansas City*, 235 Kan. 278, 680 P.2d 877 (Kan. 1984).

Firefighters and EMS personnel may also be restricted in their ability to recover for injuries caused by fellow employees. Under the Fellow Servant Rule, when an individual is injured by the acts of a co-worker in the course of their employment, the co-worker and fire service organization cannot be held liable. This rule is in effect in only a few states and, in general, has received much criticism.

The nature of the defendant's liability is often at issue in cases involving a fire service organization. Where two or more defendants bring about the harm and it is impossible to separate the portions of harm, courts often find both parties jointly liable. The parties are known as **joint tortfeasors.** In these circumstances, the tortfeasors may sue third parties who may possess some liability for contribution for the damages.

Of importance to fire and emergency service officers and others with management responsibilities is the doctrine of indemnification. This doctrine should be distinguished from contribution. In contribution, the defendants share the financial burden among the joint tortfeasors. Under **indemnification**, the fire service organization accepts responsibility for all damages resulting from the acts of its officers or employees to protect the officer or individual. The defense of immunity is normally based on public policy considerations and is designed to protect the fire and emergency service organization despite the fact that it may have committed a negligent act. There are several types of immunities including interspousal immunity, parent-child immunity, charitable immunity and a variety of types of governmental immunities. The varying degrees of governmental immunity normally exist at the federal, state, and municipal levels of government service. The governmental immunity area is one area in which fire service organizations may acquire protection.

The federal government usually may not be sued in tort except to the extent that it has consented to suit by the enactment of the Federal Tort Claims Act. Under this act, the federal government will be liable where "the United States, if a private person, would be liable to the claimant in accordance with the laws of the place where the act or omission occurred."[6] The Federal Tort Claims Act specifically states that the United States should not be liable for assault, battery, false imprisonment, and a variety of other acts.[7] The federal government is not liable under the act for performance or nonperformance of a *discretionary* function or duty on the part of one of its agencies or employees.[8] A distinction needs to be drawn between the functions at the planning level as opposed to the functions at the operations level. Liability under the Federal Tort Claims Act exists only as to the functions of the operations level.[9]

**joint tortfeasors**

two or more persons jointly or severally liable in tort for the same injury to person or property

**indemnification**

relating to corporate affairs where the directors or officers are named in litigation and the expenses incurred are paid by the corporation

---

[6]28 U.S.C.A. § 1346.
[7]28 U.S.C.A. § 1680 (h).
[8]28 U.S.C.A. § 2690 (a).
[9]*Dalehite v. U.S.,* 346 U.S. 15, 73 S.Ct. 956, 97 L.Ed. 1427 (1953).

On a state level, the doctrine of sovereign immunity has long been established. Thus, a fire or emergency service organization working as a state governmental function may be sued for torts committed by its employees and agencies only to the extent that it has, by statute or law, consented to such suits. All states, in varying degrees, have consented to negligence actions.

Municipal corporations have never enjoyed complete immunity from tort liability as with the federal and state governmental entities, because the municipality performs functions that are both governmental and proprietary in nature. Traditionally, municipalities, city governments, and fire service organizations have been held liable for torts committed by their employees in the performance of proprietary functions but have not been held liable where a tort arises out of the performance of a governmental function.

The doctrine of sovereign immunity has been dissipating rapidly. In the past, fire service organizations often relied heavily on the doctrine of sovereign immunity to avoid negligence actions being brought against their organization. Although this doctrine is still viable in many jurisdictions, fire service organizations can no longer rely solely on the sovereign immunity doctrine to exempt them from legal actions.

(This case has been edited for the purposes of this text.)

Lynne **SPENCE** and Victor Spence, Plaintiffs

v.

**ASPEN SKIING COMPANY**, a Colorado Corporation, and John Doe, Defendants.
United States District Court, D. Colorado
May 19, 1993
820 F.Supp. 542

NOTTINGHAM, Judge.

This case involves a patient who was treated by an emergency medical technician employed by the owner of a ski area. The patient brought a personal injury action against the owner. The jury returned the verdict in favor of the owner and the patient filed a motion for a new trial.

While skiing at the ski areas owned by the Aspen Skiing Company, Ms. Spence became dizzy and lightheaded. Someone called the ski patrol, and a patrolman named Steven Schreiber eventually arrived on the scene. Mr. Schreiber, an employee of the Aspen Skiing Company, was an emergency medical technician (EMT). He gave Ms. Spence an intravenous (IV) solution. At some point after the IV commenced, Ms. Spence began to experience redness, swelling, and other symptoms in the arm through which she received the IV. The symptoms were serious enough that some surgery was eventually required, and Ms. Spence lost some use of this arm.

The Plaintiff, Ms. Spence, claimed that the EMT employed by the defendant was negligent in inserting the IV and in tending to it after it was inserted.

The jury was confronted with conflicting evidence on the issues of whether the EMT's conduct was negligent, and if it was, what damages the Plaintiff sustained as a result of that negligence. The jury concluded that the defendant was negligent, that its negligence was a cause of Plaintiff's damages, and that those damages mounted to $35,500.

The jury was also informed that the Plaintiff had suffered from hypoglycemia since the mid-1970's for which she had received medical treatment. She was instructed that she follow a prescribed regimen concerning what she should eat and when she should eat it. Evidence showed that Plaintiff had deviated from this regimen at the time when she experienced the dizziness and light-headedness and that the dizziness and light-headedness were symptoms of a hypoglycemic episode. Defendant claimed that her failure to follow the regimen prescribed by her doctors constituted negligence which contributed, in large part, to any damage which she sustained.

The Court instructed the jury to consider Plaintiff's negligence. The jury found that the total amount of damages sustained by the Plaintiff was $35,500 but that 95% of Plaintiff's damage was caused by her own neglect and only 5% was chargeable to Defendant, Aspen Skiing Company. In accor-

dance with Colorado Law, the court entered judgment for the defendant and dismissed the case.

The Plaintiff filed a motion for new trial. The United States District Court, District of Colorado held that the Plaintiff's alleged failure to follow dietary regimen prescribed to treat her hypoglycemia was not the proximate cause of the arm injury she sustained but that the EMT was negligent in administering the IV solution. The Court concluded that a new trial was not required, but vacate the judgment previously entered for the defendant and enter judgment in favor of Plaintiff.

VACATED.

*Decision Explanation:*
*The motion for a new trail was denied, however, the court eliminated the earlier judgment that dismissed*

*the case because of the 95% caused by the plaintiff's own negligence and entered a judgment in favor of the plaintiff for $35,000.*

---

(This case has been edited for the purposes of this text.)

Michael K. JOHNSON, Appellant,

v.

**DISTRICT OF COLUMBIA**, et al., Appellees
District of Columbia Court of Appeals
Argued March 21, 1989
Decided Sept. 38, 1990
580 A.2d 140

STEADMAN, Associate Judge.

This case involves a tort action brought against the District of Columbia relating to the provision of the emergency ambulance services.

On March 1, 1987 Darryl Johnson telephoned the District's 911 number at the request of Ms. Sadie Tolliver, with whom he lived. Ms. Tolliver had informed Mr. Johnson that she felt weak and was in need of an ambulance. Darryl Johnson then called for emergency assistance. The emergency tele-

phone dispatcher responded that he would send an ambulance for Ms. Tolliver. Soon after Mr. Johnson placed the telephone call, Ms. Tolliver collapsed of a heart attack. When no ambulance arrived for "[a]t least ten to fifteen minutes" after the first telephone call, Mr. Johnson made a second call and was again told that an ambulance was "on the way." No ambulance arrived and Miguel Johnson, Darryl Johnson's brother made a third call "[a]pproximately twenty to thirty minutes" after the first call. Firefighters of the Emergency Ambulance Division ("EAD") finally arrived "approximately thirty or more minutes" after the first call had been made.

The only equipment the firefighters arrived with was an oxygen mask and a mouth-to-mouth resuscitation mouthpiece. They walked slowly to the house, despite Mr. Johnson's entreaties, and acted "casually and slowly" in approaching Ms. Tolliver. One firefighter lifted the victim's eyelid with the rubber antenna of his radio rather than touch her. The same firefighter made an undirected comment to "move her to the living room," which Mr. Johnson and his brother Miguel, did. The firefighters then administered cardiopulmonary resuscitation. Some time after the firefighters had arrived, certified medical technicians of the trauma unit arrived and attached the victim to a machine. These technicians rushed the victim to the hospital where she died. A doctor stated that he could have saved her if they had gotten her there a few minutes sooner, he could have saved her life.

Michael K. Johnson, Appellant, as a personal representative of the estate of Ms. Tolliver and as next friend, next of kin and natural father of Taujhis Alhazz Tolliver, a minor child and son of Ms. Tolliver, sued the District of Columbia in tort.

The trial court without opinion dismissed Appellant's amended complaint and granted summary judgment for the District. Johnson appealed.

The Appellant argued that a special relationship arose at any one of the three points in time: first, when the telephone dispatcher first assured that help would be sent; second, when the dispatcher again assured that help would be sent; and third, after the firefighting crew arrived at the scene.

The Court of Appeals stated that the only relevant issue was whether any affirmative acts of the firefighters worsened Ms. Tolliver's condition. The appellant had to show that some act of the firefighters in administering emergency medical assistance to Ms. Tolliver actually made Ms. Tolliver's condition worse than it would have been had the firefighters failed to show up at all or done nothing after their arrival.

The court concluded that the District was not entitled to summary judgment. The case was REMANDED.

*Decision Explanation:*
*The court of appeals reversed the district court's decision in granting a summary judgment for the defendants and sent the case back to district court for trial.*

(This case has been edited for the purposes of this text.)

James Quentin **TAYLOR**
v.
Terry Kenneth **ASHBURN**
Court of Appeals of North Carolina
Nov. 16, 1993
445 S.W. 2d 46

GREENE, Judge.

On September 25, 1989, plaintiff (Taylor) was driving his automobile on a road in North Carolina. Around the same time, defendant (Ashburn), a fire engineer, was operating a fire truck owned by the City of Winston-Salem (the City) and was responding to a call at a high rise housing complex. The fire truck had its emergency equipment —siren, flashing lights, and horn—in full operation. At an intersection, plaintiff's automobile and the fire truck driven by the defendant collided.

On January 21, 1992, plaintiff filed a complaint against defendant alleging that the accident was the result of defendant's negligent operation of the fire truck and caused plaintiff "substantial bodily injury, property loss, loss of income and other incidental damages." In his complaint, plaintiff did not specify whether he was suing the defendant in his individual capacity or in his capacity as a fire engineer for the City. Plaintiff alleged in his complaint that, ". . . [d]efendant was operating a . . . fire truck owned by the City of Winston-Salem and was operating said vehicle with the permission of the City . . . in connection with his employment as a fireman and was in the course and scope of his employment and agency."

On April 9, 1992, defendant filed an amended answer pleading the affirmative defenses of governmental immunity for any claims resulting in damages up to and including $1,000,000, and of immunity from liability for acts committed in the course and scope of defendant's capacity as a public officer. On October 5, 1992, based on his affirmative defenses, filed a motion for summary judgment which was denied by the trial court.

The issue on appeal is whether the plaintiff's complaint, which alleges that defendant was operating a fire truck in the course and scope of his employment as a fireman for the City when the accident occurred, constitutes suing the defendant in his official capacity so that he shares in the City's governmental immunity.

Under the doctrine of governmental immunity, a municipality is not liable for the torts of its officers and employees if the torts are committed while they are performing a governmental function, which includes the organization and operation of a fire department. Governmental immunity protects the governmental entity and its officers or employees sued in their "official capacity." Although a plaintiff generally designates in the caption of his or her complaint in what capacity a defendant is being sued, this caption is not determinative on whether or not a defendant is actually being sued in his or her individual or official capacity. The court must inspect the complaint as a whole to determine the true nature of the claim. If the plaintiff fails to advance any allegations in his or her complaint other than those relating to a defendant's official duties, the complaint does not state a claim against a defendant in his or her individual capacity, and instead, is treated as a claim against defendant in his official capacity.

In this case, plaintiff's complaint does not mention the words "individual capacity" or the words "official" or "official capacity." Plaintiff alleges that "[d]efendant was operating . . . a fire truck owned by the City . . . with the permission of the City . . . in connection with his employment . . . and was in the course and scope of his employment and agency." The allegations in plaintiff's complaint concern only defendant's actions while performing his official duties as a fire engineer of driving a fire truck for the City and responding to an emergency call. After review of this language and the complaint as a whole, we hold that plaintiff has asserted a negligence claim against defendant in his official capacity alone. Therefore, defendant shares in the City's governmental immunity. Because defendant has met his burden of showing that plaintiff cannot surmount defendant's affirmative defense of governmental immunity which bars plaintiff's claim, the trial court erred in denying defendant's motion for summary judgment. We remand for entry of summary judgment for defendant.

REVERSED AND REMANDED.

MARTIN and JOHN, JJ., concur.

*Decision Explanation:*
*The court of appeals found that the defendant was covered under the city's governmental immunity and returned the case to district court ordering the court to permit the motion for summary judgment for the fire engineer.*

## Review Questions

1. Are there any particular laws or statutes affecting emergency medical technicians in your state?

2. Does your state possess a good Samaritan statute?

3. What are the elements of a negligence action?

4. What is the decision of the court in the *Aspen* skiing case?

5. Does your department possess a 911 system? Enhanced?

6. What are the potential liabilities in a failure to respond?

7. What is the issue in the *Johnson* case?

8. Are there specific laws or regulations regarding volunteer firefighters in your state?

9. Are there specific laws or regulations regarding emergency equipment, such as sirens, in your state?

10. What is the issue in the *Taylor v. Ashburn* case?

11. What is the doctrine of governmental immunity?

12. Do you have any laws or regulations in your state addressing the attractive nuisance doctrine?

13. Provide an example where *Res Ipsa Loquitur* could apply.

# Discrimination Based on Title VII of the Civil Rights Act

Law school taught me one thing: how to take two situations that are exactly the same and show me how they are different.

*Hart Pomerantz*

## Learning Objectives

■ To understand the scope of the civil rights acts.

■ To understand the areas of potential discrimination.

■ To learn the potential liabilities for fire and emergency service organizations in this area.

**Civil Rights Act of 1964**
legislation barring discrimination based on race, color, religion, national origin, or sex in virtually all settings

The enactment of the **Civil Rights Act of 1964**,[1] which barred discrimination on race, color, religion, national origin, or sex in virtually all settings propelled workplace discrimination to the attention of the American public. Of particular importance in the employment setting is Title VII of the Civil Rights Act of 1964.[2] Title VII bars all types of discrimination based on any of the protected classes (race, religion, color, national origin, or religion). Thus, fire service organizations are required to create a nondiscriminatory work environment.

The purpose of Title VII of the Civil Rights Act of 1964 is to remove the artificial, arbitrary, and unnecessary barriers to employment when such impediments operate invidiously to discriminate against individuals on the basis of racial or other impermissible classifications.[3]

Title VII of the Civil Rights Act of 1964 was amended in 1972 by the Equal Employment Opportunity Act[4] and most recently by the Civil Rights Act of 1991. The Civil Rights Act of 1991 provides significant changes to Title VII of the Civil Rights Act of 1964 including the following:

1. Recovery of compensatory and punitive damages is now permitted and claims can be tried before a jury. Damages are capped on a sliding scale dependent on the size of the employer, and damages are expressly unavailable in disparate impact litigation.

2. The meaning of "business necessity" and other burden of proof rules in disparate impact litigation are restored to their previous status in most instances.

3. The adjustment of test scores or the use of different cutoff scores on employment-related tests on the basis of race, sex, color, religion, or national origin is prohibited.

4. Employment actions may now violate the Civil Rights Act even though the same action would have been taken absent discriminatory motive. In these cases, known as *mixed motive* cases, remedies are limited to injunctive relief, declaratory relief and attorney's fees.

5. Extraterritorial application of Title VII has been expanded and clarified.

6. Expert witness fees cannot be awarded in addition to attorney fees.

7. The time limitation for filing suits against the federal government when the federal government is the employer has been extended to ninety days.

[*Note:* The Civil Rights Act of 1991 also affects the Americans with Disabilities Act (ADA), the Rehabilitation Act of 1973, the Age Discrimination in Employment Act of 1967, and other laws.]

---

[1]42 U.S.C.A. § 2000e-2.
[2]Id.
[3]*Griggs v. Duke Power Co.*, 401 U.S. 424 (1971).
[4]P.L. 92–261 (Mar. 24, 1972) 86 Stat. 103.

**EEOC**
Equal Employment
Opportunity
Commission

The agency charged with the enforcement of the Civil Rights Act is the Equal Employment Opportunity Commission (also known as the **EEOC**). The EEOC is charged with investigating any complaints by individuals of a protected class who believe that they have been discriminated against in the workplace. The EEOC is empowered to act and to investigate employers, records, and even go so far as to represent the employee against the employer. It should be noted that the EEOC, like many government agencies, promotes the use of alternate dispute mechanisms to settle such conflicts between employers and employees.

A fire and emergency service organization falls within the jurisdiction of Title VII if:

- The organization employs fifteen or more employees;

- Employees have been employed for each working day in each of twenty or more calendar weeks in the current or preceding calendar year;

- The organization engages in an industry affecting commerce.[5]

The *employer* includes employment agencies, labor organizations, and employees of state and local governments, governmental agencies, political subdivisions, and other related divisions with certain exceptions.

As defined by Title VII, it is unlawful for an employer to:

- Fail or refuse to hire or to discharge any individual, or otherwise to discriminate against any individual with respect to his or her compensation, terms, conditions, or privileges of employment, because of such individual's race, color, religion, sex, or national origin; or

- Limit, segregate, or classify his or her employees or applicants for employment in any way that would deprive or tend to deprive any individual of employment opportunities or otherwise adversely affect his or her status as an employee, because of such individual's race, color, religion, sex, or national origin.[6]

The EEOC distinguishes employer policies that are discriminatory "on their face" (known as *disparate treatment*) from rules and policies that are facially neutral but nonetheless have a disproportionate impact on protected classes (known as *disparate impact*). Fire and emergency service organizations that enforce policies and rules that adversely impact or directly discriminate against protected class firefighters or EMS personnel may be violating Title VII.

Of particular importance to fire and emergency service organizations is the job-relatedness standard within Title VII. Title VII prohibits job requirements that have a disparate impact on individuals as a result of their protected characteristics under Title VII, unless the fire service organization can prove that the job requirement is job related. Following are some of the job requirements that have been the source of EEOC litigation:

---

[5]42 U.S.C.A. § 2000e(b).
[6]42 U.S.C.A. § 2000e-2(a).

- *Education.* As a general rule, the extent to which an employer may use education as a job requirement varies with the public's interest of health and safety in the performance of the job.[7] Given the expansive public interest in health and safety in the fire service, this requirement is normally considered job related.

- *Health Requirements.* A fire service organization is permitted to reject a prospective firefighter who fails the physical examination as long as the physical impairment or disability would prevent the applicant from performing the basic job functions. Fire service organizations must provide reasonable accommodations and meet the other requirements of the ADA (see section on the Americans with Disabilities Act).

- *Strength Requirements.* A fire service organization may reject an applicant who fails a strength test if strength is a bona fide, job-related qualification (used for measurement rather than in the abstract).[8]

- *Height and Weight Requirements.* Height and weight specifications that deny equal employment opportunities to all groups of individuals who are protected under Title VII are unlawful unless such qualifications are necessary to perform the job in question.[9]

- *Work Experience.* A fire service organization may require previous work experience as a valid and relevant job requirement so long as this requirement relates to the successful performance of the job in question[10] or is recognized as a business necessity.

- *Appearance and Dress Requirements.* As a general rule, fire service organizations cannot reject an applicant because his or her appearance is typical of minority candidates. However, fire service grooming regulations usually do not violate Title VII even though the proscribed dress standards may differ somewhat for male and female as long as the dress requirements are reasonably related to the fire service organization's needs and are commonly accepted norms in the fire service.[11]

- *Hair Requirements.* Given the nature of the activity in the fire service, the length of hair and facial hair issues are normally addressed within the realm of personal safety. (For example, beards are normally prohibited because a firefighter cannot achieve an adequate seal on the self-contained breathing apparatus. As a general rule, the EEOC generally considers it a violation of Title VII whenever an employer's grooming code requires men to have different hair styles than women.[12]

---

[7]See *Townsend v. Nassau County Medical Center*, 558 F.2d 117 (CA-2, 1977).
[8]*Dothaard v. Rawlinson*, 433 U.S. 321 (1977).
[9]*Davis v. County of Los Angeles*, 566 F.2d 1334 (CA-9, 1977).
[10]*Griggs v. Duke Power*, supra.
[11]See *Caroll v. Talman Fed. Sav. & Loan Assn.*, 604 F.2d 1028 (CA-7, 1979).
[12]See EEOC Decision No. 72-2179 (1972).

- *Credit Requirements.* Requirements that an applicant possess a good credit record are usually unlawful unless business necessity is proven.

- *Arrest and Criminal Records.* Because of the type of work performed by fire service organizations, a "clean" arrest record is usually required for safety and business reasons. Because of the disproportionate impact on black candidates, the EEOC considers a requirement that applicants have no previous arrest record as an unlawful job qualification unless the employer can show the necessity in the operations of the particular business. Additionally, the refusal to hire an applicant on the basis of a conviction of a crime may be unlawful unless the employer is able to show that it has first considered the circumstances surrounding the particular case and employment of the individual would be inconsistent with the safe and efficient operation of the business.[13]

- *Military Record.* A fire and emergency service organization that disqualifies a prospective firefighter because he or she has received a less-than-honorable discharge from the military may be guilty of unlawful discrimination unless the fire service organization can prove that the rejection is related to job performance.[14]

- *Alienage.* Citizenship requirements are unlawful whenever they have the purpose or effect of discriminating against an individual on the basis of national origin.

- *Language Requirements.* Proficiency in English as a job requirement may be unlawful unless business necessity or job relatedness is shown.

- *No Spouse Requirements.* Generally, fire and emergency service organization policies prohibiting the hiring of a spouse of a firefighter is lawful so long as the rule is neutral as to sex.

- *Marriage Status.* Policies that forbid the hiring of married women but do not forbid the hiring of married men are usually unlawful.

- *Pregnancy.* With the amendment to Title VII in 1978, employers may not discriminate against women who are affected by pregnancy, childbirth, or related medical conditions. Pregnancy discrimination is encompassed under the sex prohibition under Title VII and many states have adopted pregnancy discrimination statutes. (Also see Family and Medical Leave Act).

- *Sex Status.* It is unlawful for a fire and emergency service organization to hire a certain number of male firefighters and a certain number of female firefighters unless the fire and emergency service organization can show that this requirement is a bona fide occupational qualification.

---

[13]EEOC Decision No. 78-10 (1977).
[14]*Dozier v. Chupka,* 395 F.Supp. 836 (DC Ohio 1975).

- *Religious Convictions.* Fire and emergency service organization are required make reasonable accommodations to the religious needs of employees and may not reject candidates because of their religious needs unless such accommodations would create an undue hardship on the organization.

- *Age Requirements.* Employment discrimination on account of age between 40 and 70 violates the Age Discrimination in Employment Act. Age restrictions may also violate Title VII where the restrictions apply only to employees within a protected class.

In testing or other selection procedures for new candidates, if the test adversely affects a disproportionate number of persons protected under Title VII, it may be an unlawful selection method unless the test is related to the successful performance of the job or the test is used is necessary to the operations of the fire service organization. The EEOC has accepted three methods of validating selection tests:

- Criterion-related validations. Compares criteria with successful job performance.

- Construct validation. Compares mental and psychological traits.

- Content validation. Test closely duplicates job performance.

Fire service organizations should be aware that Title VII protections extend to work assignments, promotions, transfers, discipline, and other areas of the employment setting. Fire service organizations must maintain a work environment free of racial, sex, religious, or ethnic harassment. The fire service organization is under a duty to take reasonable measures to attempt to control or eliminate overt expressions of harassment in the workplace. For example, a fire service organization may possess a duty to remove sexually explicit photographs, calendars, and magazine centerfolds depicting females from the firehouse walls or lockers and establish policies to prevent future posting. (See discrimination in Chapter 9 for EEOC claims procedures).

A claimant under Title VII must file a claim with the EEOC within 180 days from the alleged discriminatory event or within 300 days in states with approved enforcement agencies such as the Human Rights Commission. The EEOC has 180 days to investigate the allegation and to sue the employer or issue a right-to-sue notice to the employee. The employee will have ninety days to file a civil action from the date of this notice.[15]

The original remedies provided under Title VII included reinstatement, with or without back pay, and reasonable attorney fees and costs. Title VII also provided for protection against retaliation against the employee for filing the complaint or others who may assist the employee in the investigation of the com-

---

[15]S. Rep. 101–116, 21; H. Rep. 101–485 Part 2, 51; Part 3, 28.

plaint. The remedies were designed to make the employee "whole" and to prevent future discrimination by the employer.

With the passage of the Civil Rights Act of 1991, the remedies provided under the Title VII were modified. Damages for employment discrimination, whether intentional or by practice that has a discriminatory effect, may include hiring, reinstatement, promotion, back pay, front pay, reasonable accommodation, or other action that will make an individual "whole." Payment of attorneys' fees, expert witness fees, and court costs were permitted and jury trials were allowed.

Compensatory and punitive damages were also made available where intentional discrimination is found. Damages may be available to compensate for actual monetary losses, for future monetary losses, for mental anguish, and for inconvenience. Punitive damages are also available if an employer acted with malice or reckless indifference. The total amount of punitive damages and compensatory damages for future monetary loss and emotional injury for each individual is limited, based on the size of the employer.

| Number of Employees | Damages Will Not Exceed |
|---|---|
| 15–100 | $ 50,000 |
| 101–200 | 100,000 |
| 201–500 | 200,000 |
| 500 or more | 300,000 |

Punitive damages are usually *not* available against state or local governments and thus related fire service organizations.

Although Title VII is the most visible area of potential discrimination in the fire service, other potential civil rights liabilities also exist for fire service organizations. A common federal civil rights action against governmental agencies arises under 42 U.S.C. § 1983 and involves a claim that an individual was deprived of a constitutional right by an individual or organization acting under the color of law. These actions are often based on the same conduct that forms the basis for a tort action under a state law. A § 1983 action can be brought for misuse of legal authority, even in violation of state law, and there is no general requirement that a state bring the initial action.

This statute provides that:

> Every person who, under the color of any statute, ordinance, regulation, custom, or usage, of any State or Territory, subjects, or causes to be subjected, any citizen of the United States or other person within the jurisdiction thereof to the deprivation of any rights, privileges, or immunities secured by the Constitution and laws, shall be liable to the party injured in an action at law, suit in equity, or other proper proceeding for redress.[16]

---

[16]42 U.S.C. § 1983.

Section 1983 creates no rights for the individual in and of itself but is a vehicle to redress violations of constitutional rights and certain federal statutes. Section 1983 actions have been prevalent to date against police departments and various other governmental agencies for such actions as illegal searches, use of force issues, and electronic surveillance issues.

Fire service officers should be aware that superior officers may be liable for participation in unconstitutional conduct, for grossly negligent failure to train or discipline, or for the promulgation of policies that cause constitutional violations. However, officers cannot be liable under § 1983 under the doctrine of *respondeat superior*. Additionally, a superior officer who leads or directs others who commit constitutional violations may be liable even if he or she is not at the scene.[17]

In most circumstances, states and state fire and emergency service agencies are not considered proper defendants in § 1983 actions, however state fire officers may be proper in certain circumstances. Local governments and fire or emergency services organizations attached to local governments can be liable for § 1983 actions if they adopt or tolerate policies or customs that have caused a wrong. Again, the doctrine of respondeat superior is not applicable in these types of circumstances. (See Exhibit at end of cases.)

---

[17]See *Specht v. Jensen*, 832 F.2d 1516 (10th Cir. 1987) (Superior officer who told others to "take care of it" set in motion a series of constitutional violations).

(This case has been edited for the purposes of this text.)

In re the Matter of Dominic J. **GARGANO** and Stephen S. Reffel, Complainants-Appellees
v.
**COLORADO CIVIL RIGHTS COMMISSION**, a Colorado State
Agency, Appellee,
v.
**NORTH WASHINGTON FIRE PROTECTION DISTRICT**, Respondent-Appellant
Colorado Court of Appeals
Div. II
Sept. 3, 1987
Rehearing Denied Oct. 29, 1987
754 P.2d 393

STERNBERG, Judge.
This case involved two firefighters who believed they were not hired for positions they were qualified for because of their handicaps.

The two Complainants Dominic Gargano and Stephen Reffel sought employment as entry-level firefighters for the North Washington Fire Protection District (District). Both passed an initial physical agility test, and thereby became qualified to take oral and written exams, which both also passed. They were then placed on an eligibility roster compiled by the District's Civil Service Commission.

All ten applicants on the eligibility roster were directed to take a physical examination to be conducted by the District's physician. The physician examined each applicant, and reported his findings to the District's Board of Directors.

The District found the complainants ineligible for a firefighter position because of medical reasons; accordingly, it removed their names from the eligibility list on May 8, 1981, and appointed other applicants on the roster to fill the two positions that were open. Had their names not been removed from the list, they could have been appointed.

The physical condition upon which the District based its removal of Gargano from the eligibility roster was reduced visual acuity; he had to wear eyeglasses or contact lenses for accurate vision. Reffel was disqualified because his right knee had had two surgeries performed on it as a result of a dislocation incurred while playing basketball. Although Reffel had no current trouble with his knee, after reviewing reports from several physicians who had examined Reffel, the District concluded that the condition of Reffel's knee disqualified him for the position of firefighter.

Gargano and Reffel filed complaints of alleged discrimination with the Commission, which investigated and determined that there was probable cause for finding a violation. The formal complaint filed by the Colorado Civil Rights Commission (Commission) accused the District of:

(1) Refusing to hire Gargano and Reffel, who were otherwise qualified to be firefighters, because of a handicap which did not actually disqualify them from the job and which did not actually have a significant impact on the job in violation of s 24-34-492(1)(a), C.R.S.; and

(2) Having a policy or practice of enforcing physical requirements for job positions which precludes individual consideration of job applicants in violation of s 24-34-402(1)(a).

At the request of all counsel, a hearing officer decided the case on cross-motions for summary disposition with all pleadings, depositions, and exhibits previously filed and taken to serve as evidentiary support for the cross-motions. The hearing officer held that the stipulated facts supported a prima facie case of discrimination because of the disparity in physical qualification standards used for applicants for entry level firefighter positions, as compared with standards used for current firefighters.

The hearing officer thus granted the complainants' motion for summary disposition, except as to the limited issue whether removal of Gargano and Reffel from an eligibility roster was based on business necessity or other job-related justification. The hearing officer also found that a genuine issue of fact existed whether the District had applied medical disqualifying criteria in an automatic manner, precluding individual consideration.

The Commission reviewed the hearing officer's order. It ordered the District to hire the Complainants to its next two available entry-level firefighter positions, and to pay them the value of all backpay and fringe benefits to which they would have been entitled, plus interest, had they been hired in June 1981, offset by amounts earned in the interim, stipulated to be $68,556.51 for Reffel and $1,923.07 for Gargano.

The District appealed the order of the Commission. The Colorado Court of Appeals disagreed with the rulings of the hearing officer and Commission because the complainants failed to establish that they were "handicapped" within the meaning of s 24-34-402(1)(a). Neither Gargano nor Reffel alleged that their particular physical impairments limited them in any way other than in obtaining employment as a firefighter. The Court held that the Commission's conclusions and remedies constituted an abuse of discretion and they reversed the order.

SMITH and METZGER, J.J., concur.

Decision Explanation;
The court reversed the decision of the hearing officer and commission, ordering the fire district to hire the individuals and provide back pay.

(This case has been edited for the purposes of this text.)

Gary D. **MICU**, Plaintiff-Appellant,

v.

**CITY OF WARREN**, Defendant-Appellee.

Court of Appeals of Michigan
Submitted June 11, 1985
Decided Dec. 16, 1985
382 N.W.2d 823

V.J. BRENNAN, Judge.

This case involves an action of discrimination based on height for which a potential firefighter was denied employment.

On April 5, 1979, the Plaintiff, Gary Micu, filed an employment application with the City of Warren for the position of firefighter. The city had a minimum height requirement of 5 feet 8 inches for firefighters. On the employment application, Plaintiff listed his height as 5 feet 8 inches. Actually, his height was between 5 feet 6 inches and 5 feet 6-1/2 inches. Besides completion of the employment application, to be placed on the eligibility list a firefighter applicant for the city had to meet certain "preliminary requirements," undergo written, verbal and agility tests, and be examined by the city physician. Because the Plaintiff's application did not reveal any deficiencies in the preliminary requirements, Plaintiff was allowed to complete the application process.

On July 23, 1979, during the time the Plaintiff was still involved in the application process with the City of Warren, Plaintiff accepted an unspecified position with the Farmington Hills Fire Department.

Of all the preliminary requirements and tests established by the City of Warren, Defendant, Plaintiff met and passed each with the exception of the minimum 5 feet 8 inch height requirement.

After the many tests, Plaintiff was examined by the city physician. When his height was measured and found to be less than 5 feet 8 inches, the examination ceased. In a letter allegedly dated December 18, 1979 (according to Plaintiff) or December 2, 1979 (according to Defendant), Plaintiff was informed by the Warren Police and Fire Civil Service Commission that he had not successfully completed the examination for firefighter because he did not meet the minimum height requirement. According to Defendant, Plaintiff appealed the refusal of his application to the Warren Civil Service Commission and a hearing was held on December 18, 1979, during which the denial of Plaintiff's application was affirmed.

On April 8, 1981, Plaintiff filed a two-count complaint against the City of Warren, alleging height discrimination in their employment practices, contrary to the prohibitions of the Elliott-Larsen Civil Rights Act and intentional infliction of emotional distress. Plaintiff prayed for monetary damages and attorney fees.

On May 19, 1981, in its answer, Defendant admitted that "it was an employer and Plaintiff was an applicant for the position of firefighter within the meaning of the Michigan Elliott-Larsen Act", but denied that height was not exempted as a bona fide occupational qualification. Rather, as its only affirmative defense, Defendant alleged Plaintiff was guilty of fraud by misrepresentation on his application submitted for firefighter as to his true height and is estopped to claim discrimination by reason of his fraudulent misrepresentation.

On September 11, 1981, Plaintiff's attorney filed a pretrial factual statement of Plaintiff's complaint and said that Plaintiff stated he was 5'8" tall because he believed he was 5'8" tall. On September 11, 1981, Plaintiff also moved to strike Defendant's affirmative defense and moved for partial summary judgment, contending that fraud was not an available defense to defendant "if defendant relies upon a provision contrary to law." Plaintiff, therefore, prayed that summary judgment be granted on Count I, Plaintiff's height discrimination claim.

On November 10, 1981, Defendant filed answers to interrogatories and stated that a rational basis existed for the height requirement.

The trial court denied Plaintiff's motion to strike the affirmative defense and for partial summary judgment, finding that the alleged affirmative defense (Plaintiff's misrepresentation of height) was a legally sufficient defense.

On October 11, 1983, Plaintiff filed a motion and requested that the trial court make legal findings to determine whether Elliott-Larsen supersedes the police and fire civil service act with respect to height requirements and moved for declaratory judgment. On November 4, 1983, the Defendant countered Plaintiff's motion and moved for summary judgment on both Counts I and II.

The trial court denied Plaintiff's motion and granted Defendant's motion for summary judgment on the ground that the fire and police civil service act exclusively controlled the hiring of firefighters. The court also determined that the fire and police civil service regulations, that the Defendant had adopted a rule establishing a height requirement, that height requirements "have been deemed reasonable for the proper and efficient running at a fire department," and that it was not "his duty" to find the height requirement "unenforceable."

The Court of Appeals of Michigan found that the Elliott-Larsen Act was not exclusive and the language of the Act requires that Defendant not discriminate on the basis of height unless such is established to be a bona fide occupational qualification. The Court found that the trial court erred in granting Defendant summary judgment on the ground that the Elliott-Larsen Act did not apply to the situation at bar. The Court affirmed the dismissal of Count ** of Plaintiff's complaint for failure to state a claim.

REVERSED as to Count I. AFFIRMED as to Count II.

*Decision Explanations:*
*The court of appeals reversed the granting of a summary judgment to the defendants on Count 1 and affirmed the dismissal of Count 2. Count 1 would be returned to the district court for trial.*

(This case has been edited for the purposes of this text.)

### FIREFIGHTERS LOCAL UNION
NO. 1784, Petitioner,
v.
Carl W. **STOTTS** et al.
MEMPHIS FIRE DEPARTMENT et al.
Petitioners,
v.
Carl W. **STOTTS**, etc., et al.
104 S.Ct. 2576 (1984)
The Supreme Court

WHITE, Justice.
Respondent Stotts, a black member of petitioner Memphis, Tenn., Fire Department, filed a class action in Federal District Court charging that the Department and certain city officials were engaged in a pattern or practice of making hiring and promotion decisions on the basis of race in violation of *inter alia,* Title VII of the Civil Rights Act of 1964. This action was consolidated with an action filed by respondent Jones, also a black member of the Department, who claimed that he had been denied a promotion because of his race. Thereafter, a consent decree was entered with the stated purpose of remedying the Department's hiring and promotion practices with respect to blacks. Subsequently, when the city announced that projected budget deficits required a reduction of city employees, the District Court entered an order preliminary enjoining the Department from following its seniority system in determining who would be laid off as a result of the budgetary shortfall, since the proposed layoffs would have a racially discriminatory effect and the seniority system was not a bona fide one. A modified layoff plan, aimed at protecting black employees so as to comply with the court's order, was then presented and approved, and layoffs pursuant to this plan were carried out. This resulted in white employ-

ees with more seniority than black employees being laid off when the otherwise applicable seniority system would have called for the layoff of black employees with less seniority. The Court of Appeals affirmed, holding that although the District Court was wrong in holding that the seniority system was not bona fide, it had acted properly in modifying the consent decree.

*Held:*

1. These cases are not rendered moot by the facts that the preliminary injunction purportedly applied only to 1981 layoffs, that all white employees laid off as a result of the injunction were restored to duty only one month after their layoff, and that others who were demoted have been offered back their old positions. First, the injunction is still in force and unless set aside must be complied with in connection with any future layoffs. Second, even if the injunction applied only to the 1981 layoffs, the predicate for it was the ruling that the consent decree must be modified to provide that the layoffs were not to reduce the percentage of black employees, and the lower courts' rulings that the seniority system must be disregarded for the purpose of achieving the mandated result remain undisturbed. Accordingly, the inquiry is not merely whether the injunction is still in effect, but whether the mandated modification of the consent decree continues to have an impact on the parties such that the cases remain alive. Respondents have failed to convince this Court that the modification and the *pro tanto* invalidation of the seniority system are of no real concern to the city because it will never again contemplate layoffs that if carried out in accordance with the seniority system would violate the modified decree. Finally, the judgment below will have a continuing effect on management of the Fire Department with respect to making whole the white employees who were laid off and thereby lost a month's pay and seniority, or who were demoted and thereby may have backpay claims. Unless that judgment is reversed, the layoffs and demo-

tions were in accordance with the law. The fact that not much money and seniority are involved does not determine mootness.

2. The District Court's preliminary injunction cannot be justified either as an effort to enforce the consent decree or as a valid modification thereof.

(a) The injunction does not merely enforce the agreement of the parties as reflected in the consent decree. The scope of a consent decree must be discerned within its four corners. Here, the consent decree makes no mention of layoffs or demotions nor is there any suggestion of an intention to depart from the existing seniority system or from the Department's arrangement with the union. It therefore cannot be said that the decree's express terms contemplated that such an injunction would be entered. Nor is the injunction proper as carrying out the stated purpose of the decree. The remedy outlined in the decree did not include the displacement of white employees with seniority over blacks and cannot reasonably be construed to exceed the bounds of remedies that are appropriate under Title VII. Title VII protects bona fide seniority systems, and it is inappropriate to deny an innocent employee the benefits of his seniority in order to provide a remedy in a pattern-or-practice suit such as this. Moreover, since neither the union nor the white employees were parties to the suit when the consent decree was entered, the entry of such decree cannot be said to indicate any agreement by them to any of its terms.

(b) The theory that the strong policy favoring voluntary settlement of Title VII actions permits consent decrees that encroach on seniority systems does not justify the preliminary injunction as a legitimate modification of the consent decree. That theory has no application when there is no "settlement" with respect to the disputed issue, such as here where the consent decree neither awarded competitive seniority to the minority employees not purported to depart from the existing seniority system. Nor can the injunction be so justified on the bases that is the allegations in complaint had been proved, the District court could have entered an order overriding the seniority provisions. This approach overstates a trial court's authority to disregard a seniority system in fashioning a remedy after a plaintiff had proved that an employer had followed a pattern or practice having discriminatory effect on black employees. Here, there was no finding that any of the blacks protected from layoff had been a victim of discrimination nor any award of competitive seniority to any of them. The Court of Appeals' holding that the District Court's order modifying the consent decree was permissible as a valid Title VII remedial order ignores not only the ruling in *Teamsters v. United States*, 431 U.S. 324, 97 S.Ct. 1843, 52 L.Ed.2d 396, that a court can award competitive seniority only when the beneficiary of the award had actually been a victim of illegal discrimination, but also the policy behind § 706(g) of Title VII of providing make-whole relief only to such victims. And there is no merit to the argument that the District court ordered no more than that which the city could have done by way of adopting an affirmative-action program, since the city took no such action and the modification of the decree was imposed over its objection.

679 F.2d 541, REVERSED.

## OVERVIEW OF EMPLOYMENT DISCRIMINATION LAWS

### Federal Laws

- Equal Pay Act of 1963 (section 6(d) of the Fair Labor Standard Act of 1938, 29 U.S.C. § 206 (d)0) which extends protection against sexual discrimination and requires equal pay for equal work by forbidding pay differentials predicated on gender.

- The Civil Rights Act of 1964 (Title VII) (42 U.S.C. § 2000 et seq.) which prohibits employment discrimination on the basis of race, color, religion, sex, age, or national origin by employers who employ fifteen or more employees and are engaged in industry affecting commerce.

- The Age Discrimination in Employment Act of 1967 (29 U.S.C. § 621–634) which prohibits employment discrimination against persons over the age of forty.

- Rehabilitation Act of 1973 (§ 501, 503, 504 29 U.S.C. § 791, 793, 794) which prohibits discrimination on the basis of disability by programs receiving federal funds or by federal agencies. This law is the precursor of the Americans with Disabilities Act and was created to assist individuals with disabilities in obtaining access to public buildings and enjoy-ing equal employment opportunities.

- Americans with Disabilities Act of 1990 (also known as the ADA) (42 U.S.C. § 12101) prohibits discrimination against qualified individuals with disabilities. The primary purpose of this law was to provide an esti-mated 43 million individuals with disabilities equal access to employ-ment opportunities, programs, services, and activities provided by public and private sector employers.

- Civil Rights Act of 1991 (42 U.S.C. § 1981 et seq.) This act in essence reversed a series of cases decided by the U.S. Supreme Court in 1989. The act reinstated the earlier interpretations of the Civil Rights Act of 1964. The act includes permitting full jury trials and, under certain cir-cumstances, allowing for the recovery for emotional suffering and puni-tive damages.

- Family and Medical Leave Act of 1993 (29 U.S.C. § 2601 et seq.) which requires employers with fifty or more employees to provide eligible employees up to twelve weeks unpaid, job-protected leave for family and medical reasons including activities such as the birth of a child, adop-tion, or care of a spouse, child, or parent.

- The Pregnancy Discrimination Act (42 U.S.C. § 2000 e (k)). This act pro-hibits sexual discrimination and amends the Civil Rights Act of 1964 to include pregnancy, childbirth, and pregnancy-related medical condi-tions as protected against employment discrimination.

- Vietnam Era Veterans Readjustment Assistance Act (38 U.S.C. § 2012). This act requires federal contractors with contracts over $10,000 to hire qualified veterans of any war who have disabilities and specifically qual-ified Vietnam veterans who may or may not have disabilities.

## Kentucky Specific Laws

- KRS 334.040 forbids discrimination in employment on the basis of race, color, religion, national origin, sex or age.

- KRS 207.130 prohibits discrimination in employment because of handi-cap.

- KRS 344.030 forbids discrimination based on pregnancy.

- Kentucky Public Acts, House Bill 23-X (L. 1976) prohibits an employer from attempting to deprive an employee of his or her employment because the employee receives a summons, serves as a juror, or attends court for prospective jury service.

- KRS 433.310 forbids anyone from inducing or enticing any person, who has contracted his or her employer for a specific period of time, to leave the service of the employer during the term without the employer's consent.

- KRS 121.310 forbids employers from interfering with the voting rights of their employees.

- Kentucky Public Acts, H.B. 100 (1978) and KRS 335(B)(2)(1) provide that no person shall be disqualified from public employment, nor shall a person be disqualified from pursuing, practicing, or engaging in any occupation solely because of prior conviction of a crime, unless the crime is one limited to convictions for felonies, high misdemeanors, or crimes involving moral turpitude.

- KRS 38.460 makes it unlawful for any person to willfully deprive a member of the National Guard of his or her employment.

- KRS 337.423 makes in unlawful to discriminate in the payment of wages based on sex.

## Review Questions

1. In twenty words or less, explain the parameters of Title VII of the Civil Rights Act.

2. Who are the protected classes under Title VII?

3. What are the parameters of protection provided by Title VII of the Civil Rights Act as applied to a fire department?

4. What is the issue in the *Gargano* case?

5. What is the issue in the *Micu* case?

6. Explain the concept of disparate treatment.

7. Explain the theory of a hostile work environment.

8. Explain the theory of disparate impact.

9. What is the procedure for filing a claim of discrimination with the Equal Employment Opportunity Commission?

10. Does your fire department fall within the jurisdiction of Title VII?

11. Provide examples of what may constitute unlawful employment discrimination practices.

*Chapter*

# 6

# Americans with Disabilities Act

Skill to do comes of doing.

*Ralph Waldo Emerson*

## *Learning Objectives*

- To gain a basic understanding of the ADA.
- To understand the requirements of the ADA.
- To understand the potential areas of liability under the ADA.

**Americans with Disabilities Act**
the federal statute that protects disabled workers in many environments

**The Americans with Disabilities Act** is divided into five titles, all of which have the potential to substantially impact a covered public or private sector fire service organization. Title I contains the employment provisions that protect all individuals with disabilities who are in the United States, regardless of national origin and immigration status. Title II prohibits discrimination against qualified individuals with disabilities or excluding qualified individuals with disabilities from the services, programs, or activities provided by public entities. Title II includes the transportation provisions. Title III, entitled Public Accommodations, requires that goods, services, privileges, advantages, or facilities of any public place be offered "in the most integrated setting appropriate to the needs of the individual."[1] Title III additionally covers transportation offered by private entities. Title IV addresses telecommunications. Title IV requires that telephone companies provide telecommunication relay services that and television public service announcements produced or funded with federal money include closed captioning. Title V includes the miscellaneous provisions. This title noted that the *Americans with Disabilities Act* does not limit or invalidate other federal and state laws providing equal or greater protections for the rights of individuals with disabilities and addresses related insurance, alternate dispute, and congressional coverage issues. Following is a listing of the five titles under the ADA:

TITLE I—EMPLOYMENT

TITLE II—PUBLIC ENTITIES

TITLE III—PUBLIC ACCOMMODATION

TITLE IV—TELECOMMUNICATIONS

TITLE V—MISCELLANEOUS

Title I of the ADA affects all employers and industries engaged in interstate commerce with fifteen or more employees. Just as fire service organizations were not exempt from the ADA's predecessor, the Rehabilitation Act of 1973, most fire service organizations are subject to the ADA mandates.

Title II, which applies to public services[2] and Title III, which requires public accommodations and services operated by private entities, became effective on January 26, 1992,[3] except for specific subsections of Title II, which went into effect immediately on July 26, 1990.[4] A telecommunication relay service required by Title IV must have been available by July 26, 1993[5].

Title I prohibits covered employers from discriminating against a "qualified individual with a disability" with regard to job applications, hiring, advance-

---

[1]ADA Section 305.
[2]ADA Section 204(a), 42 U.S.C. 12134.
[3]Id.
[4]ADA Section 203(a), 306(a), 42 U.S.C. 12186.
[5]ADA Section 102(a), 42 U.S.C. 12112.

ment, discharge, compensation, training, and other terms, conditions, and privileges of employment.[6]

Section 101 (8) defines a "qualified individual with a disability" as any person

who, with or without reasonable accommodation, can perform the essential functions of the employment position that such individual holds or desires . . . consideration shall be given to the employer's judgment as to what functions of a job are essential, and if an employer has prepared a written description before advertising or interviewing applicants for the job, this description shall be considered evidence of the essential function of the job.[7]

Under the ADA, an individual has a disability if he or she:

- Has a physical or mental impairment that substantially limits one or more of the major life activities of such individual;
- Has a record of such an impairment; or
- Is regarded as having such an impairment.[8]

Title I of the ADA additionally provides that if a qualified fire service organization does not make reasonable accommodation for the known limitations of a qualified individual with a disability, it may be considered to be discrimination.

Section 101 (9) defines a "reasonable accommodation" as:

(a) making existing facilities used by employees readily accessible to and usable by the qualified individual with a disability and includes:

(b) job restriction, part-time or modified work schedules, reassignment to a vacant position, acquisition or modification of equipment or devices, appropriate adjustments or modification of examinations, training materials, or policies, the provisions of qualified readers or interpreters and other similar accommodations for . . . the QID.[9]

The EEOC further defines "reasonable accommodation" as:

(1) Any modification or adjustment to a job application process that enables a qualified individual with a disability to be considered for the position such qualified individual with a disability desires, and which will not impose an undue hardship on the . . . business; OR

(2) Any modification or adjustment to the work environment, or to the manner or circumstances which the position held or desired is customarily performed, that enables the qualified individual with

---

[6]Id.

[7]ADA Section 101(8).

[8]Subtitle A, Section 3(2). The ADA departed from the Rehabilitation Act of 1973 and other legislation in using the term *disability* rather than *handicap.*

[9]ADA Section 101(9).

a disability to perform the essential functions of that position and which will not impose an undue hardship on the . . . business; OR

(3) Any modification or adjustment that enables the qualified individual with a disability to enjoy the same benefits and privileges of employment that other employees enjoy and does not impose an undue hardship on the . . . business.[10]

In essence, fire and emergency service organizations are required to make "reasonable accommodations" for any or all known physical or mental limitations of the qualified individual with a disability unless the fire and emergency service organization can demonstrate that the accommodations would impose an "undue hardship" on the business or that the particular disability directly affects the safety and health of the qualified individual with a disability or others. Included under this section is the prohibition against the use of qualification standards, employment tests, and other selection criteria that tend to screen out individuals with disabilities, unless the fire or emergency service organization can demonstrate that the procedure is directly related to the job function. In addition to the modifications to facilities, work schedules, equipment, and training programs, fire and emergency service organizations must initiate an "informal interactive (communication) process" with the qualified individual to promote voluntary disclosure of specific limitations and restrictions by the qualified individual to enable the fire service to make appropriate accommodations to compensate for the limitation.[11]

**undue hardship**
an action requiring significant difficulty or expense

Section 101 (10)(a) defines **"undue hardship"** as "an action requiring significant difficulty or expense," when considered in light of the following factors:

- Nature and cost of the accommodation
- The overall financial resources and workforce of the facility involved
- The overall financial resources, number of employees, and structure of the parent entity
- The type of operation including the composition and function of the workforce, the administration and fiscal relationship between the entity and the parent[12]

Section 102 (c)(1) of the ADA prohibits discrimination through medical screening, employment inquiries, and similar scrutiny. Fire service organizations are generally *prohibited* from conducting pre-employment physical examinations of applicants and are also prohibited from asking the prospective employee if he or she is a qualified individual with a disability. Qualified fire service organizations are further *prohibited* from inquiring as to the nature or severity of the disability even if the disability is visible or obvious. However, fire and emergency service organizations may ask any candidates for transfer or promotion

---

[10]EEOC Interpretive Guidelines.
[11]Id.
[12]ADA Section 101(10)(a).

who have a known disability whether he or she can perform the required tasks of the new position if the tasks are job related and consistent with business necessity. The fire or emergency service organization is also permitted to inquire as to the applicant's ability to perform the essential job functions prior to employment. The fire or emergency service organization should use the written job descriptions as evidence of the essential functions of the position.[13]

Fire and emergency service organizations may require medical examinations *only* if the medical examination is specifically job related and is consistent with business necessity. Medical examinations are permitted *only after* the applicant with a disability has been *offered* the job position. The medical examination may be given before the applicant *starts* the particular job and the job offer *may be conditioned* on the results of the medical examination *if* all employees are subject to the medical examinations and information obtained from the medical examination is maintained in separate confidential medical files. Fire service organizations are permitted to conduct voluntary medical examinations for current employees as part of an on-going medical health program but again the medical files must be maintained separately and in a confidential manner.[14]

The ADA does *not* prohibit fire and emergency service organizations from making inquiries or requiring medical examinations or "fit for duty" examinations when there is a need to determine whether an employee is still able to perform the essential functions of the job or where periodic physical examinations are required by medical standards of federal, state, or local law.[15]

The fire and emergency service organizations may test job applicants for alcohol and controlled substances *prior* to an offer of employment under Section 104 (d). This testing procedure for alcohol and illegal drug use is *not* considered a medical examination as defined under the ADA. Fire and emergency service organizations may additionally prohibit the use of alcohol and illegal drugs in the workplace and may require that employees not be under the influence while on the job. Fire and emergency service organizations are permitted to test for alcohol and controlled substance use by current employees in their workplace to the limits permitted by current federal and state law. The ADA requires all employers to conform to the requirements of the Drug-Free Workplace Act of 1988. Thus, most existing preemployment and postemployment alcohol and controlled substance programs that are not part of the preemployment medical examination or on-going medical screening program will be permitted in their current form.[16]

Individual employees who choose to use alcohol and illegal drugs are afforded *no protection* under the ADA, however, employees who have success-

---

[13]ADA, Title I, Section 102(C)(2).

[14]ADA Section 102(c)(2)(A).

[15]EEOC Interpretive Guidelines, 56 *Fed. Reg.* 35,751 (July 26, 1991). Federally mandated periodic examinations include such laws as the Rehabilitation Act, Occupational Safety and Health Act, Federal Coal Mine Health Act, and numerous transportation laws.

[16]ADA Section 102(c).

fully completed a supervised rehabilitation program and are no longer using or addicted are offered the protection of a qualified individual with a disability under the ADA.[17]

Title II of the ADA is designed to prohibit discrimination against disabled individuals by public entities.

Title III of the ADA builds on the foundation established by the Architectural Barriers Act and the Rehabilitation Act. This title basically extends the prohibitions that currently exist against discrimination in facility construction financed by the federal government to apply to all privately operated public accommodations. Title III focuses on the accommodations in public facilities. The ADA makes it unlawful for public accommodations not to remove architectural and communication barriers from existing facilities and transportation barriers from vehicles "where such removal is readily achievable."[18]

Title III also requires "auxiliary aids and services" be provided for the qualified individual with a disability including, but not limited to, interpreters, readers, amplifiers, and other devices to provide the qualified individual with a disability with an equal opportunity for employment or promotion.[19]

TITLE IV requires all telephone companies to provide "telecommunications relay service" to aid the hearing and speech impaired qualified individual with a disability.

Title V assures that the ADA does not limit or invalidate other federal or state laws that provide equal or greater protection for the rights of individuals with disabilities. Unique features of Title V include the miscellaneous provisions and the requirement of compliance to the ADA by all members of Congress and all federal agencies.

Congress expressed its concern that sexual preferences could be perceived as a protected characteristic under the ADA or the courts could expand ADA's coverage beyond Congress's intent. Accordingly, Congress included Section 511 (b) which contains an expansive list of conditions that are not to be considered within the ADA's definition of disability. This list includes transvestites, homosexuals, and bisexuals. Additionally, the conditions of transsexualism, pedophilia, exhibitionism, voyeurism, gender identity disorders *not resulting from physical impairment*, and other sexual behavior disorders are not considered as a qualified disability under the ADA. Compulsive gambling, kleptomania, pyromania, and psychoactive substance use disorders from *current illegal drug use* are also not afforded protection under the ADA.[20]

---

[17]ADA Section 511(b).

[18]ADA Section 302(b)(2)(A)(iv).

[19]ADA Section 3(1).

[20]ADA, Section 511(a),(b); Section 508. There is some indication that many of the conditions excluded from the disability classification under the ADA may be considered a covered handicap under the Rehabilitation Act. See *Rezza v. U.S. Dept. of Justice,* 46 Fair Empl. Prac. Cas. (BNA) 1336 (ED Pa. 1988) (compulsive gambling); *Fields v. Lyng,* 48 Fair Empl. Prac. Cas. (BNA) 1037 (D. Md. 1988) (kleptomania).

Individuals extended protection under this section of the ADA include all individuals associated with or having a relationship to the qualified individual with a disability. This *inclusion* is unlimited in nature, including family members, individuals living together, and an unspecified number of others.[21] The ADA extends coverage to all "individuals," thus the protection is provided to all individuals, legal or illegal, documented or undocumented, living within the boundaries of the United States regardless of their status.[22] Under Section 102 (b)(4), unlawful discrimination includes "excluding or otherwise denying equal jobs or benefits to a qualified individual because of the known disability of the individual with whom the qualified individual is known to have a relationship or association."[23] Thus, the protections afforded under this section are not limited to only family relationships; there appear to be no limits on the kinds of relationships or associations afforded protection. Of particular note is the *inclusion* of unmarried partners of persons with AIDS or other qualified disabilities under this section.[24]

As with most regulatory legislation, the ADA requires that employers *post* notices of the pertinent provisions of the ADA in an accessible format in a conspicuous location within the employer's facilities. A prudent organization may wish to provide additional notification on its job applications and other pertinent documents.[25]

Under the ADA, it is unlawful for an employer to "discriminate on the basis of disability against a qualified individual with a disability" in all areas including:

- Recruitment, advertising, and job application procedures
- Hiring, upgrading, promotion, award of tenure, demotion, transfer, layoff, termination, right to return from layoff, and rehiring
- Rate of pay or other forms of compensation and changes in compensation
- Job assignments, job classifications, organization structures, position descriptions, lines of progression, and seniority lists
- Leaves of absence, sick leave, or other leaves
- Fringe benefits available by virtue of employment, whether or not administered by the employer
- Selection and financial support for training including apprenticeships, professional meetings, conferences and other related activities, and selection for leave of absence to pursue training
- Activities sponsored by the employer including social and recreational programs
- Any other term, condition, or privilege of employment.[26]

---

[21]ADA Section 102(b)(4) and 302(b)(1)(E).

[22]H. Rep. 101-485, Part 2, 51.

[23]ADA Section 102(b)(4).

[24]H. Rep. 101-485, Part 2, 61–62; Part 3, 38–39.

[25]ADA Section 105.

[26]EEOC Interpretive Guidelines.

Other notable provisions of the ADA, or lack thereof, include no record-keeping requirements, no affirmative action requirements, and no preclusions or restrictions on smoking in the place of employment. The ADA possesses no retroactivity provisions.

Congress gave the ADA the same enforcement and remedies as Title VII of the Civil Rights Act of 1964 and have included the remedies provided under the Civil Rights Act of 1991.

In addition to the Americans with Disabilities Act, Section 504 of the Rehabilitation Act of 1973 provides that contractors and other recipients of federal financial assistance are prohibited from discriminating against qualified handicapped individuals solely for the reason of their handicap.[27] A handicapped individual under the Rehabilitation Act (substantially different than disabled individual under the ADA) is defined as any individual who possesses a physical or mental impairment that substantially limits one or more major life functions and either has a record of such impairment or is regarded as having such impairment.[28] The definition of handicapped individual under the Rehabilitation Act does not include alcoholics or drug abusers whose dependency upon alcohol or drugs prevent them from performing the duties of the job in question. However, the Rehabilitation Act has been found to provide protection to former drug users, individuals with former mental illnesses, epilepsy, blindness, congenital back problems, and hypersensitivities. Section 504 of the Rehabilitation Act, provides a private right of action and remedies are normally the same as afforded under Title VII of the Civil Rights Act. Under Section 503 of the Rehabilitation Act, contractors and subcontractors (with contracts in excess of $2,500) must take affirmative action to employ or advance in employment qualified handicapped individuals and not to discriminate against these individuals in employment practices. Complaints for violation of Section 503 are normally filed with the U.S. Department of Labor. The Office of Federal Contracting Compliance Programs (OFCCP) is usually involved in the conciliation of any complaint.

Under Section 501 of the Rehabilitation Act, private right of action is granted for federal government employees and applicants against federal agencies, departments, and other entities.[29]

Section 402 of the Vietnam Era Veterans Readjustment Assistance Act of 1974 additionally requires government contractors (with contracts in excess of $10,000) to take affirmative action to employ qualified disabled veterans and veterans of the Vietnam era. Additionally, Executive Order No. 11701 provides the U.S. Department of Labor with the jurisdiction to administer and enforce affirmative action programs for disabled veterans.[30] In addition to the foregoing, sev-

[27]29 U.S.C.A. § 794.
[28]29 U.S.C. § 706 (7); 45 C.F.R. § 85.31 (b) (1), (2), (3), and (4).
[29]29 U.S.C. § 791.
[30]Exec. Order No. 11701, § 2.

eral other laws, including Title VI of the Civil Rights Act of 1964,[31] the Age Discrimination Act of 1975, and Executive Order No. 11141[32] (see Chapter 14), and individual state discrimination laws may also apply to fire service organizations in the area of handicap or disability discrimination.

## Federal Laws

- Equal Pay Act of 1963 (section 6(d) of the Fair Labor Standard Act of 1938, 29 U.S.C. § 206 (d)0) which extends protection against sexual discrimination and requires equal pay for equal work by forbidding pay differentials predicated on gender.
- The Civil Rights Act of 1964 (Title VII) (42 U.S.C. § 2000 et seq.) which prohibits employment discrimination on the basis of race, color, religion, sex, age, or national origin by employers who employ fifteen or more employees and are engaged in industry affecting commerce.
- The Age Discrimination in Employment Act of 1967 (29 U.S.C. § 621–634) which prohibits employment discrimination against persons over the age of forty.
- Rehabilitation Act of 1973 (§ 501, 503, 504 29 U.S.C. § 791, 793, 794) which prohibits discrimination on the basis of disability by programs receiving federal funds or by federal agencies. This law is the precursor of the Americans with Disabilities Act and was created to assist individuals with disabilities in obtaining access to public buildings and enjoying equal employment opportunities.
- Americans with Disabilities Act of 1990 (also known as the ADA) (42 U.S.C. § 12101) prohibits discrimination against qualified individuals with disabilities. The primary purpose of this law was to provide an estimated 43 million individuals with disabilities equal access to employment opportunities, programs, services, and activities provided by public and private sector employers.
- Civil Rights Act of 1991 (42 U.S.C. § 1981 et seq.) This act in essence reversed a series of cases decided by the U.S. Supreme Court in 1989. The act reinstated the earlier interpretations of the Civil Rights Act of 1964. The act includes permitting full jury trials and, under certain circumstances, allowing for the recovery for emotional suffering and punitive damages.
- Family and Medical Leave Act of 1993 (29 U.S.C. § 2601 et seq.) which requires employers with fifty or more employees to provide eligible employees up to twelve weeks unpaid, job-protected leave for family and

---

[31] 2 U.S.C.A. § 2000 (d) et seq.
[32] 2 U.S.C.A. § 6101 et seq.

medical reasons including activities such as the birth of a child, adoption, or care of a spouse, child, or parent.

- The Pregnancy Discrimination Act (42 U.S.C. § 2000 e (k)). This act prohibits sexual discrimination and amends the Civil Rights Act of 1964 to include pregnancy, childbirth, and pregnancy-related medical conditions as protected against employment discrimination.

- Vietnam Era Veterans Readjustment Assistance Act (38 U.S.C. § 2012). This act requires federal contractors with contracts over $10,000 to hire qualified veterans of any war who have disabilities and specifically qualified Vietnam veterans who may or may not have disabilities.

## Kentucky Specific Laws

- KRS 334.040 forbids discrimination in employment on the basis of race, color, religion, national origin, sex or age.

- KRS 207.130 prohibits discrimination in employment because of handicap.

- KRS 344.030 forbids discrimination based on pregnancy.

- Kentucky Public Acts, House Bill 23-X (L. 1976) prohibits an employer from attempting to deprive an employee of his or her employment because the employee receives a summons, serves as a juror, or attends court for prospective jury service.

- KRS 433.310 forbids anyone from inducing or enticing any person, who has contracted his or her employer for a specific period of time, to leave the service of the employer during the term without the employer's consent.

- KRS 121.310 forbids employers from interfering with the voting rights of their employees.

- Kentucky Public Acts, H.B. 100 (1978) and KRS 335(B)(2)(1) provide that no person shall be disqualified from public employment, nor shall a person be disqualified from pursuing, practicing, or engaging in any occupation solely because of prior conviction of a crime, unless the crime is one limited to convictions for felonies, high misdemeanors, or crimes involving moral turpitude.

- KRS 38.460 makes it unlawful for any person to willfully deprive a member of the National Guard of his or her employment.

- KRS 337.423 makes in unlawful to discriminate in the payment of wages based on sex.

(This case has been edited for the purposes of this text.)

## U.S. EQUAL EMPLOYMENT OPPORTUNITY COMMISSION and
Charles Wessel, Plaintiff

v.

## AIC SECURITY INVESTIGATIONS. LTD., AIC
International, Ltd. and
Ruth Vrdolvak, Defendants
United States District Court, N.D. Illinois, Eastern
Division
July 28, 1994
1994 WL 395119

GUZMAN, United States Magistrate Judge

Charles Wessel was discharged on July 30, 1992 from his position as Executive Director of AIC. He retained the law firm of Lindquist & Vennum to assist him with his disability discrimination claim on that day, and he filed a charge with the EEOC on August 3, 1992. The EEOC completed its investigatory process in five weeks, and determined on September 8, 1992 that reasonable cause existed to believe that the American with Disabilities Act had been violated. After an attempted conciliation failed, Wessel, through his attorneys, moved to have his testimony preserved under Rule 27. Judge Kocoras granted his motion, and Mr. Wessel's testimony was preserved though stenographic and videotape means on November 5, 1992. On that same date, the EEOC filed a lawsuit on Wessel's behalf pursuant to Title I of the Americans with Disabilities Act of 1990, seeking back pay and compensatory damages, as well as equitable relief.

Wessel then moved to intervene in the EEOC's action, and that motion was granted. Wessel's complaint, like the EEOC's complaint, was based on Title I of the Americans with Disabilities Act of 1990. Wessel's complaint also sought punitive damages, which EEOC's complaint did not.

After an expedited discovery schedule this matter came on for trial on March 8, 1993. Following eight days of testimony, the jury found that defendants AIC and Ruth Vrdolvak, AIC's sole shareholder, discharged Wessel because of his terminal cancer, despite the fact that he remained able to perform the essential functions of his position as Executive Director of AIC.

The Supreme Court of Illinois found that Wessel's attorneys fees may be levied against Ruth Vrdolvak individually since she was the principal, if not the only, motivating force behind AIC's discriminatory actions and is the sole shareholder of the corporations. She was responsible for making the discriminatory decision to terminate Wessel and under these circumstances she is not relieved of personal liability. Wessel is hereby awarded $48,632.98 in attorney fees and $7,878.24 in costs.

---

(This case has been edited for the purposes of this text.)

Eleon **ALLEN**, individually and for others similarly situated,
Plaintiffs-Appellees,

v.

Margaret **HECKLER**, et al.,
Defendants-Appellants
United States Court of Appeals, District of Columbia
Circuit

Argued Sept. 19, 1985
Decided Dec. 31, 1985
780 F.2d 64

MIKVA, Circuit Judge.

This case involves an appeal of a class of former patients of a federal mental institution who were hired by the institution pursuant to affirmative action plan. Former patients brought suit alleging that the plan was inadequate by allowing the institution to discriminate against them on the basis of their previous institutionalization.

Eleon Allen, the named Plaintiff, is a former patient at St. Elizabeth's Hospital, a federal mental institution in Washington, D.C. After his discharge as a patient in 1978, Allen was hired by St. Elizabeth's as a housekeeping aide pursuant to 5 C.F.R. §213.3102(h) (1984) ("subsection (h)"). Subsection (h) provides that patients who have been discharged from federal mental hospitals may be given special hiring consideration at the institution where they previously received treatment. The key to subsection (h) is that ex-patients are excused from the usual competitive process for obtaining federal employment, and thus are considered "excepted" service employees. 5 U.S.C. §2103 (1982); 5 C.F.R. §213.101(a) (1984). Although the former patients must have the necessary skills for the jobs they seek, it is considered "not practical" to subject them to the competitive civil service examination. 5 C.F.R. §213.3101 (1984).

Excepted employees, such as the ex-patients here, perform exactly the same work as their "competitive" service counterparts. Their duties and responsibilities are the same, their work is judged by the same standards, and their pay is the same. The excepted workers, however, are given fewer job benefits than competitive employees.

Allen and fifty-two other former patients ("plaintiffs") working at St. Elizabeth's sued the Director of the Department of Health and Human Services and the Director of the Office Personnel Management ("the government") under the Rehabilitation Act. Section 501 of the Act requires that each executive department and agency promulgate an affirmative action plan that provides adequate "hiring, placement, and advancement" opportunities for handicapped people. 29 U.S.C. 791(b). The Plaintiffs contend, and the government does not contest, that they are "handicapped" within the meaning of the Act. Although Plaintiffs are no longer institutionalized, the Act recognizes that discrimination also occurs against those who at one time had a disabling condition.

Plaintiffs claim that granting excepted workers fewer benefits for the same work is discrimination based on their previous medical condition. They argue that subsection (h) violates the Act's affirmative action requirement because it does not provide "adequate" advancement opportunities.

The district court agreed and granted summary judgment in favor of the plaintiff class. The court ordered that all subsection (h) employees who had completed two years of satisfactory service must be allowed to convert to competitive status, with full benefits and credits. Appeal was taken.

The Court of Appeals held that the program discriminated against the former mental patients by allowing them to be given fewer job benefits than competitive employees for the same work, but that it was appropriate for District Court to reconsider remedy which allowed all such employees who had completed two or more years of satisfactory service to convert to career or career-conditional status.

Remedy VACATED and REMANDED for reconsideration; otherwise AFFIRMED.

FREIDMAN, Circuit Judge, sitting by designation, filed dissenting opinion.

Decision Explanation:
The district court's remedy of ordering the coversion of employee status was vacated and sent back to the district court for reconsideration. The remainder of the decision of the district court was affirmed.

(This case has been edited for the purposes of this text.)

Michael Emmett **WELLS**, Petitioner,

v.

**POLICE AND FIREFIGHTER'S RETIREMENT AND RELIEF BOARD**, Respondent

District of Columbia Court of Appeals

Argued En Banc Jan. 10, 1983

Decided April 6, 1983

459 A.2D 136

KELLY, Associate Judge, Retired.

This case involves a police officer who was refused disability retirement by the Police and Firemen's Retirement and Relief Board, and he sought review.

On April 24, 1978, petitioner, an officer with the Metropolitan Police Department (hereinafter "the Department") was shot while on duty. The bullet that lodged in his chest was not removed and after the wound healed, petitioner was left with residual impairment in his nerve and muscle function.

Petitioner began a rehabilitation program in June 1978. The motion and strength in his left arm and hand improved, but he continued to experience pain and parathesias (a burning or crawling sensation of the skin) in his left arm. Surgery, in March 1979, did not relieve this condition.

Dr. Alfred A. Pavot was Officer Wells' treating physician during his rehabilitation. On December 12, 1979, petitioner was presented before Dr. Pavot and the other members of the Pain Clinic at Greater Southeast Community Hospital. After the presentation, Dr. Pavot reported that petitioner had "achieved just about maximum recuperation from his injury," and that he would "be left with some permanent residuals of his left medial neuropathy which are manifesting themself (sic) with some impaired sensation in the left hand and some weakness in the muscles supplied by the left median nerve." Dr. Pavot also reported that petitioner "could return to limited duty on a full-time basis such as manning a desk in the police station and carrying out other sort of work which would not entail grappling with an assailant and similar types of activity which would necessitate normal grasp and sensory perception in both hands."

Petitioner was reexamined by specialists at the Pain Clinic in July 1980. The specialists determined that petitioner had "reached essentially a status quo position," and that he had difficulty flexing his left index finger against resistance which would prevent him from firing a revolver. They also found that although petitioner had a partial permanent disability of approximately 15% of the total body, he could return to light duty "without any significant difficulty."

In August 1980, petitioner was examined by the Board of Police and Fire Surgeons (hereinafter "Surgeons' Board"). It made findings similar to those of the Pain Clinic, but found a functional impairment of only 5%.

Petitioner filed a claim for disability retirement. At a hearing before the Board on November 6, 1980, Dr. Victor Esch, testifying for the Surgeons' Board, stated that petitioner could perform some limited duty work on a full-time basis. He recommended that petitioner not be assigned to a busy precinct where he might have to assist with prisoners, but concluded that petitioner could handle a desk job at Police Headquarters where he would have no contact with prisoners.

Petitioner contends that the Board's conclusions of fact are contrary to the weight of the evidence. Petitioner's evidence supports his claim that he could no longer perform his old job or any position entailing full police duties. Petitioner also claims that the Board erred in refusing to retire him because there was no evidence that a permanent, light-duty position was available for him. He complains that no testimony was presented at his hearing as to a specific duty he could perform and whether, if he were capable of performing, a specific position was available.

The Court of Appeals held that the Board's finding that petitioner could perform useful and efficient service for the Department was supported by substantial evidence. The Court also concluded that the petitioner failed to establish his disability from performing any job in the "grade or class of position" he last occupied.

AFFIRMED.

FERREN, Associate Judge, with whom TERRY, Associate Judge, joins, dissenting.

(This case has been edited for the purposes of this text.)

Richard D. **PETERSON**, Appellant,

v.

The **BOARD OF TRUSTEES OF the FIREMEN'S PENSION FUND OF**
the **CITY OF DES PLAINES**, Appellee.
Supreme Court of Illinois
May 21, 1973
296 N.E. 2d 721

RYAN, Justice.

This case involves a firefighter who applied to the Board of Trustees of the Firemen's Pension Fund for a disability pension.

Richard D. Peterson was a member of the Fire Department of the City of Des Plaines from August 28, 1958, until June 29, 1964. He was classified as a firefighter. On December 9, 1962, while fighting a fire, he inhaled phosgene gas produced by the use of a fire extinguisher. He became ill and was hospitalized for approximately one week. Upon leaving the hospital he returned to his duties. In June, 1964, while on duty, he again became ill and was hospitalized. His condition was diagnosed as a myocardial infarction. Following his discharge from the hospital he did not return to active duty with the fire department. In August, 1964, he applied for a leave of absence which was granted for a period of one year, from October 20, 1964, to October 20, 1965. During this period he applied for an extension of the leave of absence, but no action has been taken on this application. On April 15, 1965, he made application to the Board of Trustees for a disability pension. At the request of the Board the applicant was examined by Dr. Priest.

At the hearing before the Board on the application for a pension, Dr. Priest testified that at the time of the examination in July, 1965, Peterson had a myocardial infarction of the inferior wall of the heart and that his condition could have been caused, with reasonable medical certainty, from the occurrence on December 9, 1962. He further testified that Peterson could not perform the duties of a fire fighter but that he could perform other duties for the fire department in the field of inspecting buildings and other work in connection with the fire-prevention bureau of the department.

Dr. Slott, the applicant's physician, testified that Peterson should not perform the duties of a member of the fire-prevention bureau and that checking out fire hazards would be harmful to his health. He further testified that Peterson could perform the duties of a desk job such as typing and filing reports.

The Board found that there exist in the fire service of the City of Des Plaines necessary duties within the fire-prevention bureau of the fire department which can be performed by a person of the physical and mental capacities of Peterson. The Board also found that such a position is available and would be offered to Peterson if he seeks reinstatement.

The Board further found that the evidence failed to show that Peterson is physically or mentally permanently disabled for service in the fire department of the City of Des Plaines, so as to necessitate his retirement from fire service.

Peterson's primary argument was that since he was no longer physically capable of performing the duties of a firefighter, he can no longer be a fireman and is therefore entitled to pension.

The Supreme Court held that the fact that the applicant for firemen's pension, because of a heart condition, could no longer perform duties of a firefighter did not entitle applicant to pension on theory that he could no longer be a "fireman"; as long as applicant could perform duties of position in fire prevention bureau he was not entitled to disability pension.

AFFIRMED.

## Review Questions

1. Who is a protected individual under the Americans with Disabilities Act?

2. Describe the protections set forth in Title I of the ADA.

3. Define *reasonable accommodation*, as set forth in the ADA.

4. What is the test used to determine if an individual is qualified under the ADA?

5. What constitutes an undue hardship exception to the ADA?

6. What constitutes a safety and health exemption from the ADA?

7. Are preemployment medical exams permitted under the ADA?

8. If a firefighter is actively using illegal drugs, is the firefighter protected under the ADA? If the firefighter undergoes a qualified rehabilitation program and stops using drugs, is the firefighter protected under the ADA?

9. Who qualifies as an individual associated with a qualified individual with a disability?

10. Provide examples of at least five activities within your fire service organization in which discrimination is prohibited under the ADA.

11. Is your fire department required to comply with the ADA? Required to comply with the Rehabilitation Act?

12. What is the difference between a qualified individual with a disability under the ADA and a qualified handicapped individual under the Rehabilitation Act?

13. What is the issue in the *Heckler* case?

14. What is the issue in the *Wells* case?

15. What is the issue in the *Peterson* case?

16. What are the major differences between the Americans with Disabilities Act and the Rehabilitation Act?

# Chapter 7

# Workers' Compensation

The business that trusts to luck is a bad business.

*Publilius Syrus*

## Learning Objectives

- To acquire a general knowledge in the area of workers' compensation.
- To identify the basic principles common in most workers' compensation systems.
- To identify potential areas of liability and methods to minimize the potential risk.
- To identify other sources of information regarding your individual state's workers' compensation program.

**workers' compensation**
federal and state statutes that provide compensation to employees in the event of an employment-related injury or illness

Virtually all states, by statute or by state law, provide financial protection for work-related injuries and illnesses to paid firefighters and EMS personnel. In some states, volunteer firefighters are also afforded this protection for injuries and illnesses incurred in the course and scope of their volunteer activities. Each state possesses its own workers' compensations laws and the variation between the states may be significant. The basic concept of **workers' compensation** is to provide financial compensation to an injured employee who sustained an injury or illness that arose out of, or in the course of, his or her employment.

In most states, there are three basic components of the workers' compensation system. The first basic component consists of payment of medical bills. As a rule, all medical expenses incurred as the result of a work-related injury or illness are compensable under workers' compensation. This category includes doctor's fees, hospital fees, rehabilitation fees, and other expenses incidental to the treatment of the injury or illness. In some states, this coverage may be extended to such items as modification of vehicles, modification of home, replacement of medical devices. A second component of most workers' compensation systems consists of payment for wages while the employee is unable to work. This is often called time loss benefits or permanent partial disability (PPD) benefits. As a general rule, the injured employee would receive 66⅔% of his salary with a minimum and a maximum provided by statute during the period in which he is off work. Suppose for example, that in state X, an employee receives 66⅔% of the average weekly wage when he or she was off work. If the employee was making $500 per week, she could receive benefits of $333 per week while recuperating from the injury. If the employee was making $50 per week, the 66⅔% would provide approximately $33 per week in time loss benefits. However, in most states, a maximum amount to be received has been established by statute in addition to a minimum amount that can be received for time loss benefits. If 66⅔% of the injured employee's salary was over the maximum amount, the employee would only receive the maximum permissible amount. Conversely, if the employee was under the minimum amount, the employee would receive the minimum amount by statute. Time loss benefits or permanent partial disability benefits are usually tax free to the injured employee.

**permanent total rating**
compensation to an employee for a permanently incurred injury or illness

The third component of most workers' compensation systems includes compensation to the employee for the permanently incurred injury or illness. In some states this is called a **permanent total rating**. In this rating system, each body part is provided an amount of weeks or monies total disability due to the work-related injury or illness. When the employee has reached maximum medical recovery, the attending physician or panel of physicians rates the amount of disability the individual has incurred due to the injury or illness. This amount of disability is then calculated according to the body part and type of injury and the amount of disability the individual has incurred in accordance with the schedule established in the state. The employee is then provided a set amount, whether in lump sum or in payments, for that injury. For example, if a firefighter has his or her hand amputated at the wrist, it would constitute a 100% disability of the hand.

When the firefighter reaches maximum medical recovery, he or she would receive a set amount of money as determined by the state statute or by the state workers' compensation laws for whatever the state has determined that the hand is worth. The amounts established by state statute vary from state to state.

In most states, the fire service organization is required by law to acquire workers' compensation coverage. In some states, the fire service organization, or any employer, is provided a number of different options for acquiring this coverage. Some states offer workers compensation through a state agency whereas other states permit the employer or fire service organization to acquire workers' compensation through an established insurance company. Most states permit larger employers to become self-insured. If self-insured, the employer basically pays for all medical costs, time loss benefits, and permanent partial disability rating amounts from its bottom line. To apply for self-insurance coverage, the employer normally has to post a bond with the state, to indicate that it is able to properly administer the workers' compensation system. Many employers hire outside contractors to administer their workers' compensation program.

The compensation is normally the sole remedy for any work related injuries. To be work related, the injury or illness must have arisen out of, or in the course of the employment. Many states have established an administrative procedure through which particular forms are completed for the employee to initiate workers' compensation benefits. If workers' compensation benefits have been provided, the injured employee normally waives any rights to common law suits against the employer for the injury or illness. Under normal circumstances employees may not waive their right to workers' compensation; in some states, prior to an injury or illness, an employee may waive his right to workers' compensation by providing a written waiver to the state's workers' compensation commission. Although attorneys are not required for workers compensation coverage, most states have established a set fee for attorneys representing employees for workers' compensation claims.

In essence, workers' compensation is a method by which an employee and his or her family are relieved of the financial burden following a work-related injury or illness. The workers' compensation laws are normally liberally construed in favor of the employee. Administrative procedures have been established under most workers' compensation systems that permit the employee easy access to the benefits provided by law and quick payment of these benefits. These responsibilities are provided by statute to both the injured firefighter and the fire service organization. In most states, the fire service organization has the responsibility of providing the benefits to the injured firefighter but also has the right to evaluate medical records, investigate the particular incident, and, if found not to be work related, to deny the claim by the firefighter.

# CHART XVI—DIRECTORY OF ADMINISTRATORS

## Alabama

Workers' Compensation Division
Department of Industrial Relations
Industrial Relations Building
Montgomery, Alabama 36131
(205) 242-2868
W.F. Willett, Jr., Administrator

## Alaska

Division of Workers' Compensation
Department of Labor
P.O. Box 25512
Juneau, Alaska 99802-5512
(907) 465-2790
Paul Arnoldt, Director

Workers' Compensation Board
(*Same address as Division*)

## American Samoa

Workmen's Compensation Commission
Office of the Governor
American Samoa Government
Pago Pago, American Samoa 96799
Moaali'itele Tu'ufuli, Commissioner

## Arizona

Industrial Commission
800 West Washington
P.O. Box 19070
Phoenix, Arizona 85005-9070
(602) 542-4411
Gordon Marshall, Chairman

## Arkansas

Workers' Compensation Commission
Fourth & Spring Streets
P.O. Box 950
Little Rock, Arkansas 72203
(501) 682-3930
James Daniel, Chairman

## California

Department of Industrial Relations
Division of Workers' Compensation
455 Golden Gate Avenue, Room 5182
San Francisco, California 94102
(415) 703-3731
Casey Young, Administrative Director

Workers' Compensation Appeals Board
455 Golden Gate Avenue, Room 2181
San Francisco, California 94102
(415) 703-1700
Diana Marshall, Chairperson

## Colorado

Division of Workers' Compensation
1515 Arapahoe Street
Denver, Colorado 80202
(303) 575-8700
Barbara P. Kozelka, Director

Industrial Claims Appeals Office
1515 Arapahoe Street
Denver, Colorado 80202
(303) 620-4277
David Cain, Member
Kathy Dean, Member
Dona Halsey, Member
Robert Socolossky, Member
William Whitacre, Member

## Connecticut

Workers' Compensation Commission
1890 Dixwell Avenue
Hamden, Connecticut 06514
(203) 230-3400
Jesse M. Frankl, Chairman

## Delaware

Industrial Accident Board
820 North French Street, 6th Floor
Carvel State Office Building
Wilmington, Delaware 19801
(302) 577-2885

Karen W. Wright, Chairperson
John F. Kirk, Administrator

## District of Columbia

Department of Employment Services
Office of Workers' Compensation
1200 Upshur Street, NW
Washington, D.C. 20011
(202) 576-6265
Charles L. Green, Associate Director

## Florida

Division of Workers' Compensation
Department of Labor and Employment Security
301 Forrest Building
2728 Centerview Drive
Tallahassee, Florida 32399-0680
(904) 488-2548
Ann Clayton, Director

## Georgia

Board of Workers' Compensation
South Tower, Suite 1000
One CNN Center
Atlanta, Georgia 30303-2788
(404) 656-3875
Don L. Knowles, Director
M. Yvette Miller, Director

## Guam

Workers' Compensation Commission
Department of Labor
Government of Guam
P.O. Box 9970
Tamuning, Guam 96930-2970
(671) 647-4222
Juan M. Taijito, Director of Labor,
Ex-Officio Commissioner
Christian L. Delfin, Employment Program
  Administrator
(671) 647-4231

## Hawaii

Disability Compensation Division
Department of Labor and Industrial Relations
P.O. Box 3769

Honolulu, Hawaii 96812
(808) 586-9151
Dayton M. Nakanelua, Director

Labor and Industrial Relations Appeals Board
888 Mililani Street, Room 400
Honolulu, Hawaii 96813
(808) 548-6465
Frank Yap, Chairman

## Idaho

Industrial Commission
317 Main Street
Boise, Idaho 83720
(208) 334-6000
Rachel Gilbert, Commissioner
Jim Kerns, Commissioner
Steve Loyd, Commissioner

## Illinois

Industrial Commission
100 West Randolph Street
Suite 8-200
Chicago, Illinois 60601
(312) 814-6555
Robert J. Malooly, Chairman

## Indiana

Workers' Compensation Board
402 West Washington Street
Room W196
Indianapolis, Indiana 46204
(317) 232-3808
G. Terrence Coriden, Chairman

## Iowa

Division of Industrial Services
Department of Employment Services
1000 E. Grand Avenue
Des Moines, Iowa 50319
(515) 281-5934
Byron K. Orton, Commissioner

## Kansas

Division of Workers' Compensation
Department of Human Resources

800 SW Jackson Street, Suite 600
Topeka, Kansas 66612-1227
(913) 296-4000
George Gomez, Director

## Kentucky

Department of Workers' Claims
Perimeter Part West
1270 Louisville Road, Building C
Frankfort, Kentucky 40601
(502) 564-5550
Walter W. Turner, Commissioner

## Louisiana

Department of Labor
Office of Workers' Compensation Administration
P.O. Box 94040
Baton Rouge, Louisiana 70804-9040
(504) 342-7555
Alvin J. Walsh, Assistant Secretary of Labor

## Maine

Workers' Compensation Board
Deering Building
State House Station 27
Augusta, Maine 04333
(207) 289-3751
James McGowan, Executive Director

## Maryland

Workers' Compensation Commission
6 North Liberty Street
Baltimore, Maryland 21201
(401) 333-4700
Charles J. Krysiak, Chairman

## Massachusetts

Department of Industrial Accidents
600 Washington Street, 7th Floor
Boston, Massachusetts 02111
(617) 727-4300
James J. Campbell, Commissioner

## Michigan

Bureau of Workers' Disability Compensation
Department of Labor

201 North Washington Square
P.O. Box 30016
Lansing, Michigan 48909
(517) 322-1296
Jack F. Wheatley, Director
John P. Miron, Chief Deputy Director

Board of Magistrates
201 North Washington Square
P.O. Box 30016
Lansing, Michigan 48909
(517) 335-0642
Craig Petersen, Chief Magistrate

Workers' Compensation Appellant Commission
Department of Labor
7150 Harris Drive
P.O. Box 30015
Lansing, Michigan 48909
(517) 335-5828
J. Edward Wyszinski, Jr., Chairman

## Minnesota

Workers' Compensation Division
Department of Labor and Industry
443 Lafayette Road
St. Paul, Minnesota 55155
(612) 296-6107
John B. Lennes, Jr., Commissioner

Workers' Compensation Court of Appeals
775 Landmark Towers
345 St. Peter Street
St. Paul, Minnesota 55102
(612) 296-6526
Hon. Steven D. Wheeler, Chief Judge

## Mississippi

Workers' Compensation Commission
1428 Lakeland Drive
P.O. Box 5300
Jackson, Mississippi 39296-5300
(601) 987-4200
Claire M. Porter, Chairperson

## Missouri

Division of Workers' Compensation
Department of Labor and Industrial Relations
3315 West Truman Boulevard
P.O. Box 58
Jefferson City, Missouri 65102
(314) 751-4231
Jo Ann Karll, Director

Labor and Industrial Relations
3315 West Truman Boulevard
P.O. Box 599
Jefferson City, Missouri 65102
(314) 751-2461
Christopher S. Kelly, Chair

## Montana

State Fund Insurance Company
P.O. Box 4759
Helena, Montana 59604-4759
(406) 444-6518
Carl Swanson, President

Workers' Compensation Court
P.O. Box 537
Helena, Montana 59624
(406) 444-7794
Honorable Mike McCarter, Judge

Employment Relations Division
P.O. Box 8011
Helena, Montana 59604-8011
(406) 444-6530
Chuck Hunter, Administrator

## Nebraska

Workers' Compensation Court
State House, 12th Floor
P.O. Box 98908
Lincoln, Nebraska 68509-8908
(402) 471-2568
Hon. Ben Novicoff, Presiding Judge
Carol S. Thompson, Administrator

## Nevada

State Industrial Insurance System
515 East Musser Street

Carson City, Nevada 89714
(702) 687-5284
James J. Kropid, General Manager

Division of Industrial Relations
400 West King Street
Carson City, Nevada 89710
(702) 687-3032
Ron Swirczek, Administrator

## New Hampshire

Department of Labor
Division of Workers' Compensation
State Office Park South
95 Pleasant Court
Concord, New Hampshire 03301
(603) 271-3171
David M. Wihby, Acting Commissioner

## New Jersey

Department of Labor
Division of Workers' Compensation
CN 381
Trenton, New Jersey 08625-0381
(609) 292-2414
Hon. Paul A. Kapalko, Director & Chief Judge

## New Mexico

Workers' Compensation Administration
1820 Randolph Road, SE
P.O. Box 27198
Albuquerque, New Mexico 87125-7198
(505) 841-6000
Gerald B. Stuyvesant, Director

## New York

Workers' Compensation Board
180 Livingston Street
Brooklyn, New York 11248
(718) 802-6666
Barbara C. Deinhardt, Chairwoman

## North Carolina

Industrial Commission
Dobbs Building
430 North Salisbury Street

Raleigh, North Carolina 27611
(919) 733-4820
J. Howard Bunn, Jr., Chairman

### North Dakota

Workers' Compensation Bureau
500 East Front Avenue
Bismarck, North Dakota 58504-5685
(701) 328-3800
Randy Hoffman, Interim Exec. Director

### Ohio

Bureau of Workers' Compensation
30 West Spring Street
Columbus, Ohio 43266-0581
(614) 466-2950
Sandra H. Devery, Administrator
Michael J. Knilans, Chairman

Industrial Commission
(614) 466-3010
(*Same address as Bureau*)
Barbara A. Knapic, Chairman

### Oklahoma

Oklahoma Workers' Compensation Court
1915 N. Stiles
Oklahoma City, Oklahoma 73105
(405) 557-7600
Hon. Jerry Salyer, Presiding Judge

### Oregon

Department of Consumer and Business Services
21 Labor & Industries Building
Salem, Oregon 97310
(503) 378-4100
Kerry Barnett, Director

Workers' Compensation Board
2250 McGilchrist, S.E.
Salem, Oregon 97310
(503) 378-3308
Mary Neidig, Chairman
Dan Kennedy, Administrator

### Pennsylvania

Bureau of Workers' Compensation
Department of Labor and Industry
1171 South Cameron Street, Room 103
Harrisburg, Pennsylvania 17104-2501
(717) 783-5421
Director, Vacant

Workmen's Compensation Appeal Board
1171 South Cameron Street, Room 305
Harrisburg, Pennsylvania 17104-2511
(717) 783-7838
A. Peter Kanjorski, Chairman

### Puerto Rico

Industrial Commissioner's Office
G.P.O. Box 4466
San Juan, Puerto Rico 00936
(809) 781-0545
Basilio Torres Rivera, Esquire, President

Corporation of the State Insurance Fund
G.P.O. Box 365028
San Juan, Puerto Rico 00936-5028
Oscar L. Ramos Meléndez

### Rhode Island

Department of Labor
Division of Workers' Compensation
610 Manton Avenue
P.O. Box 3500
Providence, Rhode Island 02909
(401) 457-1800
William Tammelleo, Director

Workers' Compensation Court
One Dorrance Plaza
Providence, Rhode Island 02903
(401) 277-3097
Robert F. Arrigan, Chief Judge

### South Carolina

Workers' Compensation Commission
1612 Marion Street
P.O. Box 1715
Columbia, South Carolina 29202
(803) 737-5700

Vernon F. Dunbar, Chairman
Michael Grant LeFever, Executive Director

## South Dakota

Division of Labor and Management
Department of Labor
Kneip Building, Third Floor
700 Governors Drive
Pierre, South Dakota 57501-2277
(605) 773-3681
W.H. Engberg, Director

## Tennessee

Workers' Compensation Division
Department of Labor
710 James Robertson Parkway
Gateway Plaza, Second Floor
Nashville, Tennessee 37243-0661
(615) 741-2395
Larry B. Brinton, Jr., Director

## Texas

Workers' Compensation Commission
Southfield Building
4000 South IH 35
Austin, Texas 78704
(512) 448-7900
Jack Garey, Chairman

## Utah

Industrial Commission
P.O. Box 146600
Salt Lake City, Utah 84114-6600
(801) 530-6800
Stephen M. Hadley, Chairman

## Vermont

Department of Labor and Industry
National Life Building
Drawer 20
Montpelier, Vermont 05620-3401
(802) 828-2286
Charles D. Bond, Director

## Virgin Islands

Department of Labor
Workers' Compensation Division
3012 Vitraco Mall
Christiansted, St. Croix,
Virgin Islands 00820
(809) 692-9390
Adelbert Anduze, Director

Workers' Compensation Commission
1000 DMV Drive
P.O. Box 1794
Richmond, Virginia 23214
(804) 367-8600
Robert P. Joyner, Commissioner

## Washington

Department of Labor and Industries
Headquarters Building
7273 Linderson Way, SW, 5th Floor
Olympia, Washington 98504
(206) 956-4200
Mark O. Brown, Director

Board of Industrial Insurance Appeals
2430 Chandler Court, SW
P.O. Box 42401
Mail Stop FN 21
Olympia, Washington 98504
S. Frederick Feller, Chairperson

## West Virginia

Bureau of Employment Programs
Workers' Compensation Division
601 Morris Street
Executives Offices
Charleston, West Virginia 25332-1416
(304) 558-0475
Andrew Richardson, Commissioner

Workers' Compensation Appeal Board
601 Morris Street, Room 303
Charleston, West Virginia 25301
Richard Thompson, Chairman

## Wisconsin

Workers' Compensation Division
Department of Industry, Labor and Human
  Relations
201 East Washington Avenue, Room 161
P.O. Box 7901
Madison, Wisconsin 53707
(608) 266-9820
George Krohm, Administrator

Labor and Industry Review Commission
P.O. Box 8126
Madison, Wisconsin 53708
(608) 266-9820
Pamela Anderson, Chairperson

## Wyoming

Workers' Compensation Division
Department of Employment
122 West 25th Street, 2nd Floor
East Wing, Herschler Building
Cheyenne, Wyoming 82002-0700
(307) 777-7159
Dennis Guilford, Administrator

## United States

Department of Labor
Employment Standards Administration
Washington DC 20210
(202) 219-8305
Bernard Anderson, Assistant Secretary

Office of Workers' Compensation Programs
(202) 219-7503
Ida L. Castro, Deputy Assistant Secretary

Division of Coal Mine Workers' Compensation
(202) 219-6692
James DeMarce, Director

Division of Federal Employees' Compensation
(202) 219-7552
Thomas M. Markey, Director

Division of Longshore and Harbor Workers'
  Compensation
(202) 219-8572
Joesph F. Olimpio, Director

Division of Planning, Policy & Standards
(202) 219-7293
Diane Svenonius, Director

Benefits Review Board
800 K. Street, NW Suite 500
Washington, D.C. 20001-8001
(202) 633-7500
Nancy S. Dolder, Chief Judge

Employees' Compensation Appeals Board
300 Reporters Building
7th & D Streets, SW, Room 300
Washington, D.C. 20210
(202) 401-8600
Michael J. Walsh, Chairman

## Alberta

Workers' Compensation Board
P.O. Box 2415
9925 107th Street
Edmonton, Alberta T5J 2S5
(403) 498-4000
Dr. John Cowell, President and CEO

## British Columbia

Workers' Compensation Board
P.O. Box 5350
Vancouver, British Columbia V6B 5L5
(604) 273-2266
Jim Dorsey, Chairman of the Board of Governors

## Manitoba

Workers Compensation Board
333 Maryland Street
Winnipeg, Manitoba R3G 1M2
(204) 786-5471
Wally Fox-Decent, Chairperson
Alex Wilde, CEO

The Appeal Commission
175 Hargrave, Room 311
Winnipeg, Manitoba R3C 3R8
George Davis, Chief Appeal Commissioner
Robert MacNeil, Presiding Officer
Tom Jensen, Commissioner
Rod Frisken, Commissioner

## New Brunswick

Workplace Health, Safety and Compensation
  Commission
1 Portland Street
P.O. Box 160
Saint John, New Brunswick E2L 3X9
(506) 632-2200
John S. Roushorne, President & CEO

## Newfoundland

Workers' Compensation Commission
P.O. Box 9000
Station B
St. John's, Newfoundland A1A 3B8
(709) 778-1000
Barbara Taichman, CEO
Eric Bartlett, Executive Director of Corporate
  Services

## Northwest Territories

Workers' Compensation Board
P.O. Box 8888
Yellowknife, NW Territories X1A 2R3
(403) 920-3888
Jeffrey Gilmour, Chairman

## Nova Scotia

Workers' Compensation Board
5668 South Street
P.O. Box 1150
Halifax, Nova Scotia B3J 2Y2
(902) 424-8440
Robert Elgie, Chairman

Workers' Compensation Appeal Board
8th Floor, Lord Nelson Arcade
5675 Spring Garden Road
P.O. Box 3311
Halifiax, Nova Scotia B3J 3J1
(902) 424-4014
Linda Zambolin, Chairperson

## Ontario

Workers' Compensation Commission
200 Front Street W.
Toronto, Ontario M5V 3J1
(416) 927-6968
Kenneth B. Copeland, Vice-Chair & CEO

## Prince Edward Island

Workers' Compensation Commission
60 Belvedere Avenue, P.O. Box 757
Charlottetown, Prince Edward Island C1A 7L7
(902) 368-5680
Howard Jamieson, CEO & Board Secretary
Arthur MacDonald, Chairman

## Québec

Commission de la santé et de la sécurité du travail
524 Bourdages Street
P.O. Box 1200
Terminus postal
Québec, Québec G1K 7E2
(418) 646-4057
Pierre Shedleur, Chairman & CEO

## Saskatchewan

Workers' Compensation Board
1881 Scarth Street, #200
Regina, Saskatchewan S4P 4L1
(306) 787-4370
Stan Cameron, Chairperson & CEO

## Yukon

Workers' Compensation Board
401 Strickland Street
Whitehorse, Yukon Y1A 4N8
(403) 667-5645
William Kalssen, Chairperson

## Canada

Human Resources Development Canada
Federal Workers' Compensation Service
Ottawa, Ontario K1A 0J2
(819) 953-8001
Carol Chauvin Evans, Director

Merchant Seaman Compensation Board
Ottawa, Ontario K1A 0J2
H.P. Hansan, Chairman
(819) 953-8001

(This case has been edited for the purposes of this text.)

Brian Benson **CHARLES**, Petitioner,

v.

**WORKERS' COMPENSATION APPEALS BOARD**, and
City of Santa Ana,
Respondents
Court of Appeal, Fourth District, Division 3
June 30, 1988
Review Denied Sept. 14, 1988
202 CAL.App.3D 781

SEYMOUR, Associate Justice.

    This case involves a city paramedic who appealed an order of the Workers' Compensation Appeals Board affirming a determination that he did not qualify for special statutory treatment reserved for fire fighters and awarding temporary disability at a reduced rate.

    Petitioner Brian Charles sustained physical and emotional injuries while employed as a civilian paramedic for the City of Santa Ana. He maintained that he was statutorily entitled to a leave of absence without loss of salary in lieu of the less generous temporary disability benefits he received. Charles sustained an industrial injury to his heart and psyche and was awarded permanent disability of 14.2 percent without apportionment to nonindustrial causes. However, the Workers' Compensation judge also determined Charles did not qualify for the special statutory treatment reserved for "firefighters" and awarded total temporary disability at the reduced rate of $224 per week (rather than full salary) for the period of September 22, 1985 to June 26, 1986. The Appeals board affirmed.

    The Court of Appeals, Fourth District concluded that Charles was a firefighter within the meaning of the statute and annulled the board's order.

    The statute stated that "whenever any . . . city, county, or district firefighter . . . is disabled, whether temporarily or permanently, by injury or illness arising out of and in the course of his or her duties, he or she will become entitled . . . to leave of absence while so disabled without loss of salary in lieu of temporary disability payment, if any, which would be payable . . .

    The Workers' Compensation judge and the appeals board denied Charles' request for full salary after concluding his duties consisted "primarily of rendering medical care" and did not involve "active firefighting." Charles urged the statute should be liberally construed to cover him because he operated from the fire station, wore a fireman's "turnout" gear, had some of the same training as firefighters and had job duties which overlapped a firefighter's duties.

    The Court of Appeals concluded that those who perform some of the duties of firemen and who assume some of the physical and emotional risks firemen encounter come within the Legislature's "firefighter" rubric and are entitled to the benefits of the statute. Further, Charles lives, trains and drills with other members of the fire department, and has the same workweek and hours a fireman has. Finally, he "rolls" to the scene of every fire in a firemen's turnout gear and thereby also assumes some of the physical and emotional risks firemen encounter. The order is annulled and the cause is remanded for further proceedings. Charles shall recover his costs.

ANNULLED AND REMANDED

CROSBY, Acting P.J., and WALLIN, J., concur.

*Decision Explanation:*
*The decision of the Workers' Compensation Appeal Board denying benefits was overturned and the case was sent back to the Workers' Compensation Commission for determination of benefits.*

(This case has been edited for the purposes of this text.)

Robert **JOHNSON**, Jr. Deceased,
Employee/Respondent
v.
**CITY OF PLAINVIEW**, Minnesota, League of
Minnesota
Cities Insurance
Trust/Employee Benefit Administration,
Employer/Appellants,
and
Commissioner of Labor and Industry, as
Administrator
of Peace Officers Benefit
Fund, Relator.
Ethel **HARDEL**, spouse of Carl Hardel, Deceased,
Employee/Respondent
v.
**STATE** of Minnesota, **PEACE OFFICERS BENEFIT
FUND**, Relator
Supreme Court of Minnesota
Nov. 4, 1988
431 N.W. 2d 109

POPOVICH, Justice.

This case involves two firefighters who suffered and died from heart attacks while in the line of duty and were entitled to Workers' Compensation benefits.

Robert Johnson, Jr.

Robert Johnson had been serving as a volunteer fireman for the City of Plainview approximately four years at the time of his death. He was on call seven days a week, attended all department meetings and training sessions, and also performed other tasks such as repairing fire trucks. Johnson received $8.00 per hour for the first hour of an emergency call. $5.00 per hour for every hour thereafter, and $2.50 for attending departmental training meetings. Johnson participated in approximately 30 emergency calls each year, and received $480.50 for the six months preceding his death. In addition to being a volunteer fireman, Johnson was managing partner with his father in the Bob Johnson Sports Center, where he earned approximately $142.71 per week.

On February 10, 1985, at about 5:30 A.M., Johnson was called to a fire at the Dittrich residence about a mile and a half outside of Plainview. Johnson arrived at the fire at approximately 5:38 A.M. and was involved with three other men in setting up a 1,000 gallon portable drop tank, consisting of 5-foot high pipes holding up canvas liners and weighing approximately 80 pounds. While Johnson was attaching a 55-pound hose to the bottom of the tank, he collapsed. CPR was unsuccessful and Johnson was pronounced dead on arrival at St. Mary's Hospital.

Medical experts testified that Johnson experienced an acute rupture of an aneurysm of his ascending aorta. This rupture was due in part to the fact that Johnson had severe coronary atherosclerosis, a condition he was unaware of at the time of his death. Several experts also testified that the physical and emotional stress involved in fighting the fire was a contributing factor in Johnson's death.

Carl Hardel

Carl Hardel, a hardware store operator, was a volunteer fireman with the Brownton Fire Department for almost 30 years. In addition to performing normal firefighting duties, Hardel also took care of any electrical work required at the scene of a fire.

On July 5, 1984, shortly after 6:00 P.M., Hardel was called to a fire near his home. Before proceeding to the fire, Hardel drove to his hardware store to pick up his supply truck and then to the fire hall to retrieve some equipment. Once at the fire, Hardel could see that electrical wires were down and could hear them snapping and crackling. The water being used to fight the fire was also pooling near the downed lines.

Hardel attempted to cut the circuit breakers on the transformers near the downed lines through the use of an extension pole. After 3-5 minutes working unsuccessfully on his own, Hardel was assisted by Brownton Police Chief Mathwig, lines continued to crackle and Hardel decided to check out a second pole approximately 600 feet away. He proceeded to the second pole but on his way was rendered helpless by severe chest pains. He was transported to the Hutchinson Community Hospital where studies revealed he had suffered an extensive anterior myocardial infarction or heart attack.

Hardel was released from the hospital a week later, but on the same day he arrived home he suffered another attack while sitting at the dinner table. He was transported to the Hutchinson Community Hospital and then transferred to Methodist Hospital in St. Louis Park. Tests at both hospitals showed signs that Hardel had suffered a major stroke. Hardel died at Methodist Hospital on July 15 as the result of complications surrounding these two attacks.

The testimony of several medical experts revealed that Hardel had developed a narrowing of one or more coronary arteries as the result of atherosclerotic plaque formation in the years preceding his death. Hardel, however, was unaware of this condition. Hardel's death was linked to the first attack he suffered while fighting the fire, caused in part by the physical and emotional stress of firefighting.

The compensation judge concluded that Johnson's compensation rate was at a total weekly wage of $514.31, reflecting the combination of Johnson's imputed wage as a fireman of $371.60 plus his income of $142.71 as an employee of the Bob Johnson Sports Center. The Workers' Compensation Court of Appeals upheld this finding.

The Supreme Court of Minnesota reversed this portion of the WCCA's opinion. It stated that Johnson was not a "regular employee" of the Plainview Fire Department. By allowing Johnson to add on the salary from his regular job to the imputed wage would unfairly exaggerate his lost income. The court remanded the decision for a new calculation of Johnson's daily wage using only his imputed earnings as a firefighter.

A claim against the Fund can only be paid if the decedent was a peace officer "killed in the line of duty" whose death was not the result of "natural causes." The deaths of Johnson and Hardel qualify under this standard and entitle their dependents to benefits under the Fund.

In the cases of Johnson and Hardel, at the time of their heart attacks both men were involved in firefighting duties which exposed them to the risk of being killed. Johnson and Hardel were actively involved in fighting fires and exposed to the physical and emotional stress involved in such a dangerous activity. Johnson died at the scene of the fire as a result of his heart attack, while Hardel died a few days later as a direct result of the heart attack he suffered at the fire.

The Court concluded that Johnson and Hardel's deaths were not the result of natural causes. Although both men were suffering from atherosclerosis, neither was aware of this condition. In addition, expert testimony indicated that the physical and emotional stress associated with their firefighting activities was a contributing factor in the heart attacks each man suffered. The WCCA is affirmed on this issue.

AFFIRMED IN PART, REVERSED IN PART, AND REMANDED.

(This case has been edited for the purpose. of this text.)

Roy **ROTHELL**, Plaintiff-Appellant

v.

**CITY OF SHREVEPORT**, Defendant-Appellee

Court of Appeal of Louisiana, Second Circuit

Oct. 27, 1993

626 SO.2d 763

STEWART, Judge.

This case involves a retired Deputy Fire Chief who filed a workers' compensation suit against the City after he had a heart attack followed by a six bypass heart surgery.

Roy Rothell retired as Shreveport Fire Department (SFD) Deputy Fire Chief on March 1, 1990, after 42½ years with the department. He last performed work as a SFD employee on December 31, 1989. On July 31, 1990, he suffered a myocardial infarction (heart attack) at his residence in Eureka Springs, Arkansas. Rothell later underwent a six bypass coronary artery operation in the Springdale, Arkansas hospital.

Rothell filed suit against the City of Shreveport, alleging that his heart and lung disorders were occupational diseases as defined by the Heart and Lung Act, R.S. 33:2581, and therefore he was entitled to receive workers' compensation benefits.

At the administrative hearing, medical evidence showed that Rothell had developed heart and lung disease over a period of years while working for the SFD. Other evidence showed that Rothell had smoked for a 17-20 year period which ended in approximately 1965. Testimony was unrebutted that he had been exposed to numerous fires which involved toxic fumes or chemicals. There was detailed testimony about his duties for the various positions he held with the fire department, especially his stressful administrative responsibilities as Assistant Chief and as Deputy Chief.

The administrative hearing officer found that the defendants rebutted the R.S. 33:2581 presumption that Rothell's heart condition was related to his employment. In her "Memorandum Opinion of Judgment," the hearing officer stated the following:

Dr. Thomas Brown, an expert in adult cardiovascular disease, testified there are six accepted major risk factors in the development of a heart attack:

1. Smoking;
2. Hypertension;
3. Hypercholesterolemia;
4. Diabetes;
5. Family History; and
6. Male over the age of 55.

Of these six major risk factors there was medical evidence that the claimant suffered from five, a 20 year history of smoking, unchecked hypertension, hypercholerolemia, a family history of heart and cardiovascular disease, and the status of a male over the age of 55. Although Dr. Brown admitted that the stress of claimant's position as Deputy Fire Chief might have aggravated his predisposition to hypertension, he testified that stress itself is considered to be a minor factor in the development of a heart attack . . . . In addition to the major risk factors listed above the claimant had a history of alcohol abuse and obesity. All of the major risk factors applicable to the claimant were the result of the claimant's heredity and lifestyle, and were not related to his job as a firefighter.

Accordingly, the hearing officer denied Rothell's claim. Rothell appealed this ruling. Rothell also moved to reduce the cost of the appeal and the hearing officer denied his motion.

The applicable statute, The Heart and Lung Act 33:2581, states:

Any disease or infirmity of the heart or lungs which develops during a period of employment in the classified fire service in the state of Louisiana shall be classified as a disease or infirmity connected with employment. The employee affected, or his survivors, shall be entitled to all rights and benefits as granted by the laws of the state of Louisiana to which one suffering an occupational disease is entitled as service connected in the line of duty, regardless of whether the fireman is on duty at the time he is stricken with the disease or infirmity. Such disease or infirmity shall be presumed, prima facie, to have developed during employment and shall be presumed, prima facie, to have been caused by or to have resulted from the nature of the work performed whenever same is manifested at any time after the first five years of employment.

Rothell presented testimony by two expert witnesses, Dr. Robert C. Hernandez who specializes in emergency and internal medicine, and cardiologist Dr. Jim Haisten. Dr. Hernandez was Rothell's treating physician during the early 1980's through about 1986. Dr. Hernandez testified that, from his interactions as Rothell's physician, it was his opinion that the stress of Rothell's job accelerated the hypertension to the point that it accelerated the atherosclerosis and eventual cardiac event. Although Dr. Hernandez stated that Rothell could have had the same heart attack if he had been a secretary, it was Hernandez' opinion that there was greater potential that Rothell's job played a part in the acceleration of the hypertension than that it did not.

Both Dr. Hernandez and Dr. Haisten agreed that (1) Rothell had had heart disease for several years prior to his heart attack, (2) Rothell had several factors which contributed to the development of his heart disease, and (3) a stressful work environment could be a contributing factor in his risk profile for the development of coronary disease.

The City of Shreveport presented one expert witness, Dr. Thomas Allen Brown, a board certified cardiologist. Dr. Brown testified that he did not think that stress caused Rothell's heart attack.

The Court of Appeal of Louisiana found that the administrative hearing officer erred as a matter of law, in her determination that the City of Shreveport rebutted the presumption that Rothell's heart attack resulted from his employment as a fire fighter. The Court ordered judgment in favor of the Plaintiff, Roy Rothell, and against the Defendant, the City of Shreveport, awarding medical expenses and workers' compensation benefits at the rate of $267.00 per week to the Plaintiff, Roy J. Rothell, beginning July 31, 1990, with legal interest thereon on each past due installment from its due date until paid. Defendant-appellee, the City of Shreveport, shall pay to Plaintiff, penalties in accordance with LSA-R.S. 23:1201 E, and $3,000 in attorney's fees.

REVERSED AND RENDERED.

## Review Questions

1. Where are the statutes found that regulate workers' compensation in your state? (For example, KRS, Kentucky Revised Statute).

2. What are the benefit levels provided in your state?

3. Does your state offer vocational rehabilitation or occupational rehabilitation for injured firefighters?

4. What is the issue in the *Charles* case?

5. Does your fire service organization have a safety program?

6. What is temporary total disabilities (time loss)?

7. What is the issue in the *Johnson* case?

8. Does your state require an injury or illness to "arise out of or in the course of employment" to be covered under workers' compensation?

9. What is the issue in the *Rothell* case?

10. How is permanent partial disability determined in your state?

11. Does your state offer vocational rehabilitation?

12. What does your fire service organization do for injured firefighters who cannot return to their normal jobs?

13. Are volunteer firefighters covered under workers' compensation in your state?

14. Does your fire service organization have a workers' compensation administrator or other individual responsible for workers' compensation?

# Fair Labor Standards Act

Money is human happiness in the abstract.

*Arthur Schopenhauer*

## Learning Objectives

- To acquire a basic understanding of the FLSA.
- To acquire an understanding of the requirements of the FLSA.
- To acquire an understanding of the wage and hour provisions.

**Fair Labor Standards Act (FLSA)**

a 1938 federal act that established a standard minimum wage, and that regulates the hours and type of work performed in industries involved in interstate commerce

**Portal-to-Portal Act**

federal statute that regulates the pay for an employee's required nonproductive time

**working time**

time for which an employee is entitled to compensation, including all time an employee is required to be on duty either on the employer's premises or at a prescribed work place

The **Fair Labor Standards Act** (referred to as the FLSA) was enacted by Congress in 1938.[1] It was amended in 1947 by the Portal-to-Portal Act,[2] in 1961 (adding "enterprise" definition), and in 1977 (tightened statutory exemptions). The Equal Pay Act of 1963 also amended the FLSA by forbidding employers and unions from differentiating wages and benefits based on sex.[3] The 1974 amendment to the FLSA extended coverage to employees of federal, state, county, and municipal governments.[4]

The FLSA, as amended, generally required the payment of specified minimum wages and overtime pay for hours worked over forty in any one week period. Preliminary and postliminary activities, such as coming and going to work, are generally excluded under the **Portal-to-Portal** exclusion. **Working time** is defined as the time for which an employee is entitled to compensation and includes all time an employee is required to be on duty on the employer's premises or at a prescribed workplace.[5] In the fire service, most fire fighters are considered employees and thus covered under the FLSA. Exceptions from the requirements exist in the following areas:

- Executive Exemption
- Administrative Exemption
- Professional Exemption
- Outside Salesperson Exemption

Fire service organizations should be aware that specialized tests have been designed to evaluate whether an individual or position fits the exemption category. In general, fire service officers usually fall within one or more exemption categories.

As a general rule, all hours that a firefighter is required to provide the fire service organization are considered compensable work time. Additionally, rest and meal periods, on-duty wait time, on-call wait time, reporting time, sleep time, training time, civic and charity work (if done at the employer's request), and travel time can also be compensable. Clothes changing and washup activities can be excluded[6] as well as preemployment tests, personal medical attention, voluntary training programs, athletic events, voluntary civic or volunteer work, illness time, leave of absence, vacations, and holidays.

The Wage and Hour Administration has permitted the reasonable cost or fair value of certain noncash items to be included in the computation of wages for the purposes of satisfying the minimum wage requirements under the FLSA. These items may include meals furnished for the benefit of the firefighter, lodg-

---

[1] 29 U.S.C.A. § 201 et. seq.
[2] 29 U.S.C.A. § 251 et. seq.
[3] 29 U.S.C.A. § 206.
[4] 29 U.S.C.A. § 203 (5) (6). Also see *Garcia v. San Antonio Met. Transit Auth.*, 26 W&H Cases 65 (S.Ct. 1985).
[5] 29 CFR § 778.223.
[6] *Mitchell v. Southeastern Carbon Paper Co.*, 228 F.2d. 934 (CA 1955).

ing, merchandise, transportation, tuition, savings bonds, insurance premiums, and union dues.

In recent years, a greater importance has been placed upon personal privacies in the area of credit reporting and other laws in the area of collections. The Federal Wage Garnishment Law sets restrictions on the amount of employee earnings that can be deducted within a given period and offers protection from discharge from employment for reasons of garnishment. Garnishment is the process by which a person's debts are collected in order to fulfill a debt to a third party. The most frequent garnishments in the average workplace are for child support or alimony payments.

In the public sector, absence from work due to illness usually does not deprive an individual of his or her salary or compensation. Specifically, public sector employers usually continue to pay the officer and hold the position open through any reasonable period of time in which the individual is ill. During a period of illness, most public sector fire service organization employees are provided with sick pay as set forth by either state statute, a collective bargaining agreement, or organization policies. A fire service organization is not required by law to guarantee sick pay: There is no requirement under the Fair Labor Standards Act or any other federal legislation that requires a fire service organization to pay sick pay.

Sick pay may vary greatly depending on the fire service organization, the type of health benefits provided, and even to the type of injury involved. For long-term or catastrophic illnesses or injuries, fire service organizations often have a designated time in which the individual is maintained on salary with benefits. When an individual has incurred an injury that prevents him or her from returning to the fire service organization in their former capacity, many fire service organizations have offered alternatives such as restrictive duty programs, retraining for the firefighter, and even early retirement programs.

Section 7 (K) of the Fair Labor Standards Act provides partial exemptions of the requirements to pay a firefighter overtime of hours worked beyond forty hours per calendar week. Under this provision, a public employee (i.e., state or local government agency) can establish a work period of fewer than twenty-eight consecutive days, but not less than seven consecutive days, in lieu of the standard forty hour work week. The work period is the basis of computing any required payment for overtime. The rule provides that where fire fighters work time does not exceed 212 hours in any period of twenty-eight consecutive days, there is no requirement for payment of overtime. Work periods can stretch from seven to twenty-eight days, with each period assigned a maximum number of work hours. Fire service organizations should be very knowledgeable in the specific scheduling rules set forth under Section 7 (K).

(This case has been edited for the purposes of this text.)

Angela **HAYES**, Bruce Brouillette, Donald Gautheier, Mark Mathews, Arthur Moses, Jr., Wayne Martin, and
Silton Metoyer, Plaintiffs/Appellants

v.

The **CITY OF ALEXANDRIA**, Defendant/Appellee
Court of Appeal of Louisiana, Third Circuit
Dec. 8, 1993
629 SO.2d 435

DOMENGEAUX, Chief Judge.

This case involves current and former firefighters who brought action against the city to recover difference in wages between those actually paid by city and federal minimum wage after they began receiving state supplemental pay to raise their salaries.

The Plaintiffs were hired as firefighters for the City of Alexandria between January of 1985 and February of 1986. They alleged that they were hired at a wage of $3.04/hour, in violation of Fair Labor Standards Act. The City admitted this allegation in its answer. The petition also alleged that five of the seven Plaintiffs filed suit in federal court for unpaid minimum wage compensation, liquidated damages and attorney's fees and that the federal litigation ended in a consent order in the Plaintiffs' favor, which the City paid. The petition further alleged that the City paid back wages due under the FLSA to the two Plaintiffs who were not parties to the federal suit. The amounts paid by the City to all Plaintiffs were alleged to be the difference between $3.04/hour, the Plaintiffs' starting base salary and $3.35/hour, the minimum wage prescribed by the FLSA. These allegations were also admitted by the City.

Between June of 1985 and June of 1986, the Plaintiffs began receiving state supplemental pay pursuant to LA.R.S. 33:2002, in addition to their municipal pay. All parties agreed that the City continued to pay the Plaintiffs $3.04/hour, the prevailing minimum wage at the time they were hired. They filed the instant suit to recover the difference between the wages actually paid by the City and $3.35/hour, from the date each Plaintiff began receiving state supplemental pay.

The Court of Appeal of Louisiana found that the Plaintiffs established that they were entitled to a starting salary of $3.35 an hour. Under LA.R.S. 33:2002 and 2005, the state may not reduce the firefighters' base pay solely because of the state supplemental payments. The Court held that the City's continuation of the Plaintiffs' base pay at $3.04/hour, after actual payment to the Plaintiffs of a higher rate of pay, to be a reduction prohibited by LA.R.S. 33:2005. The Plaintiffs are entitled to the relief prayed for.

REVERSED AND REMANDED

(This case has been edited for the purposes of this text.)

William **J. SANCHEZ**, et al.

v.

**CITY OF NEW ORLEANS**, et al.

Court of Appeal of Louisiana, Fourth Circuit

Jan. 30, 1989

538 SO.2d 709

BYRNES, Judge.

This case involves an appeal of a judgment finding that new rules and regulations enacted by the New Orleans Civil Service Commission and the City of New Orleans are valid and comply with the provisions of the Fair Labor Standards Act.

On February 19, 1985 the United States Supreme Court expressly extended coverage of the Fair Labor Standards Act minimum wage provisions to state and local government employees. In 1985, the United States Congress amended the FLSA to delay compliance to be effective April 15, 1986. The 1985 amendments implemented special anti-discrimination and other monetary benefits to reduce the financial impact of the federal standard regulations. The 1985 Amendments provide that employees must assert coverage under Sec. 7 of the FLSA of 1938 and show discrimination under Sec. 8 of the 1985 Amendments.

In 1985, the Civil Service Commission of the City of New Orleans formed a committee to review options for compliance with federal regulations. Based upon its study, the committee chose a 28 work day cycle as the method to comply with federal standards. The new plan eliminated the use of a seven day week first implemented in 1979, which allowed Dutch Days or Kelly Days, a form of compensatory time off with pay. Previously, the firemen worked 60-60-48 cycles each seven day week, which computed to one day off with pay every three months (quarterly), comprising Dutch Days. By changing to the 28 day cycle, mandatory overtime was not computed until 212 hours worked. Elimination of Dutch Days decreased the amount of compensatory time off with pay by 96 hours a year, but the Commission substituted in its place, overtime

pay, in cash, for the same 96 overtime hours. Thus, the Commission opted to compensate the overtime hours in cash rather than by compensatory time. The new plan also eliminated the inclusion of sick leave and annual leave in computing the number of hours worked before overtime pay is calculated in any work period. The new plan did provide a 15% pay raise to firefighters in lieu of any lost benefits.

In its customary manner, the Civil Service Commission posted notice of its October 23, 1985 meeting. No questions were submitted by those in attendance, including Union representatives of Local 632 of the New Orleans Firefighter Association, when the explanation of the Civil Service Plan was discussed. A day or two later the City Council ratified the new Civil Services rules to be effective April 13, 1986. Between October, 1985, and April, 1986, the Union expressed concern in letters and meetings with the City and the Civil Service Commission. On April 3, 1986, William Sanchez, Union President, and others on behalf of the firefighters filed a petition for declaratory and injunctive relief. The firefighters sought a declaratory judgment decreeing that the collective bargaining agreement between the city and the union was binding and that the new Civil Service Commission rules violated provisions of the FLSA and its 1985 Amendments. Plaintiffs also claimed that the new rules produced the unlawful impairment of contractual and vested rights under LSA R.S. 33:1995 and R.S. 33:1995 and R.S. 33:1996, full back pay and an equal amount for liquidated damages, attorney's fees and costs.

After trial on June 27, 1986, Judge Connolly ruled in favor of the City and Civil Service Commission, dismissing the firefighters' petition.

On appeal the Plaintiff asserted:

(1) Coverage under Section 8 of the 1985 FLSA Amendments prohibits a unilateral reduction in wages and fringe benefits as a means of avoiding the financial impact of the FLSA and its amendments;

(2) Under 29 U.S.C. Sec. 201 of the Fair Labor Standards Act of 1938 and its 1985 Amendments, the new Civil Service rules cannot

contravene provisions of the collective bargaining agreement between the City and Union;

(3) Transitional provisions of the 1985 FLSA Amendments protected Plaintiff's contractual compensatory time benefits;

(4) Louisiana state law provides "full pay" to firefighters during sick and annual leave under LSA R.S. 33:1995 and 33:1996; and

(5) The new Civil Service Commission rules cannot retroactively apply to alter or impair obligations of the contract agreement between the City and the Union.

The Plaintiffs claimed that the Defendants discriminated against firefighters in violation of FLSA 1985 Amendment Section 8 by unilaterally reducing wages and fringe benefits for the purpose of nullifying FLSA coverage. Defendants claimed that they did not enact the new pay plan to discriminate against the Plaintiffs for the sole purpose of retaliation for asserted coverage of the Fair Labor Standards Act. Plaintiffs also allege that there was a collective bargaining agreement effective March 31, 1984 through March 31, 1988 between the City and Union which prohibited the City and Civil Service Commission from modifying compensatory time without agreement with the Union under FLSA restrictions. Under the new pay plan, Plaintiffs lost overtime compensation which they had previously received. They maintained that loss of overtime pay meant they were not receiving full pay. Plaintiffs claimed that Defendants have retroactively impaired the obligations of the contract agreement between the City and the Union.

The Court of Appeal of Louisiana found that the Plaintiffs failed to show that the Commission acted solely and directly to discriminate against the firefighters, and also failed to show that the Commission's actions resulted in any reduction of pay or benefits. The Plaintiffs failed to demonstrate that a collective bargaining agreement existed that would override the new Civil Service rules. The Court did not find that Plaintiffs, at any time, received less than full base pay.

The Court concluded that the Commission acted reasonably in balancing the interests of the City and the firefighters. The Commission did not discriminate against the firefighters nor did it contravene the City-Union collective bargaining agreement. The Civil Service Commission did not fail to comply with the provisions of the Fair Labor Standards Act.

AFFIRMED.

# Review Questions

1. What is the minimum wage in the United States as of today?

2. What constitutes working time for firefighters?

3. Are there any members of your fire service organization who are considered exempt from the FLSA?

4. Does your state have laws against garnishments?

5. What are the "exemptions" under the FLSA?

6. Please provide your thoughts as to the decision in the *Hayes v. City of Alexandria*.

7. What is the major issue in *Sanchez v. City of New Orleans*?

8. Is your fire department exempt from the FLSA?

9. How does the FLSA affect volunteer firefighters? Does the FLSA apply?

# 9

# Liabilities in Termination of Employment

Things could be worse. Suppose your errors were counted and
published every day, like those of a baseball player.

*Anonymous*

### Learning Objectives

■ To gain an understanding of the at-will doctrine.

■ To gain an understanding of the exception to this doctrine.

■ To gain knowledge of the other areas of employment protections afforded under the
law.

**88**

Every year, hundreds of firefighters and other EMS personnel are voluntarily or involuntarily terminated from their employment in the fire and emergency service due to a multitude of reasons ranging from their failure to adhere to organization policies to failure to perform adequately. Generally there are three basic ways in which someone leaves a job:

1. Voluntary termination (the employee simply quits the job)
2. Involuntary termination (the person is "fired" from the job)
3. Constructive discharge (the organization requests or forces the employee to resign from the job)

In the past, an individual who was involuntarily terminated or constructively discharged from the job possessed little, if any, recourse against the company or organization for the termination.

**employment at will**
a common law rule that is the basis on which most employment situations are analyzed

In the United States, the traditional common law rule of **employment at will** is the basis for which most employment situations are analyzed. The common law rule of employment at will, in essence, means that an employee can be terminated from employment at any time, without notice and with or without cause by the employer.[1] However, if the employee has benefit of a contractual agreement, such as a **collective bargaining agreement** between a labor organization and the company, or is protected by federal or state statute, the contractual relationship or law removes the employee from the at-will status. Additionally, many fire and emergency service organizations are under the civil service system or provided protection under specific state laws thus removing these organizations from the at-will category.

**collective bargaining agreement**
a contract that regulates the terms and conditions between an employer and the labor union

Employment at will is a well-established doctrine that has been recognized in the United States since the late 1800s. As stated in the *Payne v. Western & Atlantic Railroad* [2] case in 1884:

All (employers) may dismiss their employees at will, be they many or few, for good cause, for no cause, or even for cause normally wrong, without being thereby guilty of legal wrong.

Although the employment at-will doctrine continues to be well entrenched in the American judicial system, numerous state and federal statutes have greatly eroded the right of an employer to terminate an individual's employment without cause. Some of the federal laws that have limited the employment at-will doctrine by prohibiting retaliation or discrimination against the employee include:

- Labor Management Relations Act (includes protections for employees to form or join a union)[3]

---

[1]H.G. Wood, Law of Master and Servant, 2d Ed., § 134 (NY: J.D. Parsons, Jr. 1877, 272–273.
[2]81 Tenn. 507 (1884).
[3]29 U.S.C.A. § 151, et seq.

- Fair Labor Standards Act (includes protection for individuals who report violations)[4]
- Title VII, Civil rights Act of 1964 (discrimination to a protected class)[5]
- Age Discrimination in Employment Act of 1967 (discrimination against employees over the age of 40)[6]
- Rehabilitation Act of 1973 (disability protection)[7]
- Americans with Disabilities Act (disability protection)[8]
- Employee Retirement Income Security Act of 1974 (known as ERISA—protection for employees close to retirement)[9]
- The Vietnam Era Veterans Readjustment Assistance Act (protections to Vietnam veterans)[10]
- Occupational Safety and Health Act (protection for employees reporting violations)[11]
- Clean Air Act (protection for employees reporting violations)[12]
- Civil Service Reform Act of 1978 (special protections for individuals in the civil service system)[13]
- Consumer Credit Protection Act (protections against termination due to credit problems)[14]
- Federal Water Pollution Control Act (protection for employees reporting violations)[15]
- Judiciary and Judicial Procedure Act (protections regarding jury duty and related assistance to the courts)[16]
- and specific industry protections such as the Railway Labor Act[17] and Railroad Safety Act.[18]

In addition to the federal statutes, numerous states have made inroads into the employment at-will doctrine through the enactment of state statutes affording

---

[4]29 U.S.C.A. § 215.
[5]29 U.S.C.A. § 2000e-2 and § 2000e-3(a).
[6]29 U.S.C.A. § 621 & 734.
[7]29 U.S.C.A. § 794.
[8]Discussed infra.
[9]29 U.S.C.A. § 1140–1141.
[10]38 U.S.C.A. § 2021 (b) (1), 2024 (c).
[11]29 U.S.C. § 651–678.
[12]42 U.S.C.A. § 7622.
[13]5 U.S.C.A. § 7513 (a).
[14]15 U.S.C.A. § 1674 (c).
[15]33 U.S.C.A. § 1367.
[16]28 U.S.C.A. § 1875.
[17]45 U.S.C.A. § 152 (3).
[18]45 U.S.C.A. § 441 (a) and (b) (1).

protection to employees. As an example, the state of California has enacted the following statutes providing protection to employees:

- Antidiscrimination Laws (CGC § 3502)
- Protection of Political Freedom (CGC § 1101, et seq.)
- Jury Duty Protection (CGC § 230)
- Polygraph Restrictions (CGC § 637)
- Protection of Military Personnel (CGC § 394)
- Whistle-blowing Statute (CGC § 10545)
- Antireprisal Statute (CGC § 6400, et seq.)
- Forbidden Termination Due to Garnishment (CGC § 2929)
- Forbidden Termination for Reporting Workers' Compensation Injury or Illness (CGC § 132a)

In recent years, the state courts have additionally developed exceptions to the rigid employment at-will doctrine. Many state courts have found that when an individual is terminated from his or her employment in contravention of some substantial public policy principle, such as refusing to commit a crime, the public benefit far outweighs the employer's right to terminate the employee. Most state courts recognize exceptions from the employment at-will doctrine for the following reasons:

- Upholding the law
- Refusal to commit an unlawful act (such as manipulating air samples)
- Reporting violations of the law (whistle blowing)
- Exercising a statutory right (such as religious preference)
- Performing a public obligation (such as jury duty)

Some state courts have gone even further in finding exceptions to the employment at-will doctrine. In some states, a covenant of good faith and fair dealing between employer and employee has been found, a violation of which could sustain a private cause of action against the employer.[19] The early cases finding a covenant of good faith and fair dealing usually involved an inherent wrong, such as discrimination prior to the federal regulations[20] or terminating the employee for monetary gain.[21] Later cases suggested that an implied covenant could even be inferred from an employer's policies rather than a specific act against the employee.[22]

---

[19]*Monge v. Beebe Rubber Co.*, 114 N.H. 130, 316 A.2d 549 (1974).
[20]Id.
[21]*Fortune v. National Cash Register Co.*, 363 Mass. 96, 364 N.E. 2d 1251 (1977).
[22]*Cleary v. American Airlines, Inc.*, 111 Cal. App. 3d 443, 168 Cal. Rptr. 917 (1981).

(This case has been edited for the purposes of this text.)

REVERSED AND RENDERED

In re the Appeal by the **EAST BANK CONSOLIDATED SPECIAL SERVICE FIRE PROTECTION DISTRICT** from the Decision of the Jefferson Parish Fire Civil Service Board in the Matter of
Anthony Sambola
Court of Appeal of Louisiana, Fifth Circuit
Dec. 28, 1993
630 SO.2d 286

WICKER, Judge.

This case involves an appeal of a firefighter who was terminated for excessive leave of absence without pay.

Anthony Sambola has been a firefighter with the East Bank consolidated Special Fire Protection District of Jefferson Parish since 1986. In February 1991, Sambola began using sick leave for a non-job related illness. Sambola exhausted all of his sick and annual leave. On March 27, 1991, Sambola requested that he be placed on extended leave of absence without pay pursuant to Personnel Rule X, Section 7, Subsection 7.1 of the East Bank Consolidated Fire Department Personnel Rules. Between March 27 and September 10, 1991, Sambola returned to his duties and then again was unable to work due to his illness. By September 10, 1991 Sambola had exhausted 90 days leave of absence without pay.

Subsequently, Sambola received a letter dated September 10, 1991 from the Superintendent of Fire, Edward O'Brien, stating that Sambola had exhausted 90 days leave of absence without pay. The letter further directed Sambola to report to O'Brien's office on September 16, 1991. At the time of that meeting, the letter stated, Sambola was to provide information explaining why O'Brien should not consider Sambola resigned of his position pursuant to Rule X.

At the September 16th meeting, Sambola appeared with a letter from one of his physicians stating that Sambola would be available to return to work soon after September 20, 1991. On September 17, 1991, O'Brien sent a letter to Sambola telling him he was terminated as of that date.

Sambola appealed the Superintendent's action to the Jefferson Parish Fire Civil Service Board. The Board reversed the Superintendent's action and reinstated Sambola. The East Bank Consolidated Fire Protection District appealed the Board's decision to the District Court. The District Court affirmed the Board. The Fire Protection District appealed.

The Fire Protection District argued that it had legal cause to terminate Sambola pursuant to Personnel Rule X, once Sambola exhausted his 90 days leave without pay. The District pointed to a Policy Memorandum from the Parish President's Office. The Memorandum stated that the maximum amount of leave without pay that may be approved for any employee is 90 days. However, by delaying Sambola's termination until after the September 16th meeting, the Superintendent extended Sambola's leave without pay.

The Court of Appeal of Louisiana agreed with the Board's decision to reinstate Sambola. The Court affirmed the decision of the Jefferson Parish Fire Civil Service Board.

AFFIRMED.

(This case has been edited for the purposes of this text.)

Floyd **SAFFORD**
v.
**DEPARTMENT OF FIRE**
Court of Appeal of Louisiana, Fourth Circuit
Nov, 30, 1993
627 SO.2d 707

GULOTTA, James C., Judge Pro Tem.

This case involves a firefighter who was terminated because he refused to submit to a substance abuse test.

The Plaintiff, Floyd Safford, was a seventeen year veteran firefighter with the New Orleans Fire Department. On February 19, 1992, Plaintiff reported to his station at approximately 6:45 A.M. At approximately 8:30 A.M., the district chief received a telephone call from a woman claiming to be Plaintiff's wife. The district chief concluded from that telephone conversation that Safford was possibly intoxicated or under the influence of drugs while on duty. The chief did not know the woman's voice nor could he identify that person as being Plaintiff's wife. In order to confirm the woman's identity, the district chief called her back "at the number that she gave and also at the number she gave as her parent's house, which gave us confirmation." No other corroborating or confirmatory action was taken by the Department.

There were no complaints from Plaintiff's fellow firefighters or any evidence of any unusual conduct on Plaintiff's part on the morning in question which might indicate that he was under the influence of alcohol or drugs. Nevertheless, at 10:45 that morning, Plaintiff refused to submit to the test. Plaintiff was immediately suspended and later terminated from his job.

Plaintiff's appeal resulted in a hearing before a Civil Service Commission hearing officer who, after considering the testimony and evidence, recommended that Safford be reinstated to his position. Nonetheless, the Commission found Plaintiff guilty of violating City and Department policies in his refusal to submit to the substance abuse test and upheld Plaintiff's termination. Plaintiff appealed again.

The Court of Appeal of Louisiana concluded that there was no showing of reasonable suspicion upon which to base the order that Plaintiff submit to the test. The Civil Service Commission finding that reasonable suspicion existed was manifestly erroneous. The Court reversed and set aside the order of the Civil Service Commission terminating Plaintiff's employment with the Department. The Court ordered that Floyd Safford be reinstated with all back pay.

(This case has been edited for the purposes of this text.)

Dennis P. **WELSH**, Plaintiff and Appellant
v.
The **CITY OF GREAT FALLS**, Montana, a municipal corp. of
the State of Montana, Defendant and Respondent
Supreme Court of Montana

Submitted Dec. 8, 1983
Decided Oct. 1, 1984
Rehearing Denied Nov. 13, 1984
212 MONT. 403

SHEA, Justice.

This case involves a fireman who had been terminated for physical disability and sought relief against the city for wrongful termination.

The Plaintiff, Dennis Welsh, began his career as a Great Falls fireman when he was appointed to a six-month probationary period on March 11, 1963. After the probationary period, Welsh was nominated and appointed to serve as a full-time Great Falls fireman. In 1971, however, a benign brain tumor was discovered in Welsh's left temporal lobe. He underwent two successful operations to have the tumor removed, and was placed on medication to control seizure attacks. Welsh returned to duty after a six-month layoff. Welsh was given an office job in the fire prevention bureaus for the first year after his return, but was later returned to active firefighting duties. In 1973, Welsh passed the examination for the rank of captain and was promoted to that rank.

In 1977, Welsh suffered a severe seizure at the station house shortly after he had returned from a fire. According to the testimony of fellow fireman, Welsh passed out in the cab of the fire engine just as it was pulling into the station. After this incident, Welsh's medication dosage was increased to attempt to prevent further seizure attacks. Three years later, during a drill on July 21, 1980, Welsh was hospitalized for what was diagnosed by the emergency room doctor as "heat exhaustion." Welsh was inside a burning structure and apparently became disoriented and confused, had glazed eyes and was sweating profusely.

After the July 21 incident, Welsh had three meetings with the fire chief. The first meeting took place in the fire chief's office that day after the July 21, 1980, incident that resulted in Welsh's temporary hospitalization for either a seizure or heat exhaustion. Present were Welsh, fire chief Mike Kalovich, and Lee Bright, the operations officer. The emergency room doctor had released Welsh but the fire chief nonetheless told Welsh that he should have a physical examination. The fire chief asked and obtained permission from Welsh to speak to Welsh's doctor, Dr. Douglas Brenton. They also discussed the report from the officer in charge of the July 21 fire drill, as well as letters from other firefighters who were present at the drill and had observed Welsh's physical reactions during the drill.

The second meeting took place 10 days later (on July 31, 1980) in the fire chief's office after the fire chief had met with Welsh's doctor, Dr. Brenton. Welsh, the fire chief, and the operations officer were present again. The fire chief confronted Welsh with Dr. Brenton's letter and Welsh acknowledged that he had suffered more seizures than were known to the fire chief. When the fire chief suggested that he retire, Welsh replied that if he were in the fire chief's position he "would do the same thing."

The third and final meeting was five days later, August 5, 1980, when the fire chief handed Welsh his termination letter containing an effective resignation date of August 1, 1980. At the hearing in District Court Welsh testified that when he was given this termination letter he did not consent to his retirement and that he was never told he had a right to a hearing. The City concedes Welsh was not told he had a right to a hearing.

Finally, before the dismissal became final, Welsh was offered the traditional exit interview before the personnel board, but he declined the opportunity for an interview.

The Eighth Judicial District Court denied all relief on Welsh's complaint and he appealed.

The Supreme Court of Montana found that Welsh was improperly terminated and was entitled to full pay and benefits from the time of his termination until the final disposition of the case.

VACATED AND REMANDED.

HASWELL, C.J. and MORRISON, and SHEEHY, JJ., concur.

HARRISON, Justice, dissenting.

(This case has been edited for the purposes of this text.)

### CLEVELAND BOARD OF EDUCATION, Petitioner
v.
James **LOUDERMILL** et al.
### PARME BOARD OF EDUCATION, Petitioner
v.
Richard **DONNELLY** et al.
James **LOUDERMILL**, Petitioner
v.
### CLEVELAND BOARD OF EDUCATION et al.
Nos. 83-1362, 83-1363, and 83-6392.
105 S.Ct. 1487 (1985)
The Supreme Court

WHITE, Justice.

In No. 83-1362, petitioner Board of Education hired respondent Loudermill as a security guard. On his job application Loudermill stated that he had never been convicted of a felony. Subsequently, upon discovering that he had in fact been convicted of grand larceny, the Board dismissed him for dishonesty in filling out the job application. He was not afforded an opportunity to respond to the dishonesty charge or to challenge the dismissal. Under Ohio law, Loudermill was a "classified civil servant," and by statute, as such an employee, could be terminated only for cause and was entitled to administrative review of the dismissal. He filed an appeal with the Civil Service Commission, which, after hearings before a referee and the Commission, upheld the dismissal some nine months after the appeal had been filed. Although the Commission's decision was subject to review in the state courts, Loudermill instead filed suit in Federal District Court, alleging that the Ohio statute providing for administrative review was unconstitutional on its face because it provided no opportunity for a discharged employee to respond to charges against him prior to removal, thus depriving him of liberty and property without due process. It was also alleged that the statute was unconstitutional as applied because discharged employees were not given sufficiently prompt postremoval hearings. This District Court dismissed the suit for failure to state a claim on which relief could be

granted, holding that because the very statute that created the property right in continued employment also specified the procedures were followed, Loudermill was, by definition, afforded all the process due; that the post-termination hearings also adequately protected Loudermill's property interest; and that in light of the Commission's crowded docket the delay in processing his appeal was constitutionally acceptable. In No. 83-1363, petitioner Board of Education fired respondent Donnelly from his job as a bus mechanic because he had failed an eye examination. He appealed to the Civil Service Commission, which ordered him reinstated, but without backpay. He then filed a complaint in Federal District Court essentially identical to Loudermill's, and the court dismissed for failure to state a claim. On a consolidated appeal, the Court of Appeals reversed in part and remanded, holding that both respondents had been deprived of due process and that the compelling private interest in retaining employment, combined with the value of presenting evidence prior to dismissal, outweighed the added administrative burden of a pretermination hearing. But with regard to the alleged deprivation of liberty and Loudermill's 9-month wait for an administrative decision, finding no constitutional violation.

*Held:* All the process that is due is provided by a pretermination opportunity to respond, coupled with posttermination administrative procedures as provided by the Ohio statute; since respondents alleged that they had no chance to respond, the District Court erred in dismissing their complaints for failure to state a claim. Pp. 1491–1496.

(a) The Ohio Statute plainly supports the conclusion that respondents possess property rights in continued employment. The Due Process Clause provides that the substantive rights of life, liberty, and property cannot be deprived except pursuant to constitutionally adequate procedures. The categories of substance and procedure are distinct. "Property" cannot be defined by the procedures provided for its deprivation. Pp. 1491–1493.

(b) The principle that under the Due Process Clause an individual must be given an opportunity for a hearing *before* he is deprived of any significant property interest, required "some kind of hearing" prior to the discharge of an employee who has a constitutionally protected property interest in his employment. The need for some form of pretermination hearing is evident from a balancing of the competing interests at stake: the private interest in retaining employment, the governmental interests in expeditious removal of unsatisfactory employees and avoidance of administrative burdens, and the risk of erroneous termination. Pp. 1493–1495.

(c) The pretermination hearing need not definitively resolve the propriety of the discharge, but should be an initial check against mistaken decisions—essentially a determination of whether there are reasonable grounds to believe that the charges against the employee are true and support the proposed action. The essential requirements of due process are notice and an opportunity to respond. Pp. 1495–1496.

(d) The delay in Loudermill's administrative proceedings did not constitute a separate constitutional violation. The Due Process Clause requires provision of a hearing "at a meaningful time," and here the delay stemmed in part from the thoroughness of the procedures. P. 1496.

721 F.2d 550 (6 Cir. 1983), affirmed and remanded.

\*   \*   \*

James G. Wyman, Cleveland, Ohio, for petitioners in Nos. 83-1362 and 83-1363 and the respondents in No. 83-1363.

Robert M. Fertel, Cleveland, Ohio, for respondents in Nos. 83-1362 and 83-1363 and the petitioner in No. 83-6392.

Justice WHITE delivered the opinion of the Court.

In these cases we consider what determination process must be accorded a public employee who can be discharged only for cause.

I

In 1979 the Cleveland Board of Education, petitioner in No. 83-1362, hired respondent James Loudermill as a security guard. On his job application Loudermill stated that he had never been convicted of a felony. Eleven months later, as part of a routine examination of his employment records, the Board discovered that in fact Loudermill had been convicted of grand larceny in 1968. By letter dated November 3, 1980, the Board's Business Manager informed Loudermill that he had been dismissed because of his dishonesty in filling out the employment application. Loudermill was not afforded an opportunity to respond to the charge of dishonesty or to challenge his dismissal. On November 13, the Board adopted a resolution officially approving the discharge.

Under Ohio law, Loudermill was a "classified civil servant." Ohio Rev.Code Ann. § 124.11 (1984). Such employees can be terminated only for cause, and may obtain administrative review if discharged. § 124.34. Pursuant to this provision, Loudermill filed an appeal with the Cleveland Civil Service Commission on November 12. The Commission appointed a referee, who held a hearing on January 29, 1981. Loudermill argued that he had thought that his 1968 larceny conviction was for a misdemeanor rather than a felony. The referee recommended reinstatement. On July 20, 1981, the full Commission heard argument and orally announced that it would uphold the dismissal. Proposed findings of fact and conclusions of law followed on August 10, and Loudermill's attorneys were advised of the result by mail on August 21.

Although the Commission's decision was subject to judicial review in the state courts, Loudermill instead brought the present suit in the Federal District Court for the Northern District of Ohio. The complaint alleged that § 124.34 was unconstitutional on its face because it did not provide the employee an opportunity to respond to the charges against him prior to removal. As a result, discharged employees were deprived of liberty and property without due process. The complaint also alleged that the provision was unconstitutional as applied because discharged employees were not given sufficiently prompt post-removal hearings.

Before a responsive pleading was filed, the District Court dismissed for failure to state a claim on which relief could be granted. See Fed.Rule Civ.Proc. 12(b) (6). It held that because the very statute that created the property right in continued employment also specified the procedures for discharge, and because those procedures were followed,

Loudermill was, by definition, afforded all the process due. The post-termination hearing also adequately protected Loudermill's liberty interests. Finally, the District Court concluded that, in light of the Commission's crowded docket, the delay in processing Loudermill's administrative appeal was constitutionally acceptable. App. to Pet. for Cert. in No. 83-1362, pp. A36–A42.

The other case before us arises on similar facts and followed a similar course. Respondent Richard Donnelly was a bus mechanic for the Parma Board of Education. In August 1977, Donnelly was fired because he had failed an eye examination. He was offered a chance to retake the examination but did not do so. Like Loudermill, Donnelly appealed to the Civil Service Commission. After a year of wrangling about the timeliness of his appeal, the Commission heard the case. It ordered Donnelly reinstated, though without backpay.[1] In a complaint essentially identical to Loudermill's, Donnelly challenged the constitutionality of the dismissal procedures. The District Court dismissed for failure to state a claim, relying on its opinion in *Loudermill*.

The District Court denied a joint motion to alter or amend its judgment,[2] and the cases were consolidated for appeal. A divided panel of the Court of Appeals for the Sixth Circuit reversed in part and remanded. 721 F.2d 550 (1983). After rejecting arguments that the actions were barred by failure to

exhaust administrative remedies and by res judicata— arguments that are not renewed here—the Court of Appeals found that both respondents had been deprived of due process. It disagreed with the District Court's original rationale. Instead, it concluded that the compelling private interest in retaining employment, combined with the value of presenting evidence prior to dismissal, outweighed the added administrative burden of a pretermination hearing. *Id.*, at 561-562. With regard to the alleged deprivation of liberty, and Loudermill's 9-month wait for an administrative decision, the court affirmed the District Court, finding no constitutional violation. *Id.*, at 566.

Both employers petitioned for certiorari. Nos. 83-1362 and 83-6392. In a cross petition, Loudermill sought review of the rulings adverse to him. No. 83-6392. We granted all three petitions, 467 U.S. 1204, 104 S.Ct. 2384, 81 L.Ed.2d 343 (1984), and now affirm in all respects.

## II

[1] Respondent's federal constitutional claim depends on their having had a property right in continued employment.[3] *Board of Regents v. Roth*, 408 U.S. 564, 576–578, 92 S.Ct. 2701, 2708–2709, 33 L.Ed.2d 1162 (1901). If they did, the State could not deprive them of this property without due process. See *Memphis Light, Gas & Water Div. v. Craft*, 436 U.S. 1, 11–12, 98 S.Ct. 1554, 1561–1562, 56 L.Ed.2d 30 (1978); *Goss v. Lopez*, 429 U.S. 565, 573–574, 95 S.Ct. 729, 735–736, 42 L.Ed.2d 725 (1975).

[2] Property interests are not created by the Constitution, "they are created and their dimensions are defined by existing rules or understandings that stem from an independent source such as state law . . ." *Board of Regents v. Roth, supra*, 408 U.S. at 577, 92 S.Ct., at 2709. See also *Paul v. Davis*, 424 U.S. 693, 709, 96 S.Ct. 1155, 1164, 47 L.Ed.2d 405 (1976). The Ohio Statute plainly creates such an interest. Respondents were "classified civil service employees," Ohio Rev.Code Ann. § 124.11 (1984), entitled to

---

[1] The statute authorizes the Commission to "affirm, disaffirm, or modify the judgment of the appointing authority." Ohio Rev.Code Ann. § 124.34 (1984). Petitioner Parma Board of Education interprets this as authority to reinstate with or without backpay and views the Commission's decision as a compromise. Brief for Petitioner in No. 83-1363, p. 6, n. 3; Tr. of Oral. Arg. 14. The Court of Appeals, however, stated that the Commission lacked the power to award backpay. 721 F.2d 550, 554, n.3 (1983). As the decision of the Commission is not in the record, we are unable to determine the reasoning behind it.

[2] In denying the motion, the District Court no longer relied on the principle that the state legislature could define the necessary procedures in the course of creating the property right. Instead, it reached the same result under a balancing test based on Justice POWELL'S concurring opinion in *Arnett v. Kennedy*, 416 U.S. 134, 168–169, 94 S.Ct. 1633, 1651–1652, 40 L.Ed.2d 15 91974), and the Court's opinion in *Matthews v. Elderidge*, 424 U.S. 319, 96 S.Ct. 893, 47 L.Ed.2d 18 (1976). App. to Pet. for Cert. in No. 83-1362, pp. A54–A57.

[3] Of course, the Due Process Clause also protects interests of life and liberty. The Court of Appeals' finding of a constitutional violation was based solely on the deprivation of a property interest. We address below Loudermill's contention that he has been unconstitutionally deprived of liberty. See n. 13, *infra*.

retain their positions "during good behavior and efficient service," who could not be dismissed "except . . . for . . . misfeasance, malfeasance, or nonfeasance in office," § 124.34.[4] The statute plainly supports the conclusion, reached by lower courts, that respondents possessed property rights in continued employment. Indeed, this question does not seem to have been disputed below.[5] The Parma Board argues, however, that the property right is defined by, and conditioned on, the legislature's choice of procedures by which termination may take place.[6] The procedures were adhered to in these cases. According to petitioner, "[t]o require additional procedures would in effect expand the scope of the

property interest itself." *Id.*, at 27. See also Brief for State of Ohio et al. as *Amici Curiae* 5-10.

This argument, which was accepted by the District Court, has its genesis in the plurality opinion in *Arnett v. Kennedy*, 416 U.S. 134, 94 S.Ct. 1633, 40 L.Ed.2d 15 (1974). *Arnett* involved a challenge by a former federal employee to the procedures by which he was dismissed. The plurality reasoned that where the legislation conferring the substantive right also sets out the procedural mechanism for enforcing that right, the two cannot be separated:

> "The employee's statutorily defined right is not a guarantee against removal without cause in the abstract, but such a guarantee as enforced by the procedures which Congress has designated for the determination of cause.

> \*    \*    \*

> "[W]here the grant of a substantive right is inextricably intertwined with the limitations on the procedures which are to be employed in determining that right, a litigant in the position of appellee must take the bitter with the sweet." *Id.*, at 152–154, 94 S.Ct., at 1643–1644.

This view garnered three votes in *Arnett*, but was specifically rejected by the other six Justices. See *id.*, at 166–167, 94 S.Ct., at 1650–1651 (POWELL, J., joined by BLACKMUN, J.,); *id.*, at 177–178, 185, 94 S.Ct., at 1655–1656 (WHITE, J.,); *id.*, at 211, 94 S.Ct. at 1672 (MARSHALL, J., joined by DOUGLAS and BRENNAN, JJ). Since then, this theory has at times seemed to gather some additional support. See *Bishop v. Wood*, 426 U.S. 341, 355–361, 96 S.Ct. 2074, 2082–2085, 48 L.Ed.2d 684 (1976) (WHITE, J., dissenting); *Goss v. Lopez*, 419 U.S., at 586–587, 95 S.Ct., at 742–743 (POWELL, J., joined by BURGER, C.J., and BLACKMUN and REHNQUIST, JJ., dissenting). More recently, however, the Court has clearly rejected it. In *Vitek v. Jones*, 445 U.S. 480, 491, 100 S.Ct. 1254, 1263, 63 L.Ed.2d 552 (1980), we pointed out that "minimum [procedural] requirements [are] a matter of federal law, they are not diminished by the fact that the State may have specified its own procedures that it may deem

---

[4]The relevant portion of § 124.34 provides that no classified civil servant may be removed except "for incompetency, inefficiency, dishonesty, drunkenness, immoral conduct, insubordination, discourteous treatment of the public, neglect of duty, violation of such sections or the rules of the director of administrative services or the commission, or any other failure of good behavior, or any other acts of misfeasance, malfeasance, or nonfeasance in office."

[5]The Cleveland Board of Education now asserts that Loudermill had no property right under state law because he obtained his employment by lying on the application. It argues that had Loudermill answered truthfully he would not have been hired. He therefore lacked a "legitimate claim of entitlement" to the position. Brief for Petitioner in 83-1362, pp. 1415. For several reasons, we must reject this submission. First, it was not raised below. Second, it makes factual assumptions—that Loudermill lied, and that we would not have been hired had he not done so—that are inconsistent with the allegations of the complaint and inappropriate at this stage of litigation, which had not proceeded past the initial pleadings stage. Finally, the argument relies on a retrospective fiction inconsistent with the undisputed fact that Loudermill was hired and did hold the security guard job. The Board cannot escape its constitutional obligations by rephrasing the basis for termination as a reason why Loudermill should not have been hired in the first place.

[6]After providing for dismissal only for cause, see n. 4, *supra*, § 124.34 states that the dismissed employee is to be provided with a copy of the order of removal giving the reasons therefor. Within ten days of the filing of the order with the Director of Administrative Services, the employee may file a written appeal with the State Personnel Board of Review or the Commission. "In the event such an appeal is filed, the board or commission shall forthwith notify the appointing authority and shall hear, or appoint a trial board to hear, such appeal within thirty days from and after its filing with the board or commission, and it may affirm, disaffirm, or modify the judgment of the appointing authority." Either side may obtain review of the Commission's decision in the State Court of Common Pleas.

adequate for determining the preconditions to adverse official action." This conclusion was reiterated in *Logan v. Zimmerman Brush Co.*, 455 U.S. 422, 432, 102 S.Ct. 1148, 1155, 71 L.Ed.2d 265 (1982), where we reversed the lower court's holding that because the entitlement arose from a state statute, the legislature had the prerogative to define the procedures to be followed to protect that entitlement.

[3] In light of these holdings, it is settled that the "bitter with the sweet" approach misconceives the constitutional guarantee. If a clearer holding is needed, we provide it today. The point is straight-forward: the Due Process Clause provides that certain substantive rights—life, liberty, and property—cannot be deprived except pursuant to constitutionally adequate procedures. The categories of substance and procedure are distinct. Were the rule otherwise, the Clause would be reduced to a mere tautology. "Property" cannot be defined by the procedures provided for its deprivation any more than life or liberty. The right to due process "is conferred, not by legislature grace, but by constitutional guarantee. While the legislature may elect not to confer a property interest in [public] employment, it may not constitutionally authorize the deprivation of such an interest, once conferred, without appropriated procedural safeguards." *Arnett v. Kennedy, supra*, 416 U.S., at 167, 94 S.Ct., at 1650 (POWELL, J., concurring in part and concurring in part and dissenting in part.)

In short, once it is determined that the Due Process Clause applies, "the question remains what process is due." *Morrissey v. Brewer*, 408 U.S. 471, 481, 92 S.Ct. 2593, 2600, 33 L.Ed.2d 484 (1972). The answer to that question is not to be found in the Ohio statute.

### III

[4.5] An essential principle of due process is that a deprivation of life, liberty, or property "be preceded by notice and opportunity for hearing appropriate to the nature of the case." *Mullane v. Central Hanover Bank & Trust Co.*, 339 U.S. 306, 313, 70 S.Ct. 652, 656, 94 L.Ed. 865 (1950). We have described "the root requirement" of the Due Process Clause as being "that an individual be given an opportunity for a hearing before his is deprived of any significant property interest."[7] *Boddie v. Connecticut*, 401 U.S. 371, 379, 91 S.Ct. 780, 786, 28 L.Ed.2d 113 (1971) (emphasis in original); see *Bell v. Burson*, 402 U.S. 535, 542, 91 S.Ct. 1586, 1591, 29 L.Ed.2d 90 (1971). This principle requires "some kind of a hearing" prior to the discharge of an employee who has a constitutionally protected property interest in his employment. *Board of Regents v. Roth*, 408 U.S., at 569–570, 92 S.Ct. at 2705; *Perry v. Sinderman*, 408 U.S. 593, 599, 92 S.Ct. 2694, 2698, 33 L.Ed.2d 570 (1972). As we pointed out last Term, this rule has been settled for some time now. *Davis v. Scherer*, 468 U.S. 183, 192, n. 10, 104 S.Ct. 3012, 3018, n. 10, 82 L.Ed.2d 139 (1984); *id.*, at 200–203, 104 S.Ct., at 3022–302 (BRENNAN, J., concurring in part and dissenting in part). Even decisions finding no constitutional violation in termination procedures have relied on the existence of some pretermination opportunity to respond. For example, in *Arnett* six Justices found constitutional minima satisfied where the employee had access to the material upon which the charge was based and could respond orally and in writing and present rebuttal affidavits. See also *Barry v. Barchi*, 443 U.S. 55, 65, 99 S.Ct. 2642, 2649, 61 L.Ed.2d 365 (1979) (no due process violation where horse trainer whose license was suspended "was given more than one opportunity to present his side of the story").

The need for some form of pretermination hearing, recognized in these cases, is evident from a balancing of the competing interests at stake. These are the private interests in retaining employment, the governmental interest in the expeditious removal of unsatisfactory employees and the avoidance of administrative burdens, and the risk of erroneous termination. See *Matthews v. Eldridge*, 424 U.S. 319, 335, 96 S.Ct. 893, 903, 47 L.Ed.2d 18 (1976).

First, the significance of the private interest in retaining employment cannot be gainsaid. We have frequently recognized the severity of depriving a

---

[7]There are, of course, some situations in which a postdeprivation hearing will satisfy due process requirements. See *Ewing v. Mylinger & Casselberry, Inc.*, 339 U.S. 594, 70 S.Ct. 870, 94 L.Ed. 1088 (1950); *North American Cold Storage Co. v. Chicago*, 211 U.S. 306, 29 S.Ct. 101, 53 L.Ed. 195 (1908).

person of the means of livelihood. See *Fusari v. Steinberg*, 419 U.S. 379, 389, 95 S.Ct. 533, 539, 42 L.Ed.2d 521 (1975); *Bell v. Burson, supra*, 402 U.S., at 539, 91 S.Ct., at 1589; *Goldberg v. Kelly*, 397 U.S. 254, 264, 90 S.Ct. 101, 1018, 25 S.Ed.2d 287 (1970); *Sniadach v. Family Finance Corp.*, 395 U.S. 337, 340, 89 S.Ct. 1820, 1822, 23 L.Ed.2d 349 (1969). While a fired worker may find employment elsewhere, doing so will take some time and is likely to be burdened by the questionable circumstances under which he left his previous job. See *Lefkowitz v. Turley*, 414 U.S. 70, 83–84, 94 S.Ct. 316, 325–326, 38 L.Ed.2d 274 (1973).

Second, some opportunity for the employee to present his side of the case is recurringly of obvious value in reaching an accurate decision. Dismissals for cause will often involve factual disputes. *Cf. California v. Yamasaki*, 442 U.S. 682, 686, 99 S.Ct. 2545, 2550, 61 L.Ed.2d 176 (1979). Even where the facts are clear, the appropriateness or necessity of the discharge may not be; in such cases, the only meaningful opportunity to invoke the discretion of the decisionmaker is likely to be before the termination takes effect. See *Goss v. Lopez*, 419 U.S., at 583–584, 95 S.Ct., at 740–741; *Gagnon v. Scarpelli*, 411 U.S. 778, 784–786, 93 S.Ct. 1756, 1760–1761, 36 L.Ed.2d 656(1973).[8]

[6] The cases before us illustrate these considerations. Both respondents had plausible arguments to make that might have prevented their discharge. The fact that the Commission saw fit to reinstate Donnelly suggests that an error might have been avoided had he been provided an opportunity to make his case to the Board. As for Loudermill, given the Commission's ruling we cannot say that the discharge was mistaken. Nonetheless, in light of the referee's recommendation, neither can we say that a fully informed decisionmaker might not have exercised its discretion and decided not to dismiss him, notwithstanding its authority to do so. In any event, the termination involved arguable issues,[9] and the right to a hearing does not depend on a demonstration of certain success. *Carey v. Piphus*, 435 U.S. 247, 266, 98 S.Ct. 1042, 1053, 55 L.Ed.2d 252 (1978).

The governmental interest in immediate termination does not outweigh these interests. As we shall explain, affording the employee an opportunity to respond prior to termination would impose neither a significant administrative burden nor intolerable delays. Furthermore, the employer shares the employee's interest in avoiding disruption and erroneous decisions; and until the matter is settled, the employer would continue to receive the benefit of the employee's labors. It is preferable to keep a qualified employee on than to train a new one. A governmental employer also has an interest in keeping citizens usefully employed rather than taking the possibly erroneous and counterproductive step of forcing its employees onto the welfare rolls. Finally, in those situations where the employer perceives a significant hazard in keeping the employee on the job,[10] it can avoid the problem by suspending with pay.

---

[8]This is not to say that where state conduct is entirely discretionary the Due Process Clause is brought into play. See *Meachum v. Fano*, 427 U.S. 215, 228, 96 S.Ct. 1723, 1728, 52 L.Ed.2d 172 (1977). The point is that where there is an entitlement, a prior hearing facilitates the consideration of whether a permissible course of action is also an appropriate one. This is one way in which providing "effective notice and informal hearing permitting the [employee] to give his version of the events will provide a meaningful hedge against erroneous action. At least the [employer] will be alerted to the existence of disputes about facts and arguments about cause and effect . . . [H]is discretion will be informed and we think the risk of error substantially reduced." *Goss v. Lopez*, 419 U.S., at 583–584, 95 S.Ct. at 740–741.

[9]Loudermill's dismissal turned not on the objective fact that he was an ex-felon or the inaccuracy of his statement to the contrary, but on the subjective question whether he had lied on his application form. His explanation for the false statement is plausible in light of the fact that he received only a suspended 6-month sentence and a fine on the grand larceny conviction. Tr. of Oral Arg. 35.

[10]In the cases before us, no such danger seems to have existed. The examination Donnelly failed was related to driving school buses, not repairing them. *Id.*, at 39–40. As the Court of Appeals stated, "[n]o emergency was even conceivable with respect to Donnelly." 721 F.2d., at 562. As for Loudermill, petitioner states that "to find that we have a person who is an ex-felon as our security guard is very distressful to us." Tr. of Oral Arg. 19. But the termination was based on the presumed misrepresentation on the employment form, not on the felony conviction. In fact, Ohio law provides that an employee "shall not be disciplined for acts," including criminal convictions, occurring more than two years previously. See Ohio Admin. Code § 124-3-04 (1979). Petitioner concedes that Loudermill's job performance was fully satisfactory.

## IV

[7] The foregoing considerations indicate that the pretermination "hearing," though necessary, need not be elaborate. We have pointed out that "[t]he formality and procedural requisites for the hearing can vary, depending upon the importance of the interests involved and the nature of the subsequent proceedings." *Boddie v. Connecticut,* 401 U.S., at 378, 91 S.Ct., at 786. See *Cafeteria Workers v. McElroy,* 367 U.S. 886, 894-895, 81 S.Ct. 1743, 1748, 6 L.Ed.2d 1230 (1961). In general, "something less" than a full evidentiary hearing is sufficient prior to adverse administrative action. *Matthews v. Eldridge,* 424 U.S. at 343, 96 S.Ct., at 907. Under state law, respondents were later entitled to a full administrative hearing and judicial review. The only question is what steps were required before the termination took effect.

In only one case, *Goldberg v. Kelly,* 397 U.S. 254, 90 S.Ct. 1011, 25 L.Ed.2d 287 (1970), has the Court required a full adversarial evidentiary hearing prior to adverse governmental action. However, as the *Goldberg* Court itself pointed out, see *id.,* at 264, 90 S.Ct., at 1018, that case presented significantly different considerations than are present in the context of public employment. Here, the pretermination hearing need not definitively resolve the propriety of the discharge. It should be an initial check against mistaken decisions—essentially, a determination of whether there are reasonable grounds to believe that the charges against the employee are true and support the proposed action. See *Bell v. Burson,* 402 U.S., at 540, 91 S.Ct., at 1590.

The essential requirements of due process, and all that respondents seek or the Court of Appeals required, are notice and an opportunity to respond. The opportunity to present reasons, either in person or in writing, why proposed action should not be taken is a fundamental due process requirement. See Friendly, "Some Kind of Hearing," 123 U.Pa.L.Rev. 1267, 1281 (1975). The tenured public employee is entitled to oral or written notice of the charges against him, an explanation of the employer's evidence, and an opportunity to present his side of the story. See *Arnett v. Kennedy,* 416 U.S., at 170-171, 94 S.Ct., at 1652-1653 (opinion of POWELL, J.); *id.,* at 195-196, 94 S.Ct., at 1664-1665 (opinion of WHITE, J.); see also *Goss v. Lopez,* 419 U.S., at 581, 95 S.Ct., at 740. To require more than this prior to termination would intrude to an unwarranted extent on the government's interest in quickly removing an unsatisfactory employee.

## V

[8] Our holding rests in part on the provisions in Ohio law for a full post-termination hearing. In his cross-petition Loudermill asserts, as a separate constitutional violation, that his administrative proceedings took too long.[11] The Court of Appeals held otherwise, and we agree.[12] The Due Process Clause requires provision of a hearing "at a meaningful time." *E.g., Armstrong v. Manzo,* 380 U.S. 545, 552, 85 S.Ct. 1187, 1191, 14 L.Ed.2d 62 (1965). At some point, a delay in the post-termination hearing would become a constitutional violation. See *Barry v. Barchi,* 443 U.S., at 66, 99 S.Ct., at 2650. In the present case, however, the complaint merely recites the course of proceedings and concludes that the denial of a "speedy resolution" violated due process. App. 10. This reveals nothing about the delay except that

---

[11]Loudermill's hearing before the referee occurred two and one-half months after he filed his appeal. The Commission issued its written decision six and one-half months after that. Administrative proceedings in Donnelly's case once it was determined that they could proceed at all, were swifter. A writ of mandamus requiring the Commission to hold a hearing was issued on May 9, 1978; the hearing took place on May 30; the order of reinstatement was issued on July 6.

Section 124.34 provides that a hearing is to be held within 30 days of the appeal, though the Ohio courts have ruled that the time limit is not mandatory. *E.g., In re Bronkar,* 53 Ohio Misc. 13, 17, 372 N.E.2d 1345, 1347 (Com.Pl.1977). The statute does not provide a time limit for the actual decision.

[12]It might be argued that once we find a due process violation in the denial of pretermination hearing we need not and should not consider whether the posttermination procedures were adequate. See *Barry v. Barchi,* 443 U.S. 55, 72-74, 99 S.Ct. 2642, 2653-2654, 61 L.Ed.2d 365 (1979) (BRENNAN, J., concurring in part). We conclude that it is appropriate to consider this issue, however, for three reasons. First, the allegation of a distinct due process violation in the administrative delay is not an alternative theory supporting the same relief, but a separate claim altogether. Second, it was decided by the court below and is raised in the cross-petition. Finally, the existence of posttermination procedures is relevant to the necessary scope of pretermination procedures.

it stemmed in part from the thoroughness of the procedures. A 9-month adjudication is not, of course, unconstitutionally lengthy *per se*. Yet Loudermill offers no indication that his wait was unreasonably prolonged other than the fact that it took nine months. The chronology of the proceedings set out in the complaint, coupled with the assertion that nine months is too long to wait, does not state a claim of a constitutional deprivation.

### VI

We conclude that all the process that is due is provided by a predetermination opportunity to respond, coupled with post-termination administrative procedures as provided by the Ohio statute. Because respondents allege in their complaints that they had no chance to respond, the District Court erred in dismissing for failure to state a claim. The judgment of the Court of Appeals is affirmed, and the case is remanded for further proceedings consistent with this opinion.

*So ordered.*

Justice MARSHALL, concurring in part and concurring in the judgment.

I agree wholeheartedly with the Court's express rejection of the theory of due process, urged upon us by the petitioner Boards of Education, that a public employee who may be discharged only for cause may be discharged by whatever procedures the legislature chooses. I therefore join Part II of the opinion for the Court. I also agree that, before discharge, the respondent employees were entitled to the opportunity to respond to the charges against them (which is all they requested), and that the failure to accord them that opportunity was a violation of their constitutional rights. Because the Court holds that the respondents were due all the process they requested, I concur in the judgment of the Court.

I write separately, however, to reaffirm my belief that public employees who may be discharged only for cause are entitled, under the Due Process Clause of the Fourteenth Amendment, to more than respondents sought in this case. I continue to believe that *before the decision is made to terminate an employee's wages*, the employee is entitled to an opportunity to test the strength of the evidence "by confronting and cross-examining adverse witnesses and by presenting witnesses on his own behalf, whenever there are substantial disputes in testimonial evidence," *Arnett v. Kennedy*, 416 U.S. 134, 214, 94 S.Ct. 1633, 1674, 40 L.Ed.2d 15 (1974) (MARSHALL, J., dissenting). Because the Court suggests that even in this situation due process required no more than notice and an opportunity to be heard before wages are cut off, I am not able to join the Court's opinion in its entirety.

To my mind, the disruption caused by a loss of wages may be so devastating to an employee that, whenever there are substantial disputes about the evidence, additional pre-deprivation procedures are necessary to minimize the risk of an erroneous termination. That is, I place significantly greater weight than does the Court on the public employee's substantial interest in the accuracy of the pretermination proceeding. After wage termination, the employee often must wait months before his case is finally resolved, during which time he is without wages from his public employment. By limiting the procedures due prior to termination of wages, the Court accepts an impermissible high risk that a wrongfully discharged employee will be subjected to this often lengthy wait for vindication, and to the attendant and often traumatic disruptions to this personal and economic life.

Considerable amounts of time may pass between the termination of wages and the decision in a post-termination evidentiary hearing—indeed, in this case nine months passed before Loudermill received a decision from his postdeprivation hearing. During this period the employee is left in limbo, deprived of his livelihood and of wages on which he may well depend for basic substance. In that time, his ability to secure another job might be hindered, either because of the nature of the charges against him, or because of the prospect that he will return to his prior public employment if permitted. Similarly, his access to unemployment benefits might seriously be constrained, because many States deny unemployment compensation to workers discharged for cause. Absent an interim source of wages, the employee might be unable to meet his basic, fixed costs, such as food, rent or mortgage payments. he would be forced to spend his savings, if he had any, and to convert his possessions to cash before becoming eligible for public assistance. Even in that instance

> "[t]he substitution of a meager welfare grant for a regular paycheck may bring with it painful and irremediable personal as well as financial disloca- tion. A child's education may be interrupted, a family's home lost, a person's relationship with his friends and even his family may be irrevoca- bly affected. The costs of being forced, even temporarily, onto the welfare rolls because of wrongful discharge from tenured Government employment cannot be so easily discounted." id., at 221, 94 S.Ct., at 1677.

Moreover, it is in no respect certain that a prompt postdeprivation hearing will make the employee eco- nomically whole again, and the wrongfully discharged employee will almost inevitably suffer irreparable in- jury. Even if reinstatement is forthcoming, the same might not be true of backpay—as it was not to re- spondent Donnelly in this case—and the delay in re- ceipt of wages would thereby be transformed into a permanent deprivation. Of perhaps equal concern, the personal trauma experienced during the long months in which the employee awaits decision, dur- ing which he suffers doubt, humiliation, and the loss of an opportunity to perform work, will never be rec- ompensed, and indeed probably could not be with dollars alone.

That these disruptions might fall upon a justifi- ably discharged employee is unfortunate; that they might fall upon a wrongfully discharged employee is simply unacceptable. Yet in requiring only that the employee have an opportunity to respond before his wages are cut off, without affording him mean- ingful chance to present a defense, the Court is will- ing to accept an impermissible high risk of error with respect to a deprivation that is substantial.

Were there any guarantee that the post-depri- vation hearing and ruling would occur promptly, such as within a few days of the termination of wages, then this minimal pre-deprivation process

might suffice. But there is no such guarantee. On a practical level, if the employer had to pay the employee until the end of the proceeding, the employer obviously would have an incentive to resolve the issue expeditiously. The employer loses this incentive if the only suffering as a result of the delay is borne by the wage earner, who eagerly awaits the decision of his livelihood. Nor has this Court grounded any guarantee of this kind in the Constitution. Indeed, this Court has in the past approved, at least implicitly, an average 10 or 11- month delay in the receipt of a decision on Social Security benefits, *Matthews v. Eldridge*, 424 U.S. 319, 341–342, 96 S.Ct. 893, 905–906, 47 L.Ed.2d 18 (1976), and, in the case of respondent Loudermill, the Court gives a stamp of approval to a process that took nine months. The hardship inevitably increases as the days go by, but nevertheless the Court coun- tenances such delay. The adequacy of the predepri- vation and postdeprivation procedures are inevitably intertwined, and only a constitutional guarantee that the latter will be immediate and complete might alleviate my concern about the possibility of a wrongful termination of wages.

The opinion for the Court does not confront this reality. I cannot and will not close my eyes today—as I could not 10 years ago—to the econom- ic situation of great numbers of public employees, and to the potentially traumatic effect of a wrongful discharge on a working person. Given that so very much is at stake, I am unable to accept the Court's narrow view of the process due to a public employ- ee before his wages are terminated, and before he begins the long wait for a public agency to issue a final decision in his case.

Justice BRENNAN, concurring in part and dis- senting in part.

Today the Court puts to rest any remaining debate over whether public employers must provide meaningful notice and hearing procedures before discharging an employee for cause. As the Court convincingly demonstrates, the employee's right to fair notice and an opportunity to "present his side of the story" before discharge is not a matter of leg- islative grace, but of "constitutional guarantee."

*Ante,* at 1493, 1495. This principle, reaffirmed by the Court today, has been clearly discernible in our "repeated pronouncements" for many years. See *Davis v. Scherer,* 468 U.S. 183, 203, 104 S.Ct. 3012, 3023, 82 L.Ed.2d 139 (1984) (BRENNAN, J., concurring in part and dissenting in part).

Accordingly, I concur in Parts I–IV of the Court's opinion. I write separately to comment on two issues the Court does not resolve today, and to explain my dissent from the result in Part V of the Court's opinion.

## I

First, the Court today does not prescribe the precise form of required pretermination procedures in cases where an employee disputed the *facts* proffered to support his discharge. The cases at hand involve, as the Court recognizes, employees who did not dispute the facts but had "plausible arguments to make that might have prevented their discharge." *Ante,* at 1494. In such cases, notice and an "opportunity to present reasons," *ante,* at 1495, are sufficient to protect the important interests at stake.

As the Court correctly notes, other cases "will often involve factual disputes," *ante,* at 1494, such as allegedly erroneous records or false accusations. As Justice MARSHALL has previously noted and stresses again today, *ante* at 1497, where there exist not just plausible arguments to be made, but also "substantial disputes in testimonial evidence, " due process may well require more than a simple opportunity to argue or deny. *Arnett v. Kennedy,* 416 U.S. 134, 214, 94 S.Ct. 1633, 1674, 40 L.Ed.2d 15 (1974) (MARSHALL, J., dissenting). The Court acknowledges that what the Constitution requires prior to discharge, in general terms, is pretermination procedures sufficient to provide "an initial check against determination of whether there are reasonable grounds to believe that the charges against the employee are *true* and support the proposed action." *Ante,* at 1495 (emphasis added). When factual disputes are involved, therefore, an employee may deserve a fair opportunity before discharge to produce contrary records or testimony, or even to confront an accuser in front of a decisionmaker. Such an opportunity might not necessitate "elaborate" procedures, see *ante,* at 1395,

but the fact remains that in some cases only such an opportunity to challenge the source or produce contrary evidence will suffice to support a finding that there are "reasonable grounds" to believe accusations are "true."

Factual disputes are not involved in these cases, however, and the "very nature of due process negates any concept of inflexible procedures universally applicable to every imaginable situation." *Cafeteria Workers v. McElroy,* 367 U.S. 886, 895, 81 S.Ct. 1743, 1748, 6 L.Ed.2d 1230 (1961). I do not understand Part IV to foreclose the views expressed above or by Justice MARSHALL, *ante,* p. 1497, with respect to discharges based on disputed evidence or testimony. I therefore join Parts I–IV of the Court's opinion.

## II

The second issue not resolved today is that of administrative delay. In holding that Loudermill's administrative proceedings did not take too long, the Court plainly does *not* state a flat rule that 9-month delays in deciding discharge appeals will pass constitutional scrutiny as a matter of course. To the contrary, the Court notes that a full post-termination hearing and decision must be provided at "a meaningful time" and that "[a]t some point, a delay in the post-termination hearing would become a constitutional violation." *Ante,* at 1496. For example, in *Barry v. Barchi,* 443 U.S. 55, 99 S.Ct. 2642, 61 L.Ed.2d 365 (1979), we disapproved as "constitutionally infirm" the shorter administrative delays that resulted under a statute that required "prompt" postsuspension hearings for suspended racehorse trainers with decision to follow within 30 days of the hearing. *Id.,* at 61, 66, 99 S.Ct., at 2647, 2650. As Justice MARSHALL demonstrates, when an employee's wages are terminated pending administrative decision, "hardship inevitably increases as the days go by." *Ante,* at 1498; see also *Arnett v. Kennedy, supra,* 416 U.S., at 194, 94 S.Ct., at 1664 (WHITE, J., concurring in part and dissenting in part) ("The impact on the employee of being without a job pending a full hearing is likely to be considerable because '[m]ore than 75 percent of actions contested within employing agencies require longer to decide than the 60 days required by . . . regulations'") (citation omitted). In such cases the

Constitution itself draws a line, as the Court declares, "at some point" beyond which the State may not continue a deprivation absent decision.[13] The holding in Part V is merely that, in this particular case, Loudermill failed to allege facts sufficient to state a cause of action, and not that nine months can never exceed constitutional limits.

### III

Recognizing the limited scope of that holding in Part V, I must still dissent from its result, because the record in this case is insufficiently developed to permit an informed judgment on the issue of overlong delay. Loudermill's complaint was dismissed without answer from the respondent Cleveland Civil Service Commission. Allegations at this early stage are to be liberally construed, and "[i]t is axiomatic that a complaint should not be dismissed unless 'it appears beyond doubt that the plaintiff can prove no set of facts in support of his claim which would entitle him to relief." *Mclain v. Real Estate Bd. of New Orleans, Inc.*, 444 U.S. 232, 246, 100 S.Ct. 502, 511, 62 L.Ed.2d 441 (1980) (citation omitted). Loudermill alleged that it took the Commission over two and one-half months simply to hold a hearing in his case, over two months *more* to issue a non-binding interim decision, and more than three and one-half months after *that* to deliver a final decision. Complaint 20, 21 App. 10.[14] The

commission provided no explanation for these significant gaps in the administrative process; we do not know if they were due to an overabundance of appeals, Loudermill's own foot-dragging, bad faith on the part of the Commission, or any other variety of reasons that might affect our analysis. We do know, however, that under Ohio law the Commission is obligated to hear appeals like Loudermill's "within thirty days." Ohio Rev.Code Ann. § 124.34 (1984).[15] Although this statutory limit has been viewed only as "directory" by Ohio courts, those courts have also made it clear that when the limit is exceeded, "[t]he burden of proof [is] placed on the [Commission] to illustrate to the court that the failure to comply with the 30-day requirement . . . was reasonable." *In re Bronkar*, 53 Ohio Misc. 13, 17, 372 N.E.2d 1345, 1347 (Com.Pl. 1977). I cannot conclude on this record that Loudermill could prove "no set of facts" that might have entitled him to relief after nine months of waiting.

The Court previously has recognized that constitutional restraints on the timing, no less than the form, of a hearing and decision "will depend on appropriate accommodation of the competing interests involved." *Goss v. Lopez,* 419 U.S. 565, 579, 95 S.Ct. 729, 738–739, 42 L.Ed.2d 725 (1975). The relevant interests have generally been recognized as threefold: "the importance of the private interest and the length or finality of the deprivation, the likelihood of governmental interests involved." *Logan v. Zimmerman Brush Co,* 455 U.S. 422, 434, 102 S.Ct. 1148, 1157, 71 L.Ed.2d 265 (1982) (citations omitted); accord *Matthews v. Eldridge,* 424 U.S. 319, 334–335, 96 S.Ct. 893, 902–903, 47 L.Ed.2d 18 (1976); cf. *United States v. $8,850,* 461 U.S. 555, 564, 103 S.Ct. 2005, 2012, 76 L.Ed.2d 143 (1983) (four-factor test

---

[13]Posttermination administrative procedures designed to determine fully and accurately the correctness of discharge actions are to be encouraged. Multiple layers of administrative procedure, however, may not be created merely to smother a discharged employee with "thoroughness," effectively destroying his constitutionally protected interests by overextension. Cf. *ante,* at 1496 ("thoroughness" of procedures partially explains delay in this case).

[14]The interim decision, issued by a hearing examiner, was in Loudermill's favor and recommended his reinstatement. But Loudermill was not reinstated nor were his wages even temporarily restored; in fact there apparently exists no provision for such interim relief or restoration of backpay under Ohio's statutory scheme. See *ante,* at 1490, n. 1; cf. *Arnett v. Kennedy,* 416 U.S. 134, 196, 94 S.Ct. 1633, 1665, 40 L.Ed.2d 15 (1974) (WHITE, J., concurring in part and dissenting in part) (under federal civil service law, discharged employee's wages are only "provisionally cut off" pending appeal); *id.,* at 146 (opinion of REHNQUIST, J.) (under federal system, backpay is automatically refunded "if the [discharged] employee is reinstated on appeal"). See also N.Y.Civ.Serv.Law § 75 (3) (McKinney 1983) (suspension without pay pending determination of removal charges

may not exceed 30 days). Moreover, the final decision of the Commission to reverse the hearing examiner apparently was arrived at without any additional evidentiary development; only further argument was had before the Commission. 721 f.2d 550, 553 (CA6 1983). These undisputed facts lead me at least to question the administrative value of, and justification for, the nine-month period it took to decide Loudermill's case.

[15]A number of other States similarly have specified time limits for hearings and decisions on discharge appeals taken by tenured public employees, indicating legislative consensus that a month or two normally is sufficient time to resolve such actions. No state statutes permit administrative delays of the length alleged by Loudermill.

for evaluating constitutionally of delay between time of property seizure and initiation of forfeiture action). "Little can be said on when a delay becomes presumptively improper, for the determination necessarily depends on the facts of the particular case." *Id.*, at 565, 103 S.Ct., at 2012.

Thus the constitutional analysis of delay requires some development of the relevant factual context when a plaintiff alleges, as Loudermill has, that the administrative process had taken longer than some minimal amount of time. Indeed, all of our precedents that have considered administrative delays under the Due Process Clause, either explicitly or *sub silentio,* have been decided only after more complete proceedings in the District Courts. See *e.g. $8,850, supra;* Barry v. Barchi, 443 U.S. 55, 99 S.Ct. 2642, 61 L.Ed.2d 365 (1979); *Arnett v. Kennedy,* 416 U.S. 134, 94 S.Ct. 1633, 40 L.Ed.2d 15 (1974); *Matthews v. Eldridge, supra.* Yet in Part V. the Court summarily holds Loudermill's allegations insufficient, without advertising to any considered balancing of interests. Disposal of Loudermill's complaint without examining the competing interests involved marks an unexplained departure from the careful multifaceted analysis of the facts we consistently have employed in the past.

I previously have stated my view that "[t]o be meaningful, an opportunity for a full hearing and determination must be afforded at least at a time when the potentially irreparable and substantial harm caused by a suspension can still be avoided—*i.e.,* either before or immediately after suspension." *Barry v. Barchi, supra,* 443 U.S., at 74, 99 S.Ct., at 2654 (BRENNAN, J., concurring in part).

Loudermill's allegations of months-long administrative delay, taken together with the facially divergent results regarding length of administrative delay found in *Barchi* as compared to *Arnett,* see n. 4. *supra,* are sufficient in my mind to require further factual development. In no other way can the third *Matthews* factor—"the Government's interest, including the function involved and the fiscal and administrative burdens that the additional or substitute procedural requirement [in this case, a speedier

hearing and decision] would entail," 424 U.S., at 335, 96 S.Ct., at 903—sensibly be evaluated in this case. I therefore would remand the delay issue to the District court for further evidentiary proceedings consistent with the *Matthews* approach. I respectfully dissent from the Court's contrary decision in Part V.

Justice REHNQUIST, dissenting.

In *Arnett v. Kennedy,* 416 U.S. 134, 94 S.Ct. 1633, 40 L.Ed.2d 15 (1974), six members of this Court agreed that a public employee could be dismissed for misconduct without a full hearing prior to termination. A plurality of Justices agreed that the employee was entitled to exactly what Congress gave him, and no more. The Chief Justice, Justice STEWART, and I said:

"Here appellee did have a statutory expectancy that he not be removed other than for 'such cause as will promote the efficiency of [the] service.' But the very section of the statute which granted him that right, a right which had previously existed only by virtue of administrative regulation, expressly provided also for the procedure by which 'cause' was to be determined, and expressly omitted the procedural guarantees which appellee insists are mandated by the Constitution. Only by bifurcating the very sentence of the Act of Congress which conferred upon appellee the right not to be removed save for cause could it be said that he had an expectancy of that substantive right without the procedural limitations which Congress attached to it. In the area of federal regulation of government employees, where in the absence of statutory limitation the government employer has had virtually uncontrolled latitude in decisions as to hiring and firing, *Cafeteria Workers v. McElroy,* 367 U.S. 886, 896–897, 81 S.Ct. 1743, 1749–1750, 6 L.Ed.2d 1230 (1961), we do not believe that a statutory enactment such as the Lloyd-LaFollette Act may be parsed as discretely as appellee urges. Congress was obviously intent on according a measure of statutory job security to governmental employees which they had not previously enjoyed, but was likewise intent on excluding more elaborate procedural requirements which it felt would make the operation of the new scheme unnecessarily burdensome in practice. Where the focus of legislation was thus strongly on the procedural mechanism for enforcing the substantive right which as simultaneously conferred, we decline to conclude that the substantive right may be viewed wholly apart from the procedure provided for its enforcement. The employee's statutorily defined right is not a guarantee against removal without cause in the abstract, but such a guarantee as enforced by the procedures which Congress has designated for the determination of cause." *Id.*, at 151–152, 94 S.Ct., at 1643.

In these cases, the relevant Ohio statute provides in its first paragraph that

"[t]he tenure of every officer or employee in the classified service of the state and the counties, civil service townships, cities, city health districts, general health districts, and city school districts thereof, holding a position under this chapter of the Revised Code, shall be during good behavior and efficient service and no such officer or employee shall be reduced in pay or position, suspended, or removed, except . . . for incompetency, inefficiency, dishonesty, drunkenness, immoral conduct, insubordination, discourteous treatment of the public, neglect of duty, violation of such sections or the rules of the director of administrative services or the commission, or any other failure of good behavior, or any other acts of misfeasance, malfeasance, or nonfeasance in office." Ohio Rev.Code Ann. § 124.34 (1984).

The very next paragraph of this section of the Ohio Revised Code provides that in the event of suspension of more than three days or removal the appointing authority shall furnish the employee with the stated reasons for his removal. The next paragraph provides that within 10 days following the receipt of such a statement, the employee may

appeal in writing to the State Personnel Board of Review or the Commission, such appeal shall be heard within 30 days from the time of its filing, and the Board may affirm, disaffirm, or modify the judgment of the appointing authority.

Thus in one legislative breath Ohio had conferred upon civil service employees such as respondents in these cases a limited form of tenure during good behavior, and prescribed the procedures by which that tenure may be terminated. Here, as in *Arnett,* "[t]he employee's statutorily defined right is not a guarantee against removal without cause in the abstract, but such a guarantee as enforced by the procedures which [the Ohio Legislature] has designated for the determination of cause." 416 U.S., at 152, 94 S.Ct., at 1643 (opinion of REHNQUIST, J.) We stated in *Board of Regents v. Roth*, 408 U.S. 564, 577, 92 S.Ct. 2701, 2709, 33 L.Ed.2d 548 (1972):

"Property interests, of course, are not created by the Constitution. Rather, they are created and their dimensions are defined by existing rules or understandings that stem from an independent source such as state law—rules or understandings that secure certain benefits and that support claims of entitlement to those benefits."

We ought to recognize the totality of the State's definition of the property right in question, and not merely seize upon one on several paragraphs in a unitary statute to proclaim that in that paragraph the State has inexorably conferred upon a civil service employee something which it is powerless under the United States constitution to qualify in the next paragraph of the statute. This practice ignores our duty under *Roth* by its selective choice from among the sentences the Ohio Legislature chooses to use in establishing and qualifying a right.

Having concluded by this somewhat tortured reasoning that Ohio has created a property right in the respondents in these cases, the Court naturally proceeds to inquire what process is "due" before the respondents may be divested of that right. This customary "balancing" inquiry conducted by the Court in these cases reaches a result that is quite unobjectionable, but it seems to me that it is devoid of any principles which will either instruct or endure. The balance is simply an ad hoc weighing which depends to a great extent upon how the Court subjectively views the underlying interests at stake. The results in previous cases and in these cases have been quite unpredictable. To paraphrase Justice Black, today's balancing act requires a "pretermination opportunity to respond" but there is nothing that indicates what tomorrow's will be. *Goldberg v. Kelly,* 397 U.S. 254, 276, 90 S.Ct. 1011, 1024, 25 L.Ed.2d 287 (1970) (Black, J., dissenting). The results from today's balance certainly do not jibe with the results in *Goldberg* or *Matthews v. Eldridge,* 424 U.S. 319, 96 S.Ct. 893, 47 L.Ed.2d 18 (1976). The lack of any principled standards in this area means that these procedural due process cases will recur time and again. Every different set of facts will present a new issue on what process was due and when. One way to avoid this subjective and varying interpretation of the Due Process Clause in cases such as these is to hold that one who avails himself of government entitlements accepts the grant of tenure along with its inherent limitations.

Because I believe that the Fourteenth Amendment of the United States Constitution does not support the conclusion that Ohio's effort to confer a limited form of tenure upon respondents resulted in the creation of a "property right" in their employment, I dissent.

## Review Questions

1. Does your state provide any exceptions to the at-will doctrine? Are these exceptions created by the courts? Legislature?

2. Does your state have whistle-blower protections?

3. What is a collective bargaining agreement?

4. What is the issue in the *Safford* case?

5. Does your fire service organization test for illegal drugs?

6. What is the issue in *Welsh v. City of Great Falls*?

7. Does your fire service organization have any rules or regulations regarding physical disability?

8. Would the *Welsh* case have a different decision after the Americans with Disabilities Act?

9. What damages does your state permit in a wrongful termination action?

10. What is the at-will doctrine?

*Chapter*

# 10

# Family and Medical Leave Act

The thing that impresses me most about North America is the way parents obey their children.

*Edward, Duke of Windsor*

## *Learning Objectives*

■ To gain an understanding of the requirements of the FMLA.

■ To gain an understanding on how this law impacts your organization.

**Family and Medical Leave Act of 1993**
a law protecting employees who work for qualified employers that gives them the right to request a specified period of time away from the job for a specific circumstance

**serious health condition**
an illness, injury, impairment, or physical or mental condition involving either inpatient care or continuing treatment by a health care provider

One of the newest laws to impact many fire service organizations is the **Family and Medical Leave Act** of 1993.[1] The key provisions of this law for fire and emergency service organizations are Title I and Title II. In Title I, an *eligible employee* (i.e., individual provided coverage under this law) is defined as being an employee who has been employed at least twelve months with the fire service organization and who has provided at least 1,250 hours of service during the twelve months. Fire service organizations with fewer than fifty employees within a 75-mile radius of the worksite are excluded from coverage.

Title I also includes the broadly construed definitions of the terms *parent* and *son or daughter* (includes biological, adopted, or foster child or legal ward), but does not define *spouse*. **Serious health condition** is defined as "an illness, injury, impairment, or physical or mental condition" involving either inpatient care or continuing treatment by a health care provider.

Eligible firefighters and EMS personnel are entitled to twelve unpaid workweeks of leave during any twelve-month period for three basic reasons: (1) the birth, adoption, or placement for foster care of a child (within twelve months of the birth or placement); (2) serious health condition of a spouse, child, or parent; or (3) the firefighter's own serious health condition.

In general, the fire service organization may require that the eligible firefighter provide certification of a serious health condition of himself or herself or a family member. Certification includes the date of the serious health condition, the duration of the condition, appropriate facts regarding the condition, and a statement that the eligible firefighter needs to care for the spouse or child. For cases of intermittent leave, the dates of the leave should also be noted. The fire service organization may require that a second opinion be obtained by the firefighter at his or her own expense. The second opinion may not be provided by the health care provider employed by the fire service organization. In the event of a conflict in opinions, a third and final opinion can be required at the expense of the fire service organization.

The firefighter who completes a period of leave is to be returned either to the same position or to a position equivalent in pay, benefits, and other terms and conditions of employment. This leave cannot result in the loss of any previously accrued seniority or employment benefits but neither of these benefits is required to accrue during the leave. Health benefits are to continue throughout the leave. The fire service organization may be required to pay the health coverage premiums but may recover these premiums if the firefighter fails to return to the job following the leave.

As with most laws, the fire service organization is prohibited from discriminating against the firefighter for use of the Family and Medical Leave Act. The agency vested with the authority to enforce the FMLA is the U.S. Department of Labor. The powers to investigate, prosecute, maintain records, and enforce the

---

[1] 29 U.S.C. § 2601–2654

law are similar to the Fair Labor Standards Act. A Commission on Leave has also been established on a temporary basis to study the impact of the FMLA. The statute of limitations on filing an action under the FMLA is 2 years from the last event and 3 years if the violation is willful. For firefighters in the federal civil service system, Chapter 63 of Title 5, U.S. Code, extends coverage to federal civil service employees.

Fire service organizations should be aware that the FMLA is the foundation for coverage in this area, and states or municipalities may provide benefits equal to or greater than the FMLA. Nothing in the FMLA is meant to discourage fire service organizations from offering more generous leave policies.

## FAMILY AND MEDICAL LEAVE ACT

29 U.S.C. §§ 2601–2654

### SEC. 101. Definitions.

As used in this title:

(1) COMMERCE.—The terms "commerce" and "industry or activity affecting commerce" mean any activity, business, or industry in commerce or in which a labor dispute would hinder or obstruct commerce or the free flow of commerce, and include "commerce" and any "industry affecting commerce", as defined in paragraphs (1) and (3) of section 501 of the Labor Management Relations Act, 1947 (29 U.S.C. 142(1) and (3)).

(2) ELIGIBLE EMPLOYEE.—
(A) IN GENERAL.—The term "eligible employee" means an employee who has been employed—
(i) for at least 12 months by the employer with respect to whom leave is requested under section 102; and
(ii) for at least 1,250 hours of service with such employer during the previous 12-month period.
(B) EXCLUSIONS.—The term "eligible employee" does not include—
(i) any Federal officer or employee covered under

Subchapter V of Chapter 63 of Title 5, United States Code (as added by Title II of this Act); or
(ii) any employee of an employer who is employed at a worksite at which such employer employs less than 50 employees if the total number of employees employed by that employer within 75 miles of that worksite is less than 50.
(C) DETERMINATION.—For purposes of determining whether an employee meets the hours of service requirement specified in subparagraph (AXii), the legal standards established under section 7 of the Fair Labor Standards Act of 1938 (29 U.S.C. 207) shall apply.

(3) EMPLOY; EMPLOYEE; STATE.—The terms employ, employee, and "State" have the same meanings given such terms in subsections (c), (e), and (g) of section 3 of the Fair Labor Standards Act of 1938 (29 U.S.C. 203(c), (e), and (g)).

(4) EMPLOYER.—
(A) IN GENERAL.—The term "employer"—
(i) means any person engaged in commerce or in any industry or activity affecting commerce who employs 50 or more employees for each working day during each of 20 or more calendar workweeks in the current or preceding calendar year;

(ii) includes—
(I) any person who acts, directly or indirectly, in the interest of an employer to any of the employees of such employer; and
(II) any successor in interest of an employer; and
(iii) includes any "public agency", as defined in section 3(x) of the Fair Labor Standards Act of 1938 (29 U.S.C. 203(x)).
(B) PUBLIC AGENCY.—For purposes of subparagraph (A)(iii), a public agency shall be considered to be a person engaged in commerce or in an industry or activity affecting commerce.

(5) EMPLOYMENT BENEFITS.—The term "employment benefits" means all benefits provided or made available to employees by an employer, including group life insurance, health insurance, disability insurance, sick leave, annual leave, educational benefits, and pensions, regardless of whether such benefits are provided by a practice or written policy of an employer or through an "employee benefit plan", as defined in section 3(3) of the Employee Retirement Income Security Act of 1974 (29 U.S.C. 1002(3)).

(6) HEALTH CARE PROVIDER.—The term health care provider" means—
(A) a doctor of medicine or osteopathy who is authorized to practice medicine or surgery (as appropriate) by the State in which the doctor practices; or
(B) any other person determined by the Secretary to be capable of providing health care services.

(7) PARENT.—The term "parent" means the biological parent of an employee or an individual who stood *in loco parentis* to an employee when the employee was a son or daughter.

(8) PERSON.—The term "person" has the same meaning given such term in section 3(a) of the Fair Labor Standards Act of 1938 (29 U.S.C. 203(a)).

(9) REDUCED LEAVE SCHEDULE.—The term "reduced leave schedule" means a leave schedule that reduces the usual number of hours per workweek, or hours per workday, of an employee.

(10) SECRETARY.—The term "Secretary" means the Secretary of Labor.

(11) SERIOUS HEALTH CONDITION—The term "serious health condition" means an illness, injury, impairment, or physical or mental condition that involves—
(A) inpatient care in a hospital, hospice, or residential medical care facility; or
(B) continuing treatment by a health care provider.

(12) SON OR DAUGHTER.—The term "son or daughter" means a biological, adopted, or foster child, a stepchild, a legal ward, or a child of a person standing *in loco parentis*, who is
(A) under 18 years of age; or
(B) 18 years of age or older and incapable of self-care because of a mental or physical disability.

(13) SPOUSE.—The term "spouse" means a husband or wife, as the case may be.

## Sec. 102. Leave Requirement.

(a) IN GENERAL.—
(1) ENTITLEMENT TO LEAVE.—Subject to Section 103, an eligible employee shall be entitled to a total of 12 workweeks of leave during any 12-month period for one or more of the following:
(A) Because of the birth of a son or daughter of the employee and in order to care for such son or daughter.
(B) Because of the placement of a son or daughter with the employee for adoption or foster care.
(C) In order to care for the spouse, or a son, daughter, or parent, of the employee, if such spouse, son, daughter, or parent has a serious health condition.
(D) Because of a serious health condition that makes the employee unable to perform the functions of the position of such employee.
(2) EXPIRATION OF ENTITLEMENT.—The entitlement to leave under subparagraphs (A) and (B) of paragraph (1) for a birth or placement of a son or daughter shall expire at the end of the 12-month period beginning on the date of such birth or placement.
(b) LEAVE TAKEN INTERMITTENTLY OR ON A REDUCED LEAVE SCHEDULE.—
(1) IN GENERAL.—Leave under subparagraph (A) or (B) of subsection (aX1) shall not be taken by an employee

intermittently or on a reduced leave schedule unless the employee and the employer of the employee agree otherwise. Subject to paragraph (2), subsection (e)(2), and section 103(b)(5), leave under subparagraph (C) or (D) of subsection (a)(1) may be taken intermittently or on a reduced leave schedule when medically necessary. The taking of leave intermittently or on a reduced leave schedule pursuant to this paragraph shall not result in a reduction in the total amount of leave to which the employee is entitled under subsection (a) beyond the amount of leave actually taken.

(2) ALTERNATIVE POSITION.—If an employee requests intermittent leave, or leave on a reduced leave schedule, under subparagraph (C) or (D) of subsection (a)(1), that is foreseeable based on planned medical treatment, the employer may require such employee to transfer temporarily to an available alternative position offered by the employer for which the employee is qualified and that—

(A) has equivalent pay and benefits; and

(B) better accommodates recurring periods of leave than the regular employment position of the employee.

(c) UNPAID LEAVE PERMITTED.—Except as provided in subsection (d), leave granted under subsection (a) may consist of unpaid leave. Where an employee is otherwise exempt under regulations issued by the Secretary pursuant to section 13(a)(1) of the Fair Labor Standards Act of 1938 (29 U.S.C. 213(a)(1)), the compliance of an employer with this title by providing unpaid leave shall not affect the exempt status of the employee under such section.

(d) RELATIONSHIP TO PAID LEAVE.—

(1) UNPAID LEAVE.—If an employer provides paid leave for fewer than 12 workweeks, the additional weeks of leave necessary to attain the 12 workweeks of leave required under this title may be provided without compensation.

(2) SUBSTITUTION OF PAID LEAVE.—

(A) IN GENERAL.—An eligible employee may elect, or an employer may require the employee, to substitute any of the accrued paid vacation leave, personal leave, or family leave of the employee for leave provided under subparagraph (A), (B), or (C) of subsection (a)(1) for any part of the 12-week period of such leave under such subsection.

(B) SERIOUS HEALTH CONDITION.—An eligible employee may elect, or an employer may require the employee, to substitute any of the accrued paid vacation leave, personal leave, or medical or sick leave of the employee for leave provided under subparagraph (C) or (D) of subsection (a)(1) for any part of the 12-week period of such leave under such subsection, except that nothing in this title shall require an employer to provide paid sick leave or paid medical leave in any situation in which such employer would not normally provide any such paid leave.

(e) FORESEEABLE LEAVE.—

(1) REQUIREMENT OF NOTICE.—In any case in which the necessity for leave under subparagraph (A) or (B) of subsection (a)(1) is foreseeable based on an expected birth or placement, the employee shall provide the employer with not less than 30 days' notice, before the date the leave is to begin, of the employee's intention to take leave under such subparagraph, except that if the date of the birth or placement requires leave to begin in less than 30 days, the employee shall provide such notice as is practicable.

(2) DUTIES OF EMPLOYEE.—In any case in which the necessity for leave under subparagraph (C) or (D) of subsection (a)(1) is foreseeable based on planned medical treatment, the employee—

(A) shall make a reasonable effort to schedule the treatment so as not to disrupt unduly the operations of the employer, subject to the approval of the health care provider of the employee or the health care provider of the son, daughter, spouse, or parent of the employee, as appropriate; and

(B) shall provide the employer with not less than 30 days' notice, before the date the leave is to begin, of the employee's intention to take leave under such subparagraph, except that if the date of the treatment requires leave to begin in less than 30 days, the employee shall provide such notice as is practicable.

(f) SPOUSES EMPLOYED BY THE SAME EMPLOYER.—In any case in which a husband and wife entitled to leave under subsection (a) are employed by the same employer, the aggregate number of workweeks of leave to which both may be entitled may be limited to 12 workweeks during any 12-month period, if such leave is taken—

(1) under subparagraph (A) or (B) of subsection (a)(1);

or
(2) to care for a sick parent under subparagraph (C) of such subsection.

## Sec. 103. Certification.

(a) IN GENERAL.—An employer may require that a request for leave under subparagraph (C) or (D) of section 102(a)(1) be supported by a certification issued by the health care provider of the eligible employee or of the son, daughter, spouse, or parent of the employee, as appropriate. The employee shall provide, in a timely manner, a copy of such certification to the employer.

(b) SUFFICIENT CERTIFICATION.—Certification provided under subsection (a) shall be sufficient if it states—

(1) the date on which the serious health condition commenced:

(2) the probable duration of the condition;

(3) the appropriate medical facts within the knowledge of the health care provider regarding the condition;

(4)(A) for purposes of leave under section 102(a)(1)(C), a statement that the eligible employee is needed to care for the son, daughter, spouse, or parent and an estimate of the amount of time that such employee is needed to care for the son, daughter, spouse, or parent; and

(B) for purposes of leave under section 102(a)(1)(D), a statement that the employee is unable to perform the functions of the position of the employee;

(5) in the case of certification for intermittent leave, or leave on a reduced leave schedule, for planned medical treatment, the dates on which such treatment is expected to be given and the duration of such treatment;

(6) in the case of certification for intermittent leave, or leave on a reduced leave schedule, under section 102(a)(1)(D), a statement of the medical necessity for the intermittent leave or leave on a reduced leave schedule, and the expected duration of the intermittent leave or reduced leave schedule; and

(7) in the case of certification for intermittent leave, or leave on a reduced leave schedule, under section 102(a)(1)(C), a statement that the employee's intermittent leave or leave on a reduced leave schedule is necessary for the care of the son, daughter, parent, or spouse who has a serious health condition, or will assist in their recovery, and the expected duration and schedule of the intermittent leave or reduced leave schedule.

(C) SECOND OPINION.—

(1) IN GENERAL.—In any case in which the employer has reason to doubt the validity of the certification provided under subsection (a) for leave under subparagraph (C) or (D) of section 102(a)(1), the employer may require, at the expense of the employer, that the eligible employee obtain the opinion of a second health care provider designated or approved by the employer concerning any information certified under subsection (b) for such leave.

(2) LIMITATION.—A health care provider designated or approved under paragraph (1) shall not be employed on a regular basis by the employer.

(d) RESOLUTION OF CONFLICTING OPINIONS.—

(1) IN GENERAL.—In any case in which the second opinion described in subsection (c) differs from the opinion in the original certification provided under subsection (a), the employer may require, at the expense of the employer, that the employee obtain the opinion of a third health care provider designated or approved jointly by the employer and the employee concerning the information certified under subsection (b).

(2) FINALITY.—The opinion of the third health care provider concerning the information certified under subsection (b) shall be considered to be final and shall be binding on the employer and the employee.

(e) SUBSEQUENT RECERTIFICATION.—The employer may require that the eligible employee obtain subsequent recertification on a reasonable basis.

## Sec. 104. Employment and Benefits Protection.

(a) RESTORATION TO POSITION.—

(1) IN GENERAL.—Except as provided in subsection (b), any eligible employee who takes leave under section 102 for the intended purpose of the leave shall be entitled, on return from such leave—

(A) to be restored by the employer to the position of

employment held by the employee when the leave commenced; or

(B) to be restored to an equivalent position with equivalent employment benefits, pay, and other terms and conditions of employment.

(2) LOSS OF BENEFITS.—The taking of leave under section 102 shall not result in the loss of any employment benefit accrued prior to the date on which the leave commenced.

(3) LIMITATIONS.—Nothing in this section shall be construed to entitle any restored employee to—

(A) the accrual of any seniority or employment benefits during any period of leave; or

(B) any right, benefit, or position of employment other than any right, benefit, or position to which the employee would have been entitled had the employee not taken the leave.

(4) CERTIFICATION.—As a condition of restoration under paragraph (1) for an employee who has taken leave under section 102(a)(1)(D), the employer may have a uniformly applied practice or policy that requires each such employee to receive certification from the health care provider of the employee that the employee is able to resume work, except that nothing in this paragraph shall supersede a valid State or local law or a collective bargaining agreement that governs the return to work of such employees.

(5) CONSTRUCTION.—Nothing in this subsection shall be construed to prohibit an employer from requiring an employee on leave under section 102 to report periodically to the employer on the status and intention of the employee to return to work.

(b) EXEMPTION CONCERNING CERTAIN HIGHLY COMPENSATED EMPLOYEES.—

(1) DENIAL OF RESTORATION.—An employer may deny restoration under subsection (a) to any eligible employee described in paragraph (2) if—

(A) such denial is necessary to prevent substantial and grievous economic injury to the operations of the employer;

(B) the employer notifies the employee of the intent of the employer to deny restoration on such basis at the time the employer determines that such injury would occur; and

(C) in any case in which the leave has commenced, the employee elects not to return to employment after receiving such notice.

(2) AFFECTED EMPLOYEES.—An eligible employee described in paragraph (1) is a salaried eligible employee who is among the highest paid 10 percent of the employees employed by the employer within 75 miles of the facility at which the employee is employed.

(c) MAINTENANCE OF HEALTH BENEFITS.—

(1) COVERAGE.—Except as provided in paragraph (2), during any period that an eligible employee takes leave under section 102, the employer shall maintain coverage under any "group health plan" (as defined in section 5000(b)(1) of the Internal Revenue Code of 1986) for the duration of such leave at the level and under the conditions coverage would have been provided if the employee had continued in employment continuously for the duration of such leave.

(2) FAILURE TO RETURN FROM LEAVE.—The employer may recover the premium that the employer paid for maintaining coverage for the employee under such group health plan during any period of unpaid leave under section 102 if—

(A) the employee fails to return from leave under section 102 after the period of leave to which the employee is entitled has expired; and

(B) the employee fails to return to work for a reason other than—

(i) the continuation, recurrence, or onset of a serious health condition that entitles the employee to leave under subparagraph (C) or (D) of section 102(a)(1); or

(ii) other circumstances beyond the control of the employee.

(3) CERTIFICATION.—

(A) ISSUANCE.—An employer may require that a claim that an employee is unable to return to work because of the continuation, recurrence, or onset of the serious health condition described in paragraph (2)(B)(i) be supported by—

(i) a certification issued by the health care provider of the son, daughter, spouse, or parent of the employee, as appropriate, in the case of an employee unable to return to work because of a condition specified in section 102(a)(1)(C); or

(ii) a certification issued by the health care provider

of the eligible employee, in the case of an employee unable to return to work because of a condition specified in section 102(a)(1)(D).

(B) COPY.—The employee shall provide, in a timely manner, a copy of such certification to the employer.

(C) SUFFICIENCY OF CERTIFICATION.—

(i) LEAVE DUE TO SERIOUS HEALTH CONDITION OF EMPLOYEE.—The certification described in subparagraph (A)(ii) shall be sufficient if the certification states that a serious health condition prevented the employee from being able to perform the functions of the position of the employee on the date that the leave of the employee expired.

(ii) LEAVE DUE TO SERIOUS HEALTH CONDITION OF FAMILY MEMBER.—The certification described in subparagraph (A)(i) shall be sufficient if the certification states that the employee is needed to care for the son, daughter, spouse, or parent who has a serious health condition on the date that the leave of the employee expired.

## Sec. 105. Prohibited Acts.

(a) INTERFERENCE WITH RIGHTS—

(1) EXERCISE OF RIGHTS—It shall be unlawful for any employer to interfere with, restrain, or deny the exercise of or the attempt to exercise, any right provided under this title.

(2) DISCRIMINATION—It shall be unlawful for any employer to discharge or in any other manner discriminate against any individual for opposing any practice made unlawful by this title.

(b) INTERFERENCE WITH PROCEEDINGS OR INQUIRIES.—It shall be unlawful for any person to discharge or in any other manner discriminate against any individual because such individual—

(1) has filed any charge, or has instituted or caused to be instituted any proceeding, under or related to this title;

(2) has given, or is about to give, any information in connection with any inquiry or proceeding relating to any right provided under this title; or

(3) has testified, or is about to testify, in any inquiry or proceeding relating to any right provided under this title.

## Sec. 106. Investigative Authority.

(a) IN GENERAL—To ensure compliance with the provisions of this title, or any regulation or order issued under this title, the Secretary shall have, subject to subsection (c), the investigative authority provided under section ll(a) of the Fair Labor Standards Act of 1938 (29 U.S.C. 211(a)).

(b) OBLIGATION TO KEEP AND PRESERVE RECORDS—Any employer shall make, keep, and preserve records pertaining to compliance with this title in accordance with section 11(c) of the Fair Labor Standards Act of 1938 (29 U.S.C. 211(c)) and in accordance with regulations issued by the Secretary.

(c) REQUIRED SUBMISSIONS GENERALLY LIMITED TO AN ANNUAL BASIS.— The Secretary shall not under the authority of this section require any employer or any plan, fund, or program to submit to the Secretary any books or records more than once during any 12-month period, unless the Secretary has reasonable cause to believe there may exist a violation of this title or any regulation or order issued pursuant to this title, or is investigating a charge pursuant to section 107(b).

(d) SUBPOENA POWERS.—For the purposes of any investigation provided for in this section, the Secretary shall have the subpoena authority provided for under section 9 of the Fair Labor Standards Act of 1938 (29 U.S.C. 209).

## Sec. 10. Enforcement.

(a) CIVIL ACTIONS BY EMPLOYEES.—

(1) LIABILITY.—Any employer who violates section 105 shall be liable to any eligible employee affected—

(A) for damages equal to—

(i) the amount of—

(I) any wages, salary, employment benefits, or other compensation denied or lost to such employee by reason of the violation; or

(II) in a case in which wages, salary, employment benefits, or other compensation have not been denied or lost to the employee, any actual monetary losses sustained by the employee as a direct result of the violation, such as the cost of providing care, up to a sum equal to 12 weeks of wages or salary for the employee;

(ii) the interest on the amount described in clause (i) calculated at the prevailing rate; and

(iii) an additional amount as liquidated damages equal to the sum of the amount described in clause (i) and the interest described in clause (ii), except that if an employer who has violated section 105 proves to the satisfaction of the court that the act or omission which violated section 105 was in good faith and that the employer had reasonable grounds for believing that the act or omission was not a violation of section 105, such court may, in the discretion of the court, reduce the amount of the liability to the amount and interest determined under clauses (i) and (ii), respectively; and

(B) for such equitable relief as may be appropriate, including employment, reinstatement, and promotion.

(2) RIGHT OF ACTION.—An action to recover the damages or equitable relief prescribed in paragraph (1) may be maintained against any employer (including a public agency) in any Federal or State court of competent jurisdiction by any one or more employees for and in behalf of—

(A) the employees; or

(B) the employees and other employees similarly situated.

(3) FEES AND COSTS.—The court in such an action shall, in addition to any judgment awarded to the plaintiff, allow a reasonable attorney's fee, reasonable expert witness fees, and other costs of the action to be paid by the defendant.

(4) LIMITATIONS.—The right provided by paragraph (2) to bring an action by or on behalf of any employee shall terminate—

(A) on the filing of a complaint by the Secretary in an action under subsection (d) in which restraint is sought of any further delay in the payment of the amount described in paragraph (1)(A) to such employee by an employer responsible under paragraph (1) for the payment; or

(B) on the filing of a complaint by the Secretary in an action under subsection (b) in which a recovery is sought of the damages described in paragraph (1)(A) owing to an eligible employee by an employer liable under paragraph (1), unless the action described in subparagraph (A) or (B) is dismissed without prejudice on motion of the Secretary.

(b) ACTION BY THE SECRETARY.—

(1) ADMINISTRATIVE ACTION.—The Secretary shall receive, investigate, and attempt to resolve complaints of violations of section 106 in the same manner that the Secretary receives, investigates, and attempts to resolve complaints of violations of sections 6 and 7 of the Fair Labor Standards Act of 1938 (29 U.S.C. 206 and 207).

(2) CIVIL ACTION.—The Secretary may bring an action in any court of competent jurisdiction to recover the damages described in subsection (a)(1)(A).

(3) SUMS RECOVERED.—Any sums recovered by the Secretary pursuant to paragraph (2) shall be held in a special deposit account and shall be paid, on order of the Secretary, directly to each employee affected. Any such sums not paid to an employee because of inability to do so within a period of 3 years shall be deposited into the Treasury of the United States as miscellaneous receipts.

(c) LIMITATION.—

(1) IN GENERAL.—Except as provided in paragraph (2), an action may be brought under this section not later than 2 years after the date of the last event constituting the alleged violation for which the action is brought.

(2) WILLFUL VIOLATION.—In the case of such action brought for a willful violation of section 105, such action may be brought within 3 years of the date of the last event constituting the alleged violation for which such action is brought.

(3) COMMENCEMENT.—In determining when an action is commenced by the Secretary under this section for the purposes of this subsection, it shall be considered to be commenced on the date when the complaint is filed.

(d) ACTION FOR INJUNCTION BY SECRETARY.—The district courts of the United States shall have jurisdiction, for cause shown, in an action brought by the Secretary—

(1) to restrain violations of section 105, including the restraint of any withholding of payment of wages, salary, employment benefits, or other compensation, plus interest, found by the court to be due to eligible employees; or

(2) to award such other equitable relief as may be appropriate, including employment, reinstatement, and promotion.

(e) SOLICITOR OF LABOR.—The Solicitor of Labor may appear for and represent the Secretary on any litigation brought under this section.

## Sec. 108. Special Rules Concerning Employees of Local Educational Agencies.

(a) APPLICATION—

(1) IN GENERAL.—Except as otherwise provided in this section, the rights (including the rights under section 104, which shall extend throughout the period of leave of any employee under this section), remedies, and procedures under this title shall apply to—

(A) any "local educational agency" (as defined in section 1471(12) of the Elementary and Secondary Education Act of 1965 (20 U.S.C. 2891(12))) and an eligible employee of the agency; and

(B) any private elementary or secondary school and an eligible employee of the school.

(2) DEFINITIONS.—For purposes of the application described in paragraph (1):

(A) ELIGIBLE EMPLOYEE.—The term eligible employee means an eligible employee of an agency or school described in paragraph (1).

(B) EMPLOYER.—The term "employer" means an agency or school described in paragraph (1).

(b) LEAVE DOES NOT VIOLATE CERTAIN OTHER FEDERAL LAWS.—A local educational agency and a private elementary or secondary school shall not be in violation of the Individuals with Disabilities Education Act (20 U.S.C. 1400 *et seq.*), section 504 of the Rehabilitation Act of 1973 (29 U.S.C. 794), or Title VI of the Civil Rights Act of 1964 (42 U.S.C. 2000d *et seq.*), solely as a result of an eligible employee of such agency or school exercising the rights of such employee under this title.

(C) INTERMITTENT LEAVE OR LEAVE ON A REDUCED SCHEDULE FOR INSTRUCTIONAL EMPLOYEES.—

IN GENERAL.—Subject to paragraph (2), in any case in which an eligible employee employed principally in an instructional capacity by any such educational agency or school requests leave under subparagraph

(C) or (D) of section 102(a)(1) that is foreseeable based on planned medical treatment and the employee would be on leave for greater than 20 percent of the total number of working days in the period during which the leave would extend, the agency or school may require that such employee elect either—

(A) to take leave for periods of a particular duration, not to exceed the duration of the planned medical treatment; or

(B) to transfer temporarily to an available alternative position offered by the employer for which the employee is qualified, and that—

(i) has equivalent pay and benefits; and

(iii) better accommodates recurring periods of leave than the regular employment position of the employee.

(2) APPLICATION.—The elections described in subparagraphs (A) and (B) of paragraph (1) shall apply only with respect to an eligible employee who complies with section 102(e)(2).

(d) RULES APPLICABLE TO PERIODS NEAR THE CONCLUSION OF AN ACADEMIC TERM.—The following rules shall apply with respect to periods of leave near the conclusion of an academic term in the case of any eligible employee employed principally in an instructional capacity by any such educational agency or school:

(1) LEAVE MORE THAN 5 WEEKS PRIOR TO END OF TERM.—If the eligible employee begins leave under section 102 more than 5 weeks prior to the end of the academic term, the agency or school may require the employee to continue taking leave until the end of such term,

(A) the leave is of at least 3 weeks duration; and

(B) the return to employment would occur during the 3 week period before the end of such term.

(2) LEAVE LESS THAN 5 WEEKS PRIOR TO END OF TERM.—If the eligible employee begins leave under subparagraph (A), (B), or (C) of section 102(a)(1) during the period that commences 5 weeks prior to the end of the academic term, the agency or school may require the employee to continue taking leave until the end of such term, if—

(A) the leave is of greater than 2 weeks duration; and

(B) the return to employment would occur during the 2 week period before the end of such term.

(3) LEAVE LESS THAN 3 WEEKS PRIOR TO END OF TERM.—If the eligible employee begins leave under subparagraph (A), (B), or (C) of section 102(a)(1) during the period that commences 3 weeks prior to the end of the academic term and the duration of the leave is greater than 5 working days, the agency or school may require the employee to continue to take leave until the end of such term.

(e) RESTORATION TO EQUIVALENT EMPLOYMENT POSITION.—For purposes of determinations under section 104(a)(1)(B) (relating to the restoration of an eligible employee to an equivalent position), in the case of a local educational agency or a private elementary or secondary school, such determination shall be made on the basis of established school board policies and practices, private school policies and practices, and collective bargaining agreements.

(f) REDUCTION OF THE AMOUNT OF LIABILITY.—If a local educational agency or a private elementary or secondary school that has violated this title proves to the satisfaction of the court that the agency, school, or department had reasonable grounds for believing that the underlying act or omission was not a violation of this title, such court may, in the discretion of the court, reduce the amount of the liability provided for under section 107(a)(1)(A) to the amount and interest determined under clauses (i) and (ii), respectively, of such section.

## Sec. 109. Notice

(a) IN GENERAL.—Each employer shall post and keep posted, in conspicuous places on the premises of the employer where notices to employees and applicants for employment are customarily posted, a notice, to be prepared or approved by the Secretary, setting forth excerpts from, or summaries of, the pertinent provisions of this title and information pertaining to the filing of a charge.

(b) PENALTY.—Any employer that willfully violates this section may be assessed a civil money penalty not to exceed $100 for each separate offense.

## Sec. 401. Effect on Other Laws.

(a) FEDERAL AND STATE ANTIDISCRIMINATION LAWS.—Nothing in this Act or any amendment made by this Act shall be construed to modify or affect any Federal or State law prohibiting discrimination on the basis of race, religion, color, national origin, sex, age, or disability.

(b) STATE AND LOCAL LAWS.—Nothing in this Act or any amendment made by this Act shall be construed to supersede any provision of any State or local law that provides greater family or medical leave rights than the rights established under this Act or any amendment made by this Act.

## Sec. 402. Effect on Existing Employment Benefits.

(a) MORE PROTECTIVE—Nothing in this Act or any amendment made by this Act shall be construed to diminish the obligation of an employer to comply with any collective bargaining agreement or any employment benefit program or plan that provides greater family or medical leave rights to employees than the rights established under this Act or any amendment made by this Act.

(b) LESS PROTECTIVE.—The rights established for employees under this Act or any amendment made by this Act shall not be diminished by any collective bargaining agreement or any employment benefit program or plan.

## Sec. 403. Encouragement of More Generous Leave Policies.

Nothing in this Act or any amendment made by this Act shall be construed to discourage employers from adopting or retaining leave policies more generous than any policies that comply with the requirements under this Act or any amendment made by this Act.

## Sec. 404. Regulations.

The Secretary of Labor shall prescribe such regulations as are necessary to carry out title I and this title not later than 120 days after the date of the enactment of this Act.

## Sec. 406. Effective Dates.

(a) TITLE III.—Title III shall take effect on the date of the enactment of this Act.

(b) OTHER TITLES.—
(1) IN GENERAL.—Except as provided in paragraph (2), titles I, II, and V and this title shall take effect 6 months after the date of the enactment of this Act.
(2) COLLECTIVE BARGAINING AGREEMENTS.—In the case of a collective bargaining agreement in effect on the effective date prescribed by paragraph (1), title I shall apply on the earlier of—
(A) the date of the termination of such agreement; or
(B) the date that occurs 12 months after the date of the enactment of this Act.

## Review Questions

1. When did the Family and Medical Leave Act take effect?
2. Who is an eligible person under the FMLA?
3. What is the enforcement agency for the FMLA?
4. How long does an employee have to be with an eligible employer before requesting a leave?
5. How many employees must an employer have in order to be considered an eligible employer under the FMLA?
6. What are the benefits provided under the FMLA?
7. Under what conditions may an employee request a leave under the FMLA?

# Chapter

# 11

# OSHA and Safety Considerations

You're never wrong to do the right thing.

*Malcolm Forbes*

## *Learning Objectives*

- ■ To gain an understanding of the OSH Act.
- ■ To understand the impact of safety and health regulations on fire and emergency service organizations.
- ■ To understand your rights and responsibilities under the OSH Act or state plan program.

**Occupational Safety and Health Act**
a federal law dispensed by the Occupational Safety and Health Administration to decrease the incidents of injuries, illnesses, and death among workers as a result of the work environment

Fire and emergency service organizations generally owe a duty, under the concept of Master and Servant, to firefighters and other employees to create and maintain a safe and healthful work environment. This duty of care extends not only to the physical locations but also to equipment (inspection and repair), establishment of appropriate standard operating procedures (SOP), workplace rules and regulations, and beyond. Given a fire service organization's varied workplace, this duty to provide a safe working environment can extend to basically anywhere in which a firefighter may be assigned to work from the top of a building to below the ground.[1] Fire and emergency service organizations also normally possess an affirmative duty to create a safe and healthful work environment under the **Occupational Safety and Health Act** (hereinafter known as OSH Act) and other state and local laws (like the Public Liability Act). Failure to comply with these laws can result in monetary fines and/or penalties, and where a firefighter is injured or killed, the potential of criminal liability.

The Occupational Safety and Health Act became law in 1970 and fire service organizations were, in essence, exempt from compliance. Through incorporation of the OSH Act through state legislation and other methods, many fire and emergency service organization are now required to comply with the OSH Act and the numerous standards.

Approximately twenty states have petitioned and have been granted individualized safety and health programs for their states. These programs, known as *state plan programs*, require that the state's program possess all of the required elements and be as inclusive or more inclusive than the OSH Act requirements. Some of the states that have elected to possess their own state plan occupational safety and health program include Kentucky, Iowa, and Washington.

The OSH Act provided for the establishment of three independent branches under which the OSH Act would be managed and regulated. The Occupational Safety and Health Administration (OSHA), located in the Department of Labor, receives the most notoriety and is the enforcement branch under the OSH Act. The National Institute of Occupation Safety and Health (**NIOSH**), provides the research and testing on matters of safety and health and is located in the Department of Health and Human Services. The Occupational Safety and Health Review Commission (hereinafter **OSHRC**) is the independent judicial branch and provides one of the administrative appeal processes under the OSH Act.

**NIOSH**
National Institute of Occupational Health and Safety

**OSHRC**
Occupational Safety and Health Review Commission

Among the many requirements provided under the OSH Act and the standards, two baseline standards are of utmost importance. Under Section 5(a)(1), also known as the *general duty clause*, every employer is required to maintain its place of employment free from recognized hazards that are causing or are likely to cause death or serious physical harm to employees. The second baseline standard is Section 5(a)(2), which requires the employer to comply with all promulgated OSHA standards.

---

[1]See *Barger v. Mayor and City Council of Baltimore*, 616 F.2d 730 (1980) (Duty owed is not limited to fire apparatus and fire stations).

The OSH Act established three basic ways in which an OSHA standard could be promulgated. Under Section 6(a), the Secretary of Labor was authorized to adopt national consensus standards and establish federal standards without the rule-making procedures normally required under the OSH Act. This authority granted to OSHA ended on April 27, 1973. The second method by which OSHA may promulgate a standard is under Section 6(b), which establishes the procedures to be followed in modifying, revoking, or issuing new standards. This is the normal method by which OSHA establishes new standards today. The third method, which was provided under the OSH Act but is seldom utilized, is the emergency temporary standard under Section 6(c). Emergency Temporary Standards may be issued by the Secretary of Labor if employees are subject to grave danger from exposure to substances or agents known to be toxic or physically harmful and a standard is necessary to protect employees from harm. These standards are effective immediately upon publication in the *Federal Register* and are in effect for a period not to exceed six months.

The enforcement function under the OSH Act is provided to OSHA. OSHA compliance officers are empowered by Section 8(a) to inspect any workplace covered by the OSH Act, subject to the limitations set forth in the case law. An OSHA compliance officer is required to present his or her credentials to the owner or manager prior to proceeding on an inspection tour and the owner or manager has the right to accompany the compliance officer during the inspection tour. Employees or union representatives also have the right to accompany the compliance officer on his or her inspection tour. After the inspection, the compliance officer holds a "closing conference" with the employer and employee representatives at which time the safety and health conditions and potential citations or violations are discussed. Most compliance officers do not have the authority to issue on the spot citations but must confer with the regional or area director before issuing citations. Compliance officers do possess the authority to shut down an operation that is life threatening or that places employees in a position of imminent danger. Fire and emergency service organizations should be prepared for an OSHA inspection and develop a program to ensure all rights and responsibilities provided under the OSH Act are being complied with by all affected parties.

**citation**
a writ issued by the court requiring the person named to appear on a specific day or to show just cause as to why he or she should not

Following the closing conference, the compliance officer is required to issue a report to the area or regional director. The area or regional director usually decides whether to issue a **citation** and assesses any penalty for the alleged violation. Additionally, the area or regional director sets the date for compliance or abatement for each of the alleged violations. If a citation is issued by the area or regional director, this notice is mailed to the employer as soon as possible after the inspection, but in no event can this notification be more than six months after the alleged violation occurred. Citations must be in writing and must describe with particularity the alleged violation, the relevant standard and regulation, and the date of the alleged violation.

The OSH Act provides for a wide range of penalties from a simple notice with no fine through criminal prosecution. Violations are categorized and penalties may be assessed in the following manner:

|  | Old Penalty Schedule | New Penalty Schedule (1990) |
| --- | --- | --- |
| De Minimis Notice | $0 | $0 |
| Nonserious | $0–$1,000 | $0–$7,000 |
| Serious | $1–$1,000 | $0–$7,000 |
| Repeat | $0–$10,000 | $0–$70,000 |
| Willful | $0–$10,000 | minimum $5,000 maximum $70,000 |
| Failure to Abate Notice | $0–$1,000 per day | $0–$7,000 per day |
| New Posting Penalty (stacking permitted) |  | $0–$7,000 |

**serious violation**

an infraction in which there is substantial probability that death or serious harm could result

Each alleged violation is normally categorized and the appropriate fine issued by the area or regional director. Each citation is separate and may carry with it a monetary fine or penalty. OSHA has defined a **serious violation** as "an infraction in which there is a substantial probability that death or serious harm could result . . . unless the employer did not or could not with the exercise of reasonable diligence, know of the presence of the violation."[2]

In addition to the OSH Act, fire and emergency service organizations are also normally held to the standards developed and published by the National Fire Protection Association (known as NFPA). Although these standards developed by NFPA are advisory in nature (such as NFPA 1500), after an accident has occurred, courts often use the applicable NFPA standards as the basis for establishing the standard of care (see Chapter 12).

Labor organizations representing firefighters and EMS personnel also possess an affirmative duty to create a safe and healthful work environment under various public employee relations and labor laws. Safety and health issues are usually considered a mandatory item in collective bargaining in the fact that safety and health of firefighters often constitutes a term or condition of employment.

A relatively new phenomenon that could substantially impact the scope of potential criminal liability for fire and emergency service organizations is the use of state criminal laws by state and local prosecutors for injuries and fatalities that occur on the job. This new use of the standard state criminal laws in a workplace setting normally governed by OSHA or state plan programs is controversial but appears to be a viable method through which states can penalize officers in situations involving fatalities or serious injury. Presently this new use of state crimi-

---

[2]See 29 U.S.C. § 666.

nal laws does not appear to be preempted by the OSH Act. The use of criminal sanctions for workplace fatalities and injuries is not a new area of concern. Under the OSH Act, criminal sanctions have been available since its inception in 1970. In Europe, use of criminal sanctions for workplace fatalities are frequently used and, in the United States as far back as 1911, criminal sanctions were used in the well-known Triangle Shirt fire in New York which killed over 100 young women. In this case, the co-owners of the Triangle Shirt Company were indicted on criminal manslaughter charges although they were subsequently acquitted of these charges. Fire officers should, however, note that the sources of the potential criminal liability (i.e., state criminal codes in addition to the OSH Act) and the enforcement frequency (increased use of criminal charges under the OSH Act and state criminal codes) are recent trends.

The first degree murder conviction of the former president, plant manager, and foreman of Film Recovery Systems, Inc. for the 1983 work-related death of an employee[3], brought into the limelight the issue of whether OSHA had jurisdiction over workplace injuries and fatalities thus preempting state prosecution under state criminal statutes.

Presently, either OSHA may refer criminal matters to the U.S. Justice Department for prosecution or state prosecutors may bring criminal actions under the individual state code. Fire and emergency service organizations should also be aware that many states are adopting the OSHA standards, in whole or in part, and are requiring fire and emergency service organizations and other public sector entities to comply with these standards.

---

[3]*People v. O'Neil, et al. (Film Recovery Systems)*, Nos. 83 C 11091 & 84 C 5064 (Cir. Ct. of Cook County, Ill. June 14, 1985), rev'd 194 Ill. App.3d 79, 550 N.E.2d 1090 (1990).

---

(This case has been edited for the purposes of this text.)

### MARSHALL v. BARLOW'S, INC.
#### 436 U.S. 307 (1978)

Mr. Justice White delivered the opinion of the Court.

Section 3(a) of the Occupational Safety and Health Act of 1970 (OSHA) empowers agents of the Secretary of Labor (the Secretary) to search the work area of any employment facility within the Act's jurisdiction. The purpose of the search is to inspect for safety hazards and violations of OSHA regulations. No search warrant or other process is expressly required under the Act.

[A three-judge district court in Idaho ruled that OSHA's apparent statutory authority for warrantless inspections violated the Fourth Amendment.]

This Court has already held that warrantless searches are generally unreasonable, and that this rule applies to commercial premises as well as homes. In *Camara v. Municipal Court*, 387 U.S. 523, 528-529.(1967), we held:

"[E]xcept in certain carefully defined classes of cases, a search of private property without proper consent is 'unreasonable' unless it has been authorized by a valid search warrant."

On the same day, we also ruled:

"As we explained in *Camara*, a search of private houses is presumptively unreasonable if conducted without a warrant. The businessman, like the occupant of a residence, has a constitutional right to go about his business free from unreasonable official entries upon his private commercial property. The businessman, too, has that right placed in jeopardy if the decision to enter and inspect for violation of regulatory laws can be made and enforced by the inspector in the field without official authority evidenced by a warrant." *See v. City of Seattle*, 387 U.S. 541, 543 (1967).

These same cases also held that the Fourth Amendment prohibition against unreasonable searches protects against warrantless intrusions during civil as well as criminal investigations. The reason is found in the "basic purpose of this Amendment . . . [which] is to safeguard the privacy and security of individuals against arbitrary invasions by governmental officials." If the government intrudes on a person's property, the privacy interest suffers whether the government's motivation is to investigate violations of criminal laws or breaches of other statutory or regulatory standards. It therefore appears that unless some recognized exception to the warrant requirement applies, *See v. City of Seattle*, would require a warrant to conduct the inspection sought in this case.

The Secretary urges that an exception from the search warrant requirement has been recognized for "pervasively regulated business[es], and for "closely regulated" industries "long subject to close supervision and inspection." These cases are indeed exceptions, but they represent responses to relatively unique circumstances. Certain industries have such a history of government oversight that no reasonable expectation of privacy could exist for a proprietor over the stock of such an enterprise. Liquor and firearms are industries of this type; when an

entrepreneur embarks upon such a business, he has voluntarily chosen to subject himself to a full arsenal of governmental regulation.

Industries such as these fall within the "certain carefully defined classes of cases," referenced in *Camara*. The element that distinguished these enterprises from ordinary businesses is a long tradition of close government supervision, of which any person who chooses to enter such a business must already be aware. "A central difference between those cases and this one is that businessmen engaged in such federally licensed and regulated enterprises accept the burdens as well as the benefits of their trade, whereas the petitioner here was not engaged in any regulated or licensed business. The businessman in a regulated industry in effect consents to the restriction placed upon him."

The clear import of our cases is that the closely regulated industry is the exception. The Secretary would make it the rule. Invoking the Walsh-Healey Act of 1936, 41 U.S.C. § 35 et seq., the Secretary attempts to support a conclusion that all businesses involved in interstate commerce have long been subjected to close supervision of employee safety and health conditions. But the degree of federal involvement in employee working circumstances has never been of the order of specificity and pervasiveness that OSHA mandates. It is quite unconvincing to argue that the imposition of minimum wages and maximum hours on employers who contracted with the government under the Walsh-Healey Act prepared the entirety of American interstate commerce for regulation of working conditions to the minutest detail. Nor can any but the most fictional sense of voluntary consent to later searches be found in the single fact that one conducts a business affecting interstate commerce; under current practice and law, few businesses can be conducted without having some effect on interstate commerce.

The Secretary nevertheless stoutly argues that the enforcement scheme of the Act requires warrantless searches, and that the restrictions on search discretion contained in the Act and its regulations already protect as much privacy as a warrant would. The Secretary thereby asserts the actual reasonableness of OSHA searches, whatever the general rule against warrantless searches might be. Because "reasonableness is still the ultimate standard," the Secretary suggests that the Court decide whether a warrant is needed by arriving at a sensible balance between the administrative necessities of OSHA inspections and the incremental protection of privacy of business owners a warrant would afford. He suggests that only a decision exempting OSHA inspections from the Warrant Clause would give "full recognition to the competing public and private interests here at stake."

The Secretary submits that warrantless inspections are essential to the proper enforcement of OSHA because they afford the opportunity to inspect without prior notice and hence to preserve the advantages of surprise. While the dangerous conditions outlawed by the Act include structural defects that cannot be quickly hidden or remedied, the Act also regulates a myriad of safety details that may be amenable to speedy alteration or disguise. The risk is that during the interval between an inspector's initial request to search a plant and his procuring a warrant following the owner's refusal of permission, violations of this latter type could be corrected and thus escape the inspector's notice. To the suggestion that warrants may be issued ex parte and executed without delay and without prior notice, thereby preserving the element of surprise, the Secretary expresses concern for the administrative strain that would be experienced by the inspection system, and by the courts, should ex parte warrants issued in advance become standard practice.

We are unconvinced, however, that requiring warrants to inspect will impose serious burdens on the inspection system or the courts, will prevent inspections necessary to enforce the statute, or will make them less effective. In the first place, the great majority of businessmen can be expected in normal course to consent to inspection without warrant; the Secretary has not brought to this Court's attention any widespread pattern of refusal. In those cases where an owner does insist on a warrant, the Secretary argues that inspection efficiency will be impeded by the advance notice and delay. The Act's penalty provisions for giving advance notice of a search, and the Secretary's own regulations indicate that surprise searches are indeed contemplated.

However, the Secretary has also promulgated a regulation providing that upon refusal to permit an inspector to enter the property or to complete his inspection, the inspector shall attempt to ascertain the reasons for the refusal and report to his superior, who shall "promptly take appropriate action, including compulsory process, if necessary." The regulation represents a choice to proceed by process where entry is refused; and on the basis of evidence available from present practice, the Act's effectiveness has not been crippled by providing those owners who wish to refuse an initial requested entry with a time lapse while the inspector obtains the necessary process. Indeed, the kind of process sought in this case and apparently anticipated by the regulation provides notice to the business operator. If this safeguard endangers the efficient administration of OSHA, the Secretary should never have adopted it, particularly when the Act does not require it. Nor is it immediately apparent why the advantages of surprise would be lost if, after being refused entry, procedures were available for the Secretary to seek an ex parte warrant and to reappear at the premises without further notice to the establishment being inspected.

Whether the Secretary proceeds to secure a warrant or other process, with or without prior notice, his entitlement to inspect will not depend on his demonstrating probable cause to believe that conditions in violation of OSHA exist on the premises. Probable cause in the criminal law sense is not required. For purposes of an administrative search such as this, probable cause justifying the issuance of a warrant may be based not only on specific evidence of an existing violation but also on a showing that "reasonable legislative or administrative standards for conducting an . . . inspection are satisfied with respect to a particular [establishment]." A warrant showing that specific business has been chosen for an OSHA search on the basis of a general administrative plan for the enforcement of the Act derived from neutral sources such as, for example, dispersion of employees in various types of industries across a given area, and the desired frequency of searches in any of the lesser divisions of the area, would protect an employer's Fourth Amendment rights. We doubt that the consumption of enforcement energies in the obtaining of such warrants will exceed manageable proportions.

Finally, the Secretary urges that requiring a warrant for OSHA inspectors will mean that, as a practical matter, warrantless search provisions in other regulatory statutes are also constitutionally infirm. The reasonableness of a warrantless search, however, will depend upon the specific enforcement needs and privacy guarantees of each statute. Some of the statutes cited apply only to a single industry, where regulations might already be so pervasive that [an] exception to the warrant requirement could apply. Some statutes already envision resort to federal court enforcement when entry is refused, employing specific language in some cases and general language in others. In short, we base today's opinion on the facts and law concerned with OSHA and do not retreat from a holding appropriate to the statute because of its real or imagined effect on other, different administrative schemes.

Nor do we agree that the incremental protections afforded the employer's privacy by a warrant are so marginal that they fail to justify the administrative burdens that may be entailed. The authority to make warrantless searches devolves almost unbridled discretion upon executive and administrative officers, particularly those in the field, as to when to search and whom to search. A warrant, by contrast, would provide assurances from a neutral officer that the inspection is reasonable under the Constitution, is authorized by statute, and is pursuant to an administrative plan containing specific neutral criteria. Also, a warrant would then and there advise the owner of the scope and objects of the search, beyond which limits the inspector is not expected to proceed. These are important functions for a warrant to perform, functions which underlie the Court's prior decisions that the Warrant Clause applies to inspections for compliance with regulatory statutes. We conclude that the concerns expressed by the Secretary do not suffice to justify warrantless inspections under OSHA or vitiate the general constitutional requirement that for a search to be reasonable a warrant must be obtained.

We hold that Barlow was entitled to a declaratory judgment that the Act is unconstitutional insofar as it purports to authorize inspections without warrant or its equivalent and to an injunction enjoining the Act's enforcement to that extent. The judgment of the District Court is therefore AFFIRMED.

JUSTICE STEVENS, joined by JUSTICES BLACKMUN and REHNQUIST, dissented.

*Decision Explanation:*
*The U.S. Supreme Court decided for Barlow, finding that the employer does have a right to require OSHA to* *acquire a warrant prior to an inspection. The U.S. Supreme Court agreed with the district court's decision and overturned the appellate court's decision.*

(This case has been edited for the purposes of this text.)

In the Matter of Thomas F. **HARTNETT**, as Commissioner of
Labor of the State of New York, Appellant,

v.

**VILLAGE OF BALLSTON SPA**, Respondent
Supreme Court, Appellate Division,
Third Department
Nov. 9, 1989
152 A.D.2d 83

MERCURE, Justice.

This case involves an all-volunteer fire department which had failed to comply with the Public Employee Safety and Health Act.

The all-volunteer fire department was composed of two separate fire companies, Eagle Matt Lee Fire Company and the Union Fire Company. In 1986, an inspector from the Department of Labor conducted an inspection of respondent's fire department. Petitioner issued a notice of violation and order to comply which listed six violations of the Public Employee Safety and Health Act (PESH Act). Follow-up investigations were conducted which revealed that the Union Fire Company had corrected only two of the six violations. Petitioner sought to enforce its order to comply.

The Supreme Court of Saratoga County held that volunteer firefighters do not receive compensation in exchange for work and thus are not employees under the PESH Act and dismissed the petition. The Petitioner appealed.

The Petitioner argued that the Act's definition of "employees" was a broad definition and that, contrary to Respondent's assertion, an exchange of services for wages was not required. The Petitioner also claimed that the intent of the Act was to protect all employees who were not covered by the Federal Occupational Safety and Health Act (OSHA) and that firefighters were not excepted from coverage of the PESH Act.

The Supreme Court, Appellate Division agreed that firefighters were not excepted from coverage of the PESH Act and reversed the part of the judgment which dismissed the petition. The Court concluded that the Department's interpretation was reasonable and consistent with the remedial purpose of the PESH Act to assure safe and healthful workplaces for the State's public employees. Accordingly, the Court concluded that volunteer firefighters are included within the PESH Act definition of "employees," and if they are to be specifically excluded, it is a matter for the Legislature.

KANE, J.P,. and MIKOLL, YESAWICH, Jr. and HARVEY, JJ., concur.

(This case has been edited for the purposes of this text.)

### Barger v. **Mayor & City Council of Baltimore**
616 F.2d 730

Warren Barger and Marion Iwancio are former members of the fire department of the City of Baltimore. Barger and Iwancio worked in the Marine Division of the department as crew members of the city's fire-boats. They began to suffer from hearing loss which had stemmed from their prolonged exposure to the loud noise emitted by the engines of the diesel fire-boats. Barger and Iwancio were forced to retire because of their hearing impairments and received special disability pensions from the city.

Subsequently, Barger and Iwancio brought this suit under the Jones Act and general maritime law to recover damages for their loss of hearing, based upon the city's alleged negligence in the operation of the fireboats with the noisy engines. The jury found that the city was negligent and ruled in favor of Barger and Iwancio under both the Jones Act and general maritime law. Barger was awarded $153,000 and Iwancio received $112,500.

The city appealed principally on the basis of the district judge's instructions to the jury concerning negligence per se. The district judge instructed the jury that the city had been subject to the regulations of the Occupational Safety and Health Administration (OSHA) since 1971 and that if the city failed to comply with the OSHA noise regulation in the operation of it's fireboats, it would be guilty of negligence per se.

The city objected to the instructions on the basis that the City of Baltimore became subject to OSHA regulations only in 1973, not as early as 1971 as the trial judge instructed. The city also objected that the OSHA noise regulation does not apply to working conditions of Jones Act seamen and that therefore, the trial judge committed error by including any negligence per se instruction based on the OSHA noise standards.

The Court of Appeals concluded that while the judge did misstate the date upon which OSHA regulations began to apply to the city, this two year discrepancy was of minor significance and that the jury only applied a small part of the period during which the firemen were exposed to the loud noises. The Court also ruled that the district judge did err when he included a negligence per se instruction based upon the OSHA noise regulation but that the city could not obtain a reversal of the jury's verdict based upon this error. When the district judge presented the city with his proposed instructions, the city made no objection based upon the inapplicability of the OSHA noise regulation to Jones Act seamen. In fact, the city expressly agreed to the district judge's proposed instructions before he presented them to the jury.

AFFIRMED.

Decision Explanation:
The appellate court ruled in favor of the injured fire-fighters and agreed with the decision of the district court.

# Review Questions

1. The OSH Act became law in what year?

2. The OSH Act established three agencies. Name those agencies and the department they are aligned under.

3. What is the *general duty* clause?

4. What is the basic issue of *Marshall v. Barlow*?

5. Name the various levels and penalties provided under the OSH Act.

6. Where would you find the OSHA standards?

7. Does your fire service organization have a safety program? Does it comply with OSHA standards? Is your fire service organization required to comply with OSHA standards?

# Chapter

# 12

# Other Codes and Standards

Almost all important questions are important precisely because they are not susceptible to quantitative answers.

*Arthur Schlesinger, Jr.*

## *Learning Objectives*

- To understand the difference between a code and a standard.
- To identify the legal significance of codes and standards.
- To identify how codes and standards may impact your organization.

**code**

a system of principles or rules usually required by law

**standard**

usually minimum requirements established, often through a consensus of opinion, that may or may not be required by law

A **code** usually refers to a system of principles or rules that are usually required by law, such as, the building code. A **standard** is usually the minimum requirements established, often through a consensus of opinion, that may or may not be required by law. Standards can be performance based, specification based, or even test standards. Normally, if a standard is mandatory, such as OSHA standards, compliance is required by law. If a standard is voluntary, compliance is usually left to the judgment of the individual fire or emergency service organization. However, as the following discussion shows, this voluntary standard may be utilized as the best practice for the industry, and fire and emergency service organizations may be held to this "voluntary" standard in a court of law. Following is a listing of codes and standards:

> Code (mandatory)—Kentucky Criminal Code
>
> Mandatory Standard—29 CFR 1910, OSHA standards
>
> Mandatory (Performance) Standard—EPA Emmission Standards for Motor Vehicles
>
> Voluntary/Mandatory (Specification) Standards—NFPA 1962, Care, Use and Testing of Fire Hose including Connections and Nozzles
>
> Voluntary (Testing) Standards—NFPA test standards for aerial and ladder devices; Underwriters Laboratory testing standards
>
> Voluntary "Draft" Standard—OSHA proposed Standard on Ergonomics

In most fire and emergency service organizations, the acronyms ANSI, NFPA, BOCA, and others commonly refer to the recommendations and guidance provided by these various codes and/or organizations. Fire and emergency service organizations should be aware that these codes and rules may be advisory or compulsory, depending on the type of code and whether the particular code has been adopted by another governmental entity or jurisdiction that mandates compliance.

Probably the most common acronym used in the fire service is NFPA. NFPA stands for the National Fire Protection Association. The NFPA has published a handbook on various fire protection topics since 1896.[1] Within the NFPA Handbook, the authors cover various topics ranging from the basics of fire and fire service to fire suppression. The NFPA handbook is often considered the baseline or level of expected performance or behavior that all fire service organizations should maintain. However, the NFPA handbook and the sections contained therein, such as NFPA 1500, are advisory unless codified or adopted by the specific jurisdiction.

Another often used acronym in the fire service is ANSI. ANSI stands for the American National Standards Institute. As with the NFPA standards, ANSI standards are also advisory. However, many of the ANSI standards have been adopted within the Occupational Safety and Health standards and codified within other specific federal, state, and local codes.[2] ANSI standards tend to be broad

---

[1] *NFPA Handbook* (Seventeenth Edition) at p. 1.
[2] 29 C.F.R. § 1910.

encompassing a vast number of topics and tend to be forerunners in advanced safety and health issues such as ergonomics.

The Life Safety Code has been adopted in whole by many federal, state, and local jurisdictions. The Life Safety Code is often used by fire service organizations when evaluating construction of new or existing buildings, evaluating exits and egress, and similar specific areas.

Similar to the Life Safety Code is the BOCA codes. BOCA stands for Building Officials and Code Administrators basic fire prevention code and is primarily focused on building standards related to fire safety. As with the Life Safety Code, the entire BOCA codes have been adopted by many federal, state, and local jurisdictions.

Codes and regulations are often used in civil actions, especially negligence actions. The primary purpose of the use of the code or regulation is to show that either a duty was created for the fire service organization or its agents, or to show the standard of care that the fire service organization knew or should have known that they were required to comply with as a matter of law.

It is important for fire and emergency service personnel to maintain competency with regard to the variety of codes and standards and the modifications that are made on a periodic basis. Achieving compliance and maintaining compliance with the codes and standards that may be applicable to your fire and emergency service organization is imperative in order to avoid potential legal entanglements.

---

(This case has been edited for the purposes of this text.)

Thomas E. **BARTLES**, Kathleen M. Bartles, Judith A. Bartles
and Georgine L. Bartles, minors, by their mother and next
friend, Kathleen P. Bartles, and Kathleen P. Bartles, Widow of George E. Bartles, Respondents,

v.

**CONTINENTAL OIL COMPANY**,
a corporation, Appellant
Supreme Court of Missouri
Nov. 9, 1964
384 S.W.2d 667

BARRETT, Commissioner.

This case involves an action for wrongful death of a fireman caused when inadequately vented oil storage tank at scene of fire "rocketed" into air and engulfed him in a ball of fire.

George E. Bartles died as a result of injuries sustained while engaged in the performance of his duties as a captain in the Kansas City, Missouri, fire department. In this action to recover damages for his negligent injury and death, his widow and four minor children have recovered a judgment of $25,000.

The defendant-appellant, Continental Oil Company, operated a bulk storage plant and filling station on the northwest side of Southwest Boulevard at 31st Street in Kansas City, Kansas, abutting the Missouri-Kansas state line. On the front of the lot, facing Southwest Boulevard, there was a filling station and to the rear of the station on concrete saddles there were four 21,000 gallon capacity storage tanks. Tank number one contained 6,628 gallons of kerosene, tank number two contained 14,307 gallons of regular gasoline, tank number three contained 3,051 gallons of regular gasoline, and tank number four contained 15,555 gallons of premium gasoline.

On August 18, 1959, about eight o'clock in the morning, Fred Berry, one of the appellant's tank-truck drivers was engaged in loading his truck from storage tanks two and four. Hoses from these two tanks were open at the same time, filling truck compartments three and five. Berry was on the truck's catwalk when Jim Mitchum, another tank-wagon driver on vacation, climbed up the side of the truck to show Berry his new cigarette lighter. Almost immediately (possibly when Mitchum flicked his lighter) flames flashed from the tank-truck's fifth compartment. Berry cut off the hose from one of the storage tanks but ran from the blazing flames and did not cut off the hose from the other storage tank and throughout the ensuing fire that open hose continued pouring gasoline into the flames. Firemen from both Kansas City, Missouri, and Kansas City, Kansas, responded to the alarm and with hand lines, almost wholly from Southwest Boulevard, fought the fire with water. The streams of water were employed to confine the burning gasoline to the area of the storage tanks thereby preventing the gasoline from running down Southwest Boulevard and into the sewers of Kansas City, Missouri, endangering larger areas. About 9:15 storage tank number one ruptured at its rear and the area was engulfed in flaming kerosene, intensifying the heat in the area of the other three tanks. In the next half hour tanks two and three ruptured, and about 9:40 tank number four left its concrete cradle and "rocketed" or was catapulted 75 to 100 feet over the filling station into Southwest Boulevard and "a ball of fire" engulfed the several crews of fire fighters in the street, killing one bystander and five firemen, including Captain Bartles, and injuring twenty-three people.

Continental's four storage tanks, constructed of quarter-inch steel, were installed in 1924. They were 30 feet long, eleven feet in diameter, with flat ends, and when installed, as well as 35 years later, "were fitted with two-inch pressure valves." Originally the tanks were installed perpendicularly but in 1958 Continental Oil Company dismantled the storage facilities, removed the two-inch vents, and installed the four tanks horizontally on the concrete saddles. But in changing or replacing the installation, Continental again placed two-inch vents on the tops of the storage tanks. And in response to interrogatories and "requests for admissions" Continental stated that the "four tanks was (sic) vented only by a single A.Y. McDonald 2-inch Plate 925 combination gauge hatch and P & P valve, known as a 'breather valve.' . . . that these said four tanks had no vent or mechanism designed to relieve pressure (such as might occur during a fire) other than whatever relief was provided by the single 2-inch diameter breather valve on each tank."

In addition to these admissions the Chief of the Fire Prevention Division of the Kansas City, Missouri, Fire Department, who was present while the fire was in progress and when tank number four rocketed, and afterwards made an investigation, testified that the reason the tanks ruptured was that "(t) the vents were too small." He testified that the Flammable Liquids Code required "5-1/2 inches of emergency venting." An expert witness, "a protection engineer specializing in petroleum hazards," referred to all the literature connected with the petroleum industry (with which Continental was familiar) and in connection with his years of experience in the industry, testified that for petroleum storage tanks of 18,000 to 25,000 gallon capacity "a free circular opening of 5-1/2 inches in diameter" was required. As he testified on both direct and cross-examination this witness gave formulas and made the computations as to the "normal safety operating pressures of a tank" as well as the pressures built up in a storage tank during a fire. He testified that tank number four rocketed because the two-inch vent was not large enough to take care of the vapor generated by the fire. And, he said, "I can say without any worry at all that no tank that has been equipped with the vent of the size specified in these suggested standards (5-1/2 inches) has ever rocketed."

The tanks were old-fashioned and the two-inch vents in the tanks were smaller than was recommended by the literature, by the National Fire Protection Association and the Association of Petroleum Industries.

The Circuit Court of Jackson County gave judgment to fireman's heirs. The Supreme Court held that finding of existence of hidden danger known to owner and failure to warn fireman of danger which

he was not bound to accept as a usual peril of his profession was supported by the evidence.

AFFIRMED.

---

(This case has been edited for the purposes of this text.)

Allen **FRAZIER**, a minor by Ferrol Frazier, next friend, Plaintiff-Appellant, v. **CONTINENTAL OIL COMPANY**, et al., Defendants-Appellees. United States Court of Appeals, Fifth Circuit Feb. 24, 1978 568 F.2d 378

GEWIN, Circuit Judge.

This case involves an action for damages against an oil company and others, by a child who was injured in a flash fire at a service station.

On the morning of February 28, 1972, the appellant, Allen Frazier, along with five other children, was injured in a flash fire at a Continental Oil Company (Conoco) gasoline station located across the street from the school where the appellant was a student. Before classes began, the appellant accompanied another student, Mark Buckley, to the station. Buckley went inside the building to purchase some candy, and the appellant walked around to the back of the station to use the restroom. While the boys were at the station, a gasoline truck was filling underground storage tanks on the premises. Each tank was vented by a pipe which opened above and between the entrances to the men's and women's restrooms. As the appellant passed beneath these pipes, an

explosion occurred, and he sustained burns on his face and arms.

Under Mississippi law negligence may be proved entirely by circumstantial evidence. There was testimony by the appellant's expert witness, Mr. Brown, that the vent pipes running from the tanks were equipped with "T" vents which would cause the fumes to be discharged downward in violation of industry standards set forth in the National Fire Protection Association Code (NFPA). Brown also testified that one pipe was 10 feet and the other 10 feet 6 inches above the adjacent ground level, in violation of the industry standard of 12 feet. He further testified that the Code required upward discharge of gasoline vapor, which is gasoline in its most combustible form, in order to "disperse the gasoline vapor into an atmospheric condition and away from a point where people might congregate or be," and to prevent the vapors from collecting under an eave or some part of a building.

Brown testified that the industry standards governing service station facilities were found in NFPA Pamphlet 30, and that the Conoco station violated the standards set forth in that publication for venting gasoline storage tanks.

The United States District Court granted oil company's motion for directed verdict, and child appealed. The Court of Appeals held that (1) trial court erred in excluding testimony of expert witness that ventilation pipes running from gasoline storage tank violated industry standards set forth in published fire safety code; (2) evidence presented jury question as to whether oil company could be held liable under theories of negligence or strict liability, and (3) evidence presented jury question concerning oil company's control of service station equipment.

REVERSED AND REMANDED.

*Decision Explanation:*
*The district court's granting of a directed verdict for* the oil company was reversed and the case was sent back to district court for hearing.

## Review Questions

1. Does your fire service organization have a copy of the NFPA standards?

2. Does your fire service organization inspect for violations of BOCA, (National Electrical Code) NEC, and other standards?

3. Does your firehouse or other structures comply with the various codes?

4. What is the issue in the *Bartles* case?

5. How are the standards applicable in the *Bartles* case?

6. How are NFPA codes used in the *Frazier* case?

7. What are the specific requirements under NFPA 1500?

8. How can noncompliance with an NFPA standard or other standard be used against a fire service organization? How can compliance with an NFPA standard or other standard be used as a defense for a fire service organization?

# 13

# Products Liability in the Fire Service

The biggest problem in the world could have been solved when it was small.

*Ralph J. Bunche*

## *Learning Objectives*

- To gain a basic understanding of product liability.
- To understand the basic concept of the laws involving defective products.
- To gain a knowledge base regarding selection of products to avoid defects.

Products liability, as related to the fire and emergency service, usually involves the purchase of particular equipment that ultimately is found to be defective in design, manufacturing, or other areas, such as warnings and instructions. Given today's changing technology, this chapter was developed so that fire and emergency service organizations can be aware of the possibility of dangerous products when purchasing new equipment or replacing existing equipment. Fire service organizations should pay special attention to the equipment purchased for use by firefighters and other personnel to ensure that the product is free of all hazards and defects.

Products liability can be divided into three basic areas:

1. Negligence
2. Misrepresentation
3. Strict liability in tort

These principles normally apply to the manufacturer; however, suppliers and distributors may also acquire liability dependent upon their actions or inactions.

Negligence usually applies when the manufacturer fails in its duty in the actual design of manufactured goods or in supplying warranty or instructions for the product. The second area, tortious misrepresentation, is, in essence, the defendant's state of mind in making such representations as to the defective product. The three primary areas of tortious misrepresentation today are fraud, deceit, or negligent misrepresentation. Additionally, a manufacturer or other defendant may also be held strictly liable in tort for certain public misrepresentations with regard to its product, regardless of whether the company acted in good faith or exercised reasonable care.

"It has been said that the concept of implied warranty rests upon the foundation of business ethics and constitutes an exception to the maxim 'let the buyer beware,' and self encompassing the idea that their being no warranty implied with respect to the quality of goods being sold."[1] Selling for a sound price raises implied warranty that a thing sold is free from defects, known or unknown (to the seller).[2] Most purchasers believe that the product they are purchasing is free of defects and is essentially safe to use for its specified purpose. As can be seen from the courts, and as specified in the Uniform Commercial Code,[3] the courts have found that, under the theory of misrepresentation, there is an implied warranty with products that the product is, in fact, safe. As long as the purchaser uses the product for the particular purpose for which the product was made, the manufacturer is normally liable under implied warranty of fitness or merchantability to ensure that the product is free of any and all defects. In the area of warranties, it should be noted that not only is the individual who purchased the particular

---

[1]*Lambert v. Sistrunck*, 58 So. 2d 434 (Fla. 1952).
[2]*Southern Iron Equipment Co. v. Railroad Company*, 151 S.C. 506, 149 S.E. 271 (1929).
[3]Section 2-314, Implied Warranty: Merchantability; Usage of Trade.

product protected, but third parties also may benefit through the privity of contract or other relationships regarding protections from unwarranted defects.

With regard to fire service equipment, the most often used theory is that of strict liability in tort. The Restatement (Second) of Torts § 402 (A), Special Liability of Seller of Products from Physical Harm to User Consumer, states:

> (1) One who sells any product in a defective condition unreasonably dangerous to the user consumer or to his property is subject to liability for physical harm thereby caused to the ultimate user consumer or to his property if: a) the seller is engaged in the business of selling such a product; and b) it is expected to or does reach the user consumer without substantial change in the condition in which it was sold.
>
> (2) The rules stated above apply although (a) the seller has exercised all possible care in the preparation and sale of his product and (b) the user consumer has not bought the product from or entered into any contractual relationship with the seller.

In most fire and emergency service organization situations, the manufacturer of the particular product is held within the strict liability theory because of the particular design and use of the specialized product. As a general rule, most new products purchased by fire service organizations are protected under warranty or other contractual arrangements. With used fire service equipment, **caveat emptor** or "Let the buyer beware" normally applies. The manufacturer in many jurisdictions can use the defenses of contributory or comparative negligence, assumption of the risk, or misuse of the particular product. Damages available in products liability actions can include the basic personal injury and death benefits, emotional distress damages, economic and property damage losses, and punitive damages.

Special issues in the area of products liability often affect fire and emergency service organizations. Products liability litigation involving claims against manufacturers of cars, motorcycles, and other vehicles alleging defects in the vehicle makes up much of the total products liability litigation. For fire service personnel, the issue is whether the pumper, ladder truck, EMS vehicle, or other vehicles were properly designed, crash worthy,[4] or possessed other defects that caused a particular accident. Issues such as the failure to use seatbelts by fire service personnel, the design of the fire service vehicle, the design of the particular apparatus on the fire service vehicle, postsale warnings on used equipment, and the issue of recall by the manufacturers also are considered within this area of products liability. The second special area of products liability involves toxic substance litigation or what is often referred to as *toxic torts*. Toxic substance litigation cases in the area of products liability has involved cigarette litigation and individuals being exposed to various chemicals or chemical mixtures after a fire, various diseases, and other agents, such as Agent Orange,[5] or nuclear radiation,

**caveat emptor**
buyer beware

---

[4]Motor vehicle Information and Cost Savings Act, 15 U.S.C.A. § 1901 (14).
[5]*In re Agent Orange Product Liability Litigation,* 611 f. Supp. 1267.

are potential areas in which litigation has arisen in the area of product liability. In these types of cases, the scientific uncertainty in tracing the cause and effect between the substance and the injury have prevented recovery in many of the toxic substance or toxic tort cases. Cases such as the asbestos cases,[6] in which there are several manufacturers, the damages, if proven, can be distributed among the various manufacturers of the defective product.

For a prudent fire and emergency service organization to avoid the potential of acquiring a defective product that ultimately could harm a firefighter, fire service organizations should be diligent in the acquisition of appropriate equipment for use by fire service personnel. Fire and emergency service organizations should do their homework to ensure that the products purchased from a manufacturer meet all the specifications and are being used in the capacity the manufacturer recommends. Caveat Emptor or "let the buyer beware" should be foremost in thinking of volunteer or other fire service organizations that are purchasing used equipment. All used equipment should be carefully evaluated to ensure that the equipment is functioning and necessary inspection and repairs made prior to placing the equipment in service.

With the expanding technology being used by fire service organizations today, the potential of encountering a defective product that could injure a firefighter or other personnel has increased significantly. Today, virtually all states have adopted the ruling in *MacPherson v. Buick Motor Company*,[7] which held the manufacturer of a defective product liable in negligence to a person injured by the product, despite the absence of privity between the parties. Additionally, most jurisdictions have adopted the warranty theory that manufacturers of defective products can be held strictly liable in tort.[8] Many states have developed their own laws with regard to defective products[9] usually based on the Restatement (Second) of Torts § 402A.

Manufacturers, suppliers, assemblers, and component parts manufacturers can all be liable for a defective product. Individuals protected by most product liability laws include the actual user of the product, consumers, bystanders, and even rescuers if the defective product threatened or caused injury during the rescue of the imperiled individual.

Under the strict liability laws, foreseeability of harm by the manufacturer is not required. Strict liability can apply even if the defective product was being misused at the time of the injury, provided the misuse was foreseeable.[10] Additionally, where the manufacturer improperly designed the product, there is no defense that the design defect was obvious to the consumer.[11] However, where a

---

[6]*Martin v. Owens Corning Fiberglass Corporation*, 515 Pa. 377, 528 A.2d 947 (Supreme Court of Penn. 1987).

[7]Supra.

[8]See *Greenman v. Yuba Power Products, Inc.*, 59 Cal.2d 67, 27 Cal. Rptr. 697, 377 P.2d 897 (1963).

[9]See, as example, Kentucky Product Liability Act, KRS 411.300 et seq.

[10]*Moran v. Fabrege, Inc.*, 273 Md. 538, 332 A.2d 11 (1975).

[11]*Palmer v. Massey-Ferguson, Inc.*, 3 Wash. App. 508 (1970) ("The law, we think, ought to discourage misdesign rather than encourage it in its obvious form.").

product possesses unavoidably unsafe characteristics, such as certain drugs, a warning may be required. For example, the Pasteur treatment for rabies can lead to serious injury and damaging consequences when injected. Since the disease itself possesses the potential for a painful death, both the marketing and use of the vaccine was justified notwithstanding the high degree of risk.[12] Under this principle, most courts hold that manufacturers can rely on warnings, such as labels for drugs, identifying the risks of the product. However, failure to communicate the risks by third parties, such as physicians with prescription drugs, may result in strict liability.[13]

Strict liability may be applied if the product is unreasonably dangerous and the manufacturer fails to provide proper warnings of the dangers or proper directions as to the product's use.[14] The duty to warn may be present even where the use by the consumer of the product is abnormal if the abnormal use is foreseeable.[15]

Circumstantial evidence may be used to prove strict liability just as it can be used to prove negligence. The circumstantial evidence must be sufficient to warrant the inference that the defective condition existed and was linked to the individual's injury.[16] For example, if a firefighter was injured when his turnout gear melted and caught fire, the proof that the manufacturer produced the turnout gear and the turnout gear melted at a temperature in which the gear was supposed to provide protection should be sufficient to achieve this causal nexus.

Two of the often-used defenses to a strict liability claim by firefighters are assumption of risk and contributory negligence. Assumption of the risk generally bars recovery when an individual who knew the danger in the situation nevertheless voluntarily accepted the risk. An Assumption of the risk defense is basically an argument that the individual unreasonably accepted the risk and this acceptance of the risk should protect the defendant. The general test for this defense is that the injured individual must have been aware of the defect in the product and not merely that a reasonable person should have been aware of the defect.[17] This test is subjective and the jury is not required to accept the injured individual's testimony. Other factors such as age, experience, knowledge, understanding, obviousness of defect, and danger the defect poses can also be taken into consideration.

---

[12]See also *Hines v. St. Joesph's Hospital,* 86 N.M. 763, 527 P.2d 1075 (1974) (no process to determine hepatitis virus in blood supply at that time).

[13]See *Davis v. Wyeth Labs., Inc.,* 399 F.2d 121 (9th Cir. 1968) (live polio vaccine—strict liability found).

[14]Id.

[15]See *Spruill v. Boyle-Midway, Inc.,* 308 F.2d 79 (4th Cir. 1962) (14-month old infant died from ingesting furniture polish—failure to warn of toxic nature).

[16]See *Elmore v. American Motors Corp,* supra.; *Briner v. General Motors Corp.,* 461 S.W.2d 99 (Ky. 1970); *Perkins v. Trailco Mfg. & Sales Co.,* 613 S.W.2d 855 (Ky. 1981).

[17]See *Williams v. Brown Mfg. Co.,* 45 Ill.2d 418, 261 N.E.2d 305 (1970).

(This case has been edited for the purposes of this text.)

Sam **HORNBECK**, Appellant,

v.

**WESTERN STATES FIRE APPARATUS, INC.,**
an Oregon Corporation,
Respondent,
Supreme Court of Oregon, In Banc.
Argued and Submitted Sept. 8, 1977
Decided Dec. 20, 1977
280 OR. 647

LINDE, Justice.

This case involves an injured firefighter who sued a fire truck designer-manufacturer for negligence.

The Plaintiff, Sam Hornbeck, a firefighter for the City of Albany Fire Department, was injured in a fall from a fire truck designed and manufactured by Defendant, Western States Fire Apparatus. He brought action for damages against the Defendant, alleging, first, that defects of design and manufacture made the truck unreasonably dangerous for its intended use because the horizontal bar furnished as a handhold for firemen riding the rear footboard was placed too low, and second, that the Defendant was negligent in failing to design and provide a more adequate handhold.

Defendant pleaded that Plaintiff was contributorily negligent in a number of respects and that "(P)laintiff assumed the risk of falling from the rear of said fire apparatus." Plaintiff made a motion to withdraw Defendant's plea of assumption of risk from the jury and the Court denied and instructed the jury on the defense.

A jury trial resulted in a verdict for Defendant on both counts. On appeal, Plaintiff assigned as error a number of rulings of the trial court. The most important was the submission of a defense of assumption of risk to the jury.

The Defendant claimed that the jury verdict in its favor should nevertheless be sustained because it implicitly found no liability irrespective of the defense of assumption of risk.

The Supreme Court of Oregon held that in the absence of any evidence of an express assumption of risk by firefighter, it was reversible error to submit defense of assumption of risk on the strict liability claim; and the fact that the jury found for manufacturer on negligence claim as well as strict liability claim did not necessarily mean that jury found that the handhold was not unreasonably dangerous, and thus did not preclude necessity for new trial on the products liability cause of action.

*Decision Explanation*
*The Oregon Supreme Court found that the lower court had errored and returned this case to the lower court for rehearing.*

(This case has been edited for the purposes of this text.)

REVERSED AND REMANDED.
**JACKSON** et al., Appellants and Cross-Appellees
v.
**ALERT FIRE AND SAFETY EQUIPMENT, INC**. et al.
Appellees: South Akron Awning et al.
Appellees and Cross-Appellant
Supreme Court of Ohio
Submitted Nov. 19, 1990
Decided March 6, 1991
58 OHIO ST.3d 48

This case involves two firefighters who were severely burned while fighting a structural fire and filed products liability action against sellers of firefighting gloves and pants, and company which lengthened firefighting coat.

On March 3, 1986, Appellants and Cross-Appellees, Michael D. Jackson ("Jackson") and Jerry S. Kelley ("Kelley"), firefighters for the Akron Fire Department, were severely burned over various parts of their bodies while fighting a structural fire in Akron.

In June 1988, Jackson and Kelley filed separate amended complaints in the Court of Common Pleas of Summit County. Jackson, in his amended complaint, named various defendants including his employer, the City of Akron ("City"), and Defendant-Appellee Alert Fire & Safety Equipment, Inc. ("Alert"). Kelley, in his amended complaint, named the same Defendants as did Jackson, and included, among others, Defendant-Appellee M.F. Murdock Company ("Awning"), and Levinson's Co. ("Levinson's").

Jackson and Kelley advanced virtually the same theories of recovery against the Defendants: strict liability, negligence, implied and express warranty, and negligent and intentional infliction of emotional distress.

Alert's and Murdock's alleged liability arose from being retailer/suppliers of gloves. Jackson alleged that he sustained burns to his hands and wrists in the structural fire as a result of wearing defective "Polar Bear" gloves sold to him by Alert. Jackson's purchase of the gloves was approved by the City and

the City reimbursed him. Kelley alleged that he received burns to his hands and wrists as a result of wearing defective "Nitty-Gritty 93 NFW" gloves purchased by the City from Murdock, Alert or another supplier and issued by the City to Kelley. Nitty-Gritty 93 NFW gloves are coated with natural rubber. Kelley, in his deposition, testified that firefighters could let the City issue the gloves directly to them. Kelley stated that even though a different type of glove (leather) would have been better, he opted to let the City issue him the rubber Nitty-Gritty 93 NFW gloves.

Levinson's alleged liability emanated from being a retailer/supplier of "Mr. Two-Ply" pants worn by Kelley at the time of fighting the structural fire. Kelley charged that the pants were defective and, as a result, he received burns to his legs, thighs and body. The pants were part of the uniform that the City mandated that firefighters wear. Kelley purchased the pants directly from Levinson's.

Awning's alleged liability stemmed from adding approximately ten inches of material to the bottom of Kelley's "Morning Pride" fire coat. Kelley claimed he sustained burns to his legs, thighs, and body as a result of wearing a defective Morning Pride coat "remanufactured" by Awning. Kelley stated that the added material "shriveled up" as a result of the fire. The coat was lengthened at the request of the City to keep water from entering Kelley's boots and the City paid for the lengthening.

All Defendants filed summary judgment motions. Alert and Murdock filed jointly, while Levinson's and Awning filed separately.

On October 24, 1988, the trial court, after reviewing the pleadings and evidence submitted in support, sustained all summary judgment motions.

Jackson and Kelley appealed to the Court of Appeals for Summit County. The Court of Appeals found that summary judgment was properly granted as to all claims in favor of Alert and Murdock. The court further determined that the trial court improperly granted the motions of Levinson's and Awning as to the claims of strict liability, implied warranty in tort and negligence, and properly granted their motions as to the claim of intentional infliction of emotional distress. Jackson and Kelley appealed again.

The applicable statute in this case is R.C. 2305.33. This statute provided that a seller, pursuant to Civ. R. 56 was entitled to summary judgment if the seller proves that: (1) it did not alter or fail to maintain the product while it was in the seller's possession; (2) the manufacturer is amenable to suit and not financially insolvent; (3) the seller, upon request of the injured party, provides the person making the request with the name and address of manufacturer; and (4) the seller did not have actual knowledge of the alleged defect in the product and, based upon facts available to him, could not have been expected to have had knowledge of the alleged defect.

Jackson and Kelley did not contend that Alert and Murdock failed to comply with R.C. 2305.33 (B) (1) through (3). Rather, Jackson and Kelley asserted

Alert and Murdock failed to offer probative evidence sufficient to satisfy the prerequisites of subsection (B) (4) of the statute.

The Supreme Court of Ohio found that summary judgment was properly granted to Alert and Murdock on Jackson's and Kelley's claims of strict liability and implied warranty in tort. Material issues of fact existed, precluding summary judgment as to whether firefighting pants were defective and whether seller had duty to warn purchasers; and the company which lengthened firefighter's coat was providing a product, not a service, under products liability law, and was not entitled to summary judgment.

AFFIRMED AND REMANDED.

HOLMES, J., filed an opinion concurring in part and dissenting in part in which MOYER, C.J., joined.

*Decision Explanation:*
*The Supreme Court of Ohio agreed with the district* *and appellate courts that the granting of a summary* *judgment to the defendants was appropriate.*

# Review Questions

1. Explain, in your own words, the theory of strict liability in tort as related to a fire service product.

2. What are the four required elements of a negligence action?

3. Does your fire service organization identify potentially dangerous products prior to purchase? If yes, what is the procedure?

4. What is the theory of caveat emptor?

5. Provide at least one example of a potential misrepresentation regarding a fire service product.

6. Could a firefighter's failure to wear seatbelts affect a product liability case?

7. Does your fire service organization possess Material Safety Data Sheets (MSDS) for all industrial sites in your area?

8. Can the industrial site or the manufacturer be held liable for a firefighter's injuries as a result of failure to provide the required information to the fire service organization?

9. What is the issue involved in the *Hornbeck* case?

10. What type of action was utilized in the *Jackson* case?

*Chapter*

# 14

# Age Discrimination Employment Act

Everything subject to time is liable to change.

*Joseph Albo*

Don't ever take a fence down until you know the reason why it was put up.

*Gilbert Keith Chesterton*

## Learning Objectives

■ To understand the parameters of the ADEA.

■ To understand the requirements of the ADEA.

■ To learn the prohibitions set forth in the ADEA and the methods of avoiding discrimination.

The Age Discrimination Employment Act of 1967[1] (ADEA) was enacted by Congress for the purpose of promoting employment of individuals over the age of forty and prohibiting arbitrary age discrimination by employers. In essence, ADEA protects individuals who are between the ages of forty and seventy from being discriminated against or discharged because of their age.[2] The ADEA covers virtually all fire service organizations, private sector employers, employment agencies, and labor organizations.[3] An employer is defined under the ADEA as "one engaged in an industry affecting commerce and having twenty or more employees for each working day in each of the twenty or more calendar weeks during the current or preceding calendar year."[4] State or political subdivisions of a state, as well as federal agencies, are included in this definition of an employer. However, federal employees are not entitled to a right to a jury trial as specified under the ADEA.[5] The ADEA prohibits covered employers and other entities from discriminating against employees, members, referrals, applicants, or other covered individuals as a means of retaliation because the employee has made a charge of discrimination, has testified, has assisted, or has participated in any manner in any investigation, proceeding, or litigation under the ADEA.[6] Additionally, covered employers are prohibited from printing or publishing or causing to be printed or published any notice or advertisement related to employment that indicates a preference, limitation, specification, or discrimination based upon age.[7] The ADEA prohibits an employer from:

- Failing, refusing to hire, or discharging any individual or otherwise discriminating against any individual with respect to the employee's compensations, terms, conditions, or privileges of employment because of such individual's age;
- Limiting, segregating, or classifying its employees in any way that would deprive or tend to deprive any individual of employment opportunities or otherwise adversely affect the status of the employee because of age.
- Reducing the wage rate of any other employee or employees in order to comply with the ADEA.[8]

The ADEA prohibits labor organizations from:

- Excluding or expelling from its membership or otherwise discriminating against any individual because of his or her age.
- Limiting, segregating, or classifying its membership, or classifying or failing to refuse to refer for employment any individual in a way that would

[1] 29 U.S.C.A. § 621, et seq.
[2] Id.
[3] Id.
[4] 29 U.S.C.A. § 630 (b).
[5] *Lehman v. Nakshain,* 453 U.S. 156, (1981).
[6] 29 U.S.C. § 623 (d).
[7] 29 U.S.C. § 623 (e).
[8] 29 U.S.C.A. § 623 (a).

tend to deprive that individual of employment opportunities, or that would limit such employment opportunities or otherwise adversely affect his or her status as an employee or as an applicant for employment because of the individual's age.

- Causing or attempting to cause an employer to discriminate against an individual in violation of the ADEA.[9]

Originally the responsibility for enforcing the ADEA was with the Wage and Hour Division of the Department of Labor. However, in 1978, the enforcement and administration of the ADEA was transferred from the Wage and Hour Division to the Equal Employment Opportunity Commission (EEOC).[10] The EEOC has established guidelines for investigating and conciliating ADEA claims and the procedures mirror those prescribed for Title VII of the Civil Rights Act (see Chapter 5 for details).

The EEOC regulations regarding the ADEA require employers to keep records that identify employee's names, addresses, dates of birth, occupations, rates of pay, and additional information for a period of three years.[11] The EEOC regulations also require that personnel records be maintained by an employer for each employee for a period of one year. These records include:

- Job applications, resumes, or other replies to job advertisements, including applications for temporary positions and records pertaining to failure to hire

- Records pertaining to promotion, demotion, transfer, selection of training, layoffs, recall, or discharge

- Job orders submitted to employment agencies or unions

- Test papers in connection with employer administered aptitude tests or other employment related tests

- Records pertaining to the results of any physical examinations

- Job advertisements or notices to employees regarding openings, promotions, training programs, or other opportunities for overtime work.[12]

Covered employers are also required to maintain records pertaining to employee benefit plans, written seniority or merit rate systems, and preemployment records of applicants for temporary positions. Additionally, the EEOC has prepared a standardized poster, as with OSHA and other agencies, that is required to be posted in a conspicuous place on the premises at all times.[13] Failure to post the required EEOC notice may be used to toll the statute of limitations for filing

---

[9]29 U.S.C. § 623 (c).

[10]Organization Plan No. 1 of 1978, 43 *Fed. Reg.* 19807.

[11]29 U.S.C.A. § 1627.3.

[12]29 U.S.C.A. § 1627.3 (b) (1).

[13]29 U.S.C.A. § 627; 29 C.F.R. § 1627.10.

an ADEA claim.[14] The ADEA provides certain exceptions to the requirements of this law. Employers who have adopted a bona fide employee benefit plan that provides lower benefits to older employees as long as the plan is not a subterfuge to evade the purpose of the act is permitted.[15] Qualified employee benefit plans must be the type or similar to those prescribed by statute (i.e., retirement plans, pension plans, or insurance plans) and must be able to justify the age-related cost considerations.[16] Additionally, the Tax Equity and Fiscal Responsibility Act of 1982, as well as the ADEA, may require that employers provide employees between the ages of sixty-five and seventy and their dependents with the same group health insurance benefits as employees under the age of sixty-five.[17]

The ADEA prohibits an employer from involuntarily retiring any individual between the ages of forty and sixty-nine because of the individual's age. An exception has been provided under the 1978 amendments which preserves this provision for bona fide executives.[18] To apply for a bona fide executive exception, the employee must be employed as an executive or in a high policy-making position for two years prior to his or her retirement. He or she must be entitled to retirement benefits of at least $44,000, exclusive of the employee's own contributions, and the employee must be between the ages of sixty-five and sixty-nine.[19]

The statute of limitations for an ADEA claim are relatively confusing. An individual who has been discriminated against can file an EEOC charge, no later than 180 days from the alleged unlawful acts, unless the misconduct occurred in a state where the age antidiscrimination group or agency permits a 300-day statute of limitations.[20] The ADEA charge must be filed in writing with the state agency or the EEOC. An investigative procedure and conciliation would mirror that as set forth under Title VII of the Civil Rights Act (see Chapter 5).

A person may seek to bring a civil action against the employer in any court of competent jurisdiction if the following conditions are met prior to the filing of the action:

- The employee has waited sixty days after filing the charge with the EEOC or state agency before initiating his or her private lawsuit.
- The Age Discrimination Charge is filed within the specified statute of limitations.
- The lawsuit by the employee is filed within the two- or three-year statute of limitations.

The EEOC may bring suit against the employer in any court of competent jurisdiction on behalf of the employee to seek remedies for violation of the ADEA.

---

[14] *Vance v. Whirlpool Corporation*, 32 FEP Cases 1391 (Ca.–4, 1983).
[15] 19 U.S.C.A. § 623 (f) (2).
[16] 29 U.S.C.A. § 860.120 (b).
[17] See 29 U.S.C.A. § 623 (g) (1); § 116 (a) of Pub. L. 97-248, 96 Stat. 353, September 3, 1982.
[18] 29 U.S.C.A. § 631 (c).
[19] 29 U.S.C.A. § 631 (c) (as amended).
[20] 29 U.S.C.A. § 626 (d) (2).

The ADEA provides that an individual is entitled to a trial by jury for most remedies except for injunctive relief.[21] Types of relief permitted under an ADEA action include permanent injunctions, reinstatement or promotion, judgments compelling employment, monetary damages, and, in willful situations, liquidated damages.[22] Attorney's fees can be awarded to the prevailing party.

The defenses to an ADEA action against an employer include providing deference based on reasonable factors other than age, the bona fide occupational qualification (BFOQ) defense, a bona fide seniority system, the employer's good faith reliance on a written administrative order, or procedural or technical deficiencies in the ADEA claim itself.

---

[21]29 U.S.C.A. § 217.
[22]29 U.S.C.A. § 216(b) and (c).

(This case has been edited for the purposes of this text.)

Edwin **BOYLAN**, John E. Butler, Dominic G. Ferlauto, Thomas Fitzpatrick, John W. Gardner, Edward J. Gray, William P. Hayes, John Hussey, James N. Karas, William N. Luck, Thomas W. Martin, William R. Nixon, and Albert J. Shleeh, Plaintiffs-Respondents,

v.

**STATE OF NEW JERSEY**, City of Jersey City, City of Irvington, Town of Montclair, and City of Newark, Defendant-Appellants

William J. **COMER**, Fire Chief of the City of Paterson; Kenneth Peterson, Fire Chief of the City of Passaic; George Sbarra, Fire Chief of the Township of Belleville; James Houn, Fire Chief of the City of Hoboken; Edward Woods, Fire Chief of the City of Margate City; John W. Gardner, Fire Chief of the Town of Montclair; and New Jersey Paid Fire Chiefs' Association, Plaintiffs-Respondents,

v.

The **CITY OF PATERSON**, the City of Passaic, the Township of Belleville, the City of Hoboken, the City of Margate City, the Town of Montclair, and all Municipalities of the

State of New Jersey, Defendants-Appellants.

Alexander M. **BEATTIE**, Jr., Appellant,

v.

**CITY OF CLIFTON, N.J.** (a Municipal Corporation of the State of New Jersey); and the State of New Jersey, N.J. Division of Pensions, Respondents. Gioacchino **FIORENTINO**, Firemen's Mutual Benevolent Association— Local 21; David Slaughter; and City of Ocean City, Respondent,

v.

**NEW JERSEY DEPARTMENT OF the TREASURY— DIVISION OF PENSIONS;**

Hon. Douglas R. Forrester, Director of the Division of Pensions; and the Police and Firemen's Retirement System of New Jersey, Appellants

Supreme Court of New Jersey

Argued Jan. 31, 1989

Decided Aug. 2, 1989

116 N.J. 236

POLLOCK, J.

This case involves challenges that were made to enforcement of early mandatory retirement provision in state statutes for law enforcement officers and firefighters.

All Plaintiffs were advised by a February 19, 1987 directive issued by the Division of Pensions (Division) in the State Department of Treasury, based on legal advice from the Attorney General, that they had to retire because they were over age 65.

In the appeal brought by Edward Boylan and others, all of the Plaintiffs served as officers in various positions in police and fire departments. The Plaintiffs included one police chief, four deputy chiefs, three police captains and one detective. They also included one fire chief, two battalion chiefs and two captains from fire departments. An additional Plaintiff named in the complaint was one superintendent Nixon. Each Plaintiff was between the age of 65 and 70 years and was a member of either the P & FRS or the PERS.

In the Comer appeal, Plaintiffs, paid fire chiefs in their respective municipalities, were between the ages of 65 and 70, and were members of the P & FRS. In the third appeal, Beattie, a deputy fire chief, was 68 years of age. In the Fiorentino appeal, the Plaintiffs included a fire captain, the City of Ocean City, a collective bargaining representative and Firemen's Mutual Benevolent Association, Local 27.

The Plaintiffs in each of these matters, with the exception of the Beattie appeal, sought injunctive relief against the enforcement of the mandatory retirement requirement. The matters were transferred to the Appellate Division in three of the appeals on orders entered by the trial court. Beattie took a direct appeal from the mandatory retirement directive issued by the Division of Pensions and sought and obtained injunctive relief pending disposition of the appeal.

As that summary indicated, Plaintiffs held super-visory positions in county or municipal, police, and fire departments in New Jersey. Each was a member of either the P & FRS or PERS pension system and each was over sixty-five but under seventy.

In the latter part of 1983, the Attorney General of New Jersey issued a formal opinion concluding that the pension provisions requiring mandatory retirement before age seventy for uniformed police officers and firefighters were unenforceable under the ADEA, unless those provisions were a bona fide occupational qualification. The amendments to the ADEA took effect in 1986, in which, the Attorney General issued a formal opinion concluding that the 1986 amendments allowed the mandatory retirement provisions in the New Jersey pension plans.

The ADEA was enacted in 1967 and its purpose was to ban discrimination in the hiring or discharging of employees on the basis of age. However, it permitted employers to make employment decisions on the basis of age when it is a "bona fide occupational qualification reasonably necessary to the normal operation of the particular business."

Against the background of the Attorney General's opinion, the Appellate Division held that the 1986 amendment allowed the mandatory retirement of only those Plaintiff's involved in "active" law-enforcement or firefighting efforts.

The Supreme Court of New Jersey reversed the decision of the Appellate Division. The Court held that the 1986 amendments to the federal Age Discrimination in Employment Act (ADEA) permits the early mandatory retirement of such officers and firefighters who serve primarily in an administrative or supervisory capacity.

REVERSED.

*Decision Explanation:*
*The Supreme Court of New Jersey agreed with the district court's decision and reversed the appellate court's decision permitting early mandatory retirement.*

(This case has been edited for the purposes of this text.)
Dale Duane **LINK**, Plaintiff and Appellant,
v.
**CITY OF LEWISTOWN**, Defendant and Respondent
Supreme Court of Montana
Submitted on Brief May 21, 1992
Decided June 11, 1992
253 MONT. 451

TURNAGE, Chief Justice.

This case involves a part-paid firefighter who brought suit against City claiming discrimination on basis of age, political beliefs and retaliation.

Dale Duane Link began working as a part-paid firefighter for the City of Lewistown in 1981. Under the City's personnel rules, regulations, and requirements, part-paid firefighters are appointed by the fire chief and must serve a six-month probationary period. They receive a monetary fee for hours or fractions thereof served in fighting fires and in training.

In 1988, Link applied for a position as a full-time firefighter for the City. The position was not initially publicly advertised because the Fire Chief viewed the pool of qualified applicants to be the part-paid firefighters. Appointment to this position required approval by the Mayor and City Council. After testing and interviews, the Fire Chief sought approval of Link for the full-time position at the regular Lewistown City Council meeting of September 19, 1988.

At the meeting, Council members expressed concern with the interpretation of s 7-33-4107, MCA, which states that firefighters "shall not be more than 34 years of age at the time of original appointment." Link served as a part-paid firefighter since he was thirty-one years old, but he was thirty-nine when he applied for the full-time position. Link met all the other qualifications for the job. After discussion, the City Council directed that the full-time position be publicly advertised.

Link contacted an attorney who advised the City Council that a complaint was being filed with the Montana Human Rights Commission. The matter came before the City Council again at its October 17, 1988 meeting. The Fire Chief informed the Council that he had publicly advertised the position and that he still recommended that Link be approved for the position.

After discussion, the Council tabled the matter pending review by the City Attorney. At a meeting the following month, Council members agreed to seek the opinion of the Montana Attorney General on the applicability of the age limitation in s 7-33-4107, MCA.

The effort to obtain an Attorney General opinion was abandoned when Link filed a complaint before the Montana Human Rights Commission in November 1988. In January 1990, that complaint was dismissed and Link was authorized to file a complaint in District Court pursuant to ss 49-2-509 and 49-3-312, MCA.

Link's complaint in District Court alleged that the City discriminated against him on the basis of age, political beliefs, and retaliation. The court originally denied cross-motions for summary judgment. However, on the parties' joint request for reconsideration, it granted summary judgment to the City on the age discrimination issue. It relied on the City's argument concerning the effect of the following language in s 7-33-4106, MCA:

The mayor . . . shall nominate and, with the consent of the council or commission, appoint . . . all firefighters.

The Court reasoned that because the Mayor and City Council did not have input into Link's appointment as a part-paid firefighter, his appointment to that position was not an "original appointment" under s 7-33-4107, MCA.

The Supreme Court of Montana held that Link's appointment as a part-paid firefighter was his "original appointment" under s 7-7-7-33-4107, MCA and that he did not have to be under maximum hiring age in order to be subsequently appointed to full-time position. Therefore, the Court reversed the partial summary judgment of the District Court.

REVERSED AND REMANDED.                    GRAY, Justice, dissenting.

Decision Explanation:                     summary judgment granted by the district court and
The Supreme Court of Montana reversed the partial    sent the case back to the district court for hearing.

## Review Questions

1. The ADEA protects individuals over what age?

2. An ADEA claim is filed with what agency?

3. What prohibitions are provided for under the ADEA?

4. What defenses are available to covered organizations when an ADEA claim is filed?

5. What is the statute of limitations for an ADEA claim?

6. What is the issue in the *Boylan* case?

7. What are the issues involved in the *Link* case?

8. What is the decision of the Montana Human Rights Commission in the *Link* case?

9. What is a right-to-sue notice by the EEOC?

*Chapter*

# 15

# Labor and Employment Laws

Labor disgraces no man; unfortunately you occasionally find men disgrace labor.

*Ulysses S. Grant*

I believe in the dignity of labor, whether with head or hand; that the world owes no man a living but that it owes every man an opportunity to make a living.

*John D. Rockefeller, Jr.*

## Learning Objectives

■ To gain an understanding of the various federal and state labor and employment laws federal and state.

■ To gain an understanding of the scope of the various laws that may impact your fire and emergency service organization.

■ To gain an understanding of the various rights that may apply to the employment of firefighters or EMS personnel.

In addition to the specific employment issues addressed in Chapter 8, a myriad of other labor and employment-related laws may directly or indirectly affect a fire and emergency service organization. At the federal level, and additionally in most states, laws have been adopted in the labor and employment area addressing specific employees working within their jurisdiction. Fire and emergency service organizations should take careful note of the laws applicable to their organization on a federal, state, and even a local level to ensure compliance.

As a general rule, most American workforce employees are considered "at will" and can be terminated for good cause, bad cause, or no cause at all. The general exceptions to this rule are if an employee is governed by a collective bargaining agreement (union contract) or the individual possesses a personal contract for employment. The courts have generally found that these contractual relationships remove the particular party from the at-will doctrine. In addition to the above exceptions, many state courts throughout the United States have developed additional exceptions from the at-will doctrine including, but not limited to, the following:

- Implied Contract Exception. Many courts have recognized a cause of action where the employer has promised through a personnel handbook or other document that the employee would not be discharged until given a fair chance to perform. The court recognizes this exception through a handbook as constituting an implied in fact employment contract.[1]

- Covenant of Good Faith and Fair Dealing. The courts have recognized that the employer may have an implied obligation to deal fairly with an employee and not act arbitrarily in a discharge situation. Most of these cases involve employees who have spent an extended period of time with the employer and then are terminated under the at-will doctrine.[2]

- Independent Consideration Exception. This exception involves an employer's oral promise not to terminate the employee except for good cause and when services are performed in addition to regular required employment services.[3]

- Public Policy Exceptions. Most states recognize that an employer may not terminate an employee for pursuing a workers' compensation claim, complaint to OSHA, EEOC claim, or other public policy rights.[4]

- Presumption of Term Exception. In many states an oral contract for a yearly salary raises the inference of a year-to-year oral employment contract in which the employee cannot be lawfully discharged except for cause.[5]

---

[1]See *Walker v. Northern San Diego County Hospital District,* 135 Cal. App. 3d. 896, 185 Cal. Rep. 617 (1982).
[2]See *Pugh v. See's Candies, Inc.,* 116 Cal. App. 3d. 311, 171 Cal. Rep. 917 (1981); *Cancellier v. Federated Department Stores,* 672 F.2d. 1312 (Ca-9, 1982).
[3]See *Alvarez v. Dart Industries, Inc.,* 55 Cal. App. 3d. 91, 127 Cal. Rep. 222 (1976).
[4]See *Firestone Textile Company Divisions v. Meadows,* 666 S.W. 2d 730 (Ky. 1983); *Tammany v. Atlantic Richfield Co.,* 27 Cal. 3d 167, 610 P. 2d. 1330, 164 Cal. Rep. 839 (1980).
[5]*Moore v. YWCA,* Case No. 84-CA-1508-MR (Ky. Court of Appeals, 1985).

- Promissory Estoppel Exception. Some courts have recognized a promissory estoppel exception to the at-will rule holding that an employee may recover moving expenses where he or she detrimentally relied upon prospective employer's promise of a job.[6]
- Intentional Infliction of Emotional Distress Exception. Some courts have found that where an employer's conduct is so outrageous as to go beyond all possible bounds of decency resulting in severe emotional distress to an employee, an action will be permitted to address this wrongdoing.[7]
- Refusal to Commit Unlawful Acts. Most states recognize that if an employee refuses to commit an unlawful act and the employer discharges the employee, it is an exception to the at-will doctrine.[8]

The at-will employment doctrine in the United States is deteriorating through the numerous exceptions being applied by the courts. Additionally, there is a movement in the United States to protect employee's rights in the workplace given the substantial deterioration of union representation of employees in the American workplace. The at-will doctrine is still judicially sound in most jurisdictions however, in addition to the recognized court exceptions, many federal and state laws regarding discrimination or retaliation also limit this doctrine.

In addition to the at-will doctrine, numerous labor and employment-related laws have been enacted on the federal and state levels that directly impact fire and emergency service organizations. The Welsh-Healey Act, 41 U.S.C. § 35, enacted in 1936, addresses the situation of government contracts in excess of $10,000. Employees working for contractors governed by this act must pay the minimum wage and overtime benefits as specified by law. There are numerous exceptions to the Welsh-Healey Act[9]. The Department of Labor enforces the Welsh-Healey Act and failure to comply may permit the government to blacklist the contractor from further U.S. government contracts.[10]

The Davis-Bacon Act, enacted in 1931, requires the payment of minimum wages and fringe benefits to laborers on federal public works contracts in excess of $2,000.[11] The provisions of the act normally apply to construction activities as distinguished from manufacturing activities and require the employer to comply with wage and hour provisions. Violations of the Davis-Bacon Act are grounds for criminal prosecution.[12] The Department of Labor is responsible for the administration and enforcement of the Davis-Bacon Act.

---

[6]*Lorson v. Falcon Coach, Inc.,* 522 P. 2d. 449 (Kansas 1974).

[7]*Harris v. First Federal Savings & Loan Assoc. of Chicago,* 473 N.E. 2d. 457 (Ill. Court of Appeals, 1984).

[8]See *Peterson v. International Brotherhood of Teamsters Local 396,* 174 Cal. App. 2d 184, 344 P. 2d 25 (1959).

[9]Exceptions include transportations by common carriers, rentals, foreign goods, executive and administrative professional employees. CFR §§ 50-201.603.

[10]41 U.S.C.A. § 37.

[11]40 U.S.C.A. § 276 (a).

[12]29 C.F.R. § 5.10 (b).

In addition to the foregoing laws, the Service Contract Act provides that every contract entered into by the U.S. government in excess of $2,500 contain stipulations with regard to minimum wage and fringe benefits.[13] Additionally, employers covered under the Service Contract Act must maintain certain employee records and post notices of compensation required under the Act in prominent places. Failure to comply with this law can permit the U.S. government to cancel the employer's contract or blacklist it from future government contracts. The Department of Labor is responsible for administering and enforcing the Service Contract Act.

The Copeland Act, sometime known as the antikickback act, prohibits government contractors from compelling or inducing the employee to give back part or all of his compensation.[14] Any employer who, by force, intimidation, or threat of discharge, or any other manner whatsoever, who induces an employee to provide part of his or her compensation shall be subject to fines up to $5,000 or imprisonment up to five years, or both.[15]

The Miller Act provides that government contractors working on public works contracts in excess of $2,000 must furnish a bond to protect the payment of wages to all employees.[16] The Contract Work Hours and Safety Standards Act provides that employers pay time and one-half for hours worked in excess of eight hours in any one day or forty hours in any one week and additionally requires that the employer provide a safe and healthy work environment for employees.[17]

In recent years, the issue of garnishment and credit has woven its way into the working environment. Having an effect on both the employer and the employee, Congress has enacted several laws addressing the issue of credit and garnishment in the workplace. Under the Consumer Credit Protection Act (Title III), restrictions are set on the amount of an individual's earning that may be deducted in any one week as a result of a garnishment proceeding.[18] The Consumer Credit Protection Act establishes a maximum of the aggregate disposable income of any individual that may be subject to garnishment be no more than 25% of the employee's disposable earnings in any one week, or the amount by which the disposable earnings for that week exceeds thirty times the federal minimum hourly wage as set forth under the Fair Labor Standards Act.[19] Exceptions can be made from a court order on any bankruptcy as set up under the Bankruptcy Act or any debt due for the payment of state or federal taxes.

The Consumer Credit Protection Act further provides that no employer may discharge any employee by reason of the fact that his or her earnings are subject

---

[13]41 U.S.C.A. § 315(a).

[14]18 U.S.C.A. § 874.

[15]18 U.S.C.A. § 874.

[16]40 U.S.C.A. § 270(a); 32 C.F.R. § 10.101–110.

[17]40 U.S.C.A. § 329.

[18]15 U.S.C.A. § 1671, et seq.

[19]Id.

to garnishment for any one indebtedness.[20] "Any one indebtedness" refers to a single debt regardless of the numbers of levies made or the number of creditors seeking satisfaction.[21] Employers willfully violating the statute may be subject to a fine up to $1,000 or imprisonment of not more than one year, or both.[22]

Numerous other federal and state labor and employment acts address specific industries and specific situations. For example, the railroad system is governed by the Railway Labor Act and, in the event of an impasse in labor negotiations, the Federal Service Labor Management Relations Act established Federal Service Impasse Panels to assist in labor negotiations. Most states have enacted labor laws that stress such situations as child labor,[23] voting times,[24] payment of wages,[25] payment of wages upon termination,[26] garnishment of wages,[27] and medical insurance conversion.[28] Some states have also enacted regulations with regard to employment practices and with regard to strikes, picketing, and boycotting during a labor strike. In the area of employment practices, many states have adopted laws concerning antidiscrimination, protections of political freedom, jury duty, investigating consumer reports, arrest record, access to personnel files, confidentiality of medical information, employment under false pretense, polygraph restrictions, protection of military personnel, blacklisting, fingerprinting, whistle-blowing statutes, alcohol rehabilitation, plant closure, equal pay, and reprisal. In the area of strikes, many states have developed particular statutes addressing firefighting organizations such as California Labor Code § 1960, which permits firefighters within the state of California the right of self-organization and the right to join and assist labor organizations (also see Chapter 16). Other concerns that may be addressed under state labor laws include collective bargaining rights of public employees, higher education employees, non-right to work policies, yellow dog contract laws, strike replacement laws, and anti-injunction statutes. Prudent fire and emergency service organizations should become familiar with the particular laws of their jurisdiction as well as the federal laws that apply to their fire and emergency service organization.

---

[20]15 U.S.C.A. § 1674.
[21]15 U.S.C.A. § 1671 et seq.
[22]15 U.S.C. § 1674.
[23]See, for example, KRS § 339.280.
[24]KRS § 118.035.
[25]KRS § 337.020.
[26]KRS § 337.055.
[27]KRS § 427.140.
[28]KRS § 304.18–110.

(This case has been edited for the purposes of this text.)

Martha Anne **BOYLE**, Executrix of the Estate of John T. Boyle, Deceased: et al, Appellants,

v.

**CITY OF ANDERSON**: Anderson Fire Fighters Association

Local 1262, AFL-CIO, an unincorporated association: Professional Fire Fighters Union of Indiana, AFL-CIO: and

Individual Fire Fighters and Members and Officers of Said Unions, et al, Appellees.

Supreme Court of Indiana

March 3, 1989

534 N.E. 2d 1083

DeBRULER, Justice, dissenting to Denial of Transfer.

This case involves a consolidated appeal involving four cases, in which Plaintiffs sought damages for loss of property from fire in the business district of downtown Anderson, Indiana.

On August 30, 1978, at 4:30 A.M., a fire broke out in a lounge on a downtown street in Anderson, Indiana. The fire spread to adjoining buildings and ultimately a half block of older commercial buildings was destroyed. At the time, the City of Anderson was negotiating with the local firemen and their Unions, and the firemen had been on strike for several days. The main station was manned by the fire chief, a handful of firemen who were ordinarily administrators and a handful of probationary firemen. The City had mutual assistance agreements with the fire departments in some small surrounding communities. The chief and a few of his men responded to the fire as did several of the departments of the surrounding communities. Some fire equipment reached the site in a short time. The striking Anderson firemen refused to fight the fire, and two of the incoming community departments were held up by strikers for a short time, before being passed through to join the firefighting. Representatives of local, state, and international firefighters' Unions involved in local meetings on or about this time, and decisions were reached at those meetings to withhold services in the event of a fire. There was no loss of life, but the buildings and their contents were lost. The City had not sued in court for an order requiring the firemen to return to work, at the time of the fire.

The Plaintiffs asserted that the firemen engaged in unlawful conduct when they intentionally refused to fulfill their obligations as city firemen to fight the fire and when they engaged in efforts to obstruct others from fighting the fire as well.

The trial court concluded that the conduct of firemen in engaging in an illegal strike, and any assistance or encouragement of that conduct by the state and national unions did not give rise to a liability against private persons such as the Plaintiffs for loss of property. The trial court also concluded that the matters before it on summary judgment would show only that the asserted intentional obstructive conduct was that of unnamed persons, that it lasted for only three or four minutes, and that those briefly detained were passed on and did proceed to the fire scene.

The Court of Appeals agreed with the trial court that the claims asserted against the City of Anderson could not properly be maintained. However, the Court of Appeals disagreed with the trial court wherein it concluded that the claims against the individual striking firemen and the unions could not properly be maintained, and consequently ordered those claims to be put to trial. All parties, except the City of Anderson filed petitions to transfer to the Supreme Court of Indiana. The Supreme Court denied all transfer petitions.

*Decision Explanation:*
*The Court of Appeals permitted the summary judgment for the city and permitted the actions against the* *firefighters and union. The petitions to move the actions to the Supreme Court were denied.*

(This case has been edited for the purposes of this text.)

Loretta **WHITE**, et al., Appellants,

v.

**INTERNATIONAL ASSOCIATION OF FIREFIGHTERS**, Local 42, et al., Respondents

Missouri Court of Appeals, Western District

Oct. 27, 1987

738 S.W. 2d 933

BERREY, Judge.

This case involves a citizen whose home was destroyed by fire during firefighters strike. Plaintiff sued union in intentional tort.

The Plaintiffs, Loretta White and Charles A. White, alleged that on March 17, 1980, the Defendants began strike and refused to respond to fire alarms or calls; established picket lines to hinder the City's normal operations; engaged in acts of sabotage against the City's fire fighting equipment and facilities; and that those actions were in violation of s 105.530 RSMo 1978, the prohibition against public employee strikes. Plaintiffs further alleged that on March 20, 1980, a fire was discovered on Plaintiff's property located at 4003 Prospect and with the inexperienced personnel and inadequate equipment available, the fire was not brought under control and totally destroyed their real and personal property.

In Plaintiffs petition, they contend that had the Defendants "been performing the duties for which they were employed, the fire could have been brought under control with inconsequential damages to the premises and its contents." Plaintiffs also asserted that Defendants' actions established the tort of outrage.

The Defendants filed a motion to dismiss for failure to state a claim on which relief could be granted. The trial court granted the Defendant's motion.

Plaintiffs contend the trial court erred in granting Defendants' dismissal because private citizen may maintain a cause of action under a theory of intentional tort against the firefighters' union for damages occurring during an illegal strike by public employees.

The Missouri Court of Appeals held that private citizens whose home was destroyed during strike by fire fighters' union could not maintain cause of action against union in intentional tort. Therefore, the Court of Appeals affirmed the trial court's order dismissing the Plaintiffs' petition.

ALL CONCUR.

# Review Questions

1. Give an example of how the implied contract exception may be used.

2. Provide an example of a public policy exception to the at-will doctrine.

3. Provide an example of an intentional infliction of emotional distress exception in an employment setting.

4. What are the limitations of the Bacon-Davis Act of 1931?

5. In your own words, describe, the protections provided under the Consumer Credit Protection Act.

6. What is the difference between a lockout and a strike?

7. What is the major issue of the *Boyle* case?

8. Why do you think that the Anderson Firefighters Association, Local 1262, AFL-CIO, the Professional Firefighters Union of Indiana, AFL-CIO, and the International Association of Firefighters, AFL-CIO, were named defendants in this case?

9. Why do you think that the court dismissed the plaintiff's petition in the *White* case?

*Chapter*

# 16

# Collective Bargaining in the Fire Service

Men are more important than tools. If you don't believe so, put a good tool into the hands of a poor workman.

*John J. Bernet*

## *Learning Objectives*

- To gain a basic understanding of the collective bargaining process.
- To gain a basic understanding of the laws and regulations governing the collective bargaining process.
- To gain an understanding of individual rights and responsibilities in the collective bargaining process.

The rights of firefighters and EMS personnel to form or join a labor organization is governed by state law in most states. State labor laws are usually modeled after the federal labor relations laws which can be found in the Norris-LaGuardia Act, the National Labor Relations Act (NLRA), the Taft-Hartley Act (also known as the Labor Management Relations Act, or LMRA), and the Landrum-Griffin Act (also known as the Labor Management Reporting and Disclosure Act, or LMRDA). The federal labor relations laws were developed in the early 1900s to govern the activities and actions of both management and labor.

Given the extensive and complicated nature of this subject area, this text addresses only limited topics significant to firefighter and EMS personnel that are common between the federal and most state laws in order to provide a framework for further research by the reader. It is recommended that firefighters and EMS personnel interested in further information contact your state's labor cabinet and/or agency, or legal counsel.

**National Labor Relations Board (NLRB)**
the governing body established to administer and enforce the laws in the labor relations area

First, the governing body established on the federal level to administer and enforce the laws in the labor relations area is the **National Labor Relations Board** (NLRB). The NLRB was originally established in 1935 as part of the Wagner Act and modified in 1947 through the Labor Management Relations Act. The NLRB consists of five members appointed by the president with Senate approval for five-year terms, as well as an office of General Counsel. The Board's primary function is primarily judicial, however it does possess limited powers to investigate and prosecute unfair labor practices. The Board has complete authority over matters related to Section 9, as well as rule-making powers under Section 6 of the LMRA .

The National Labor Relations Board usually has jurisdiction when a labor dispute exists and the dispute affects commerce and meets the minimum dollar standard (for retail industries, the gross business volume in sales and taxes must exceed $500,000). Additionally, only employers, their agents and employees (being broadly construed) fall within the jurisdiction of the LMRA. Several employers, including governmental agencies, railroads, and airlines, as well as several categories of employees including supervisors, independent contractors, agricultural employees, and others are excluded from coverage.

For many fire and emergency service organizations, specific organizations modeled after the NLRB have been established to address labor disputes. Additionally, many states have enacted specific labor-related laws that set forth the parameters of permitted labor activity for fire, police, and EMS personnel who perform essential functions.[1]

The most often evoked rights for employees under the LMRA are specified in Section 7 of the Act and correlating state statutes.[2] Section 7 provides protec-

---

[1]For example, the Commonwealth of Kentucky has enacted laws that limit labor organization for fire service personnel to cities containing a population of at least 300,000 (KRS § 345.010, et seq.) and prohibits strikes by firefighters who have formed a union (Kentucky Public Act, H.B. 217, L. 1972).
[2]For example, Kentucky Revised Statute § 336.130 provides that Kentucky employees shall have the right to self-organization and to form, join, or assist labor unions, and such employees shall also have the right to bargain collectively through representatives of their own choosing.

tion for three specific rights of employees covered under the LMRA, namely the right to "self-organization, to form, join, or assist labor organizations, [and] to bargain collectively through representatives of their own choosing . . . ;" to protect employees' activities for "other mutual aid or protection;" and the right to refrain from union or concerted activities."[3] Many states have adopted the language from Section 7 of the Act directly into the individual state labor laws.

A violation of Section 7 or any of the requirements under the LMRA and other labor laws is called an *unfair labor practice*. The action normally filed against the employer or union as a result of this violation is called an *unfair labor practice charge*. Under the LMRA and other federal laws, this type of charge is normally filed with the National Labor Relations Board (See NLRB Charge form). On the state level, this type of charge is normally filed with the state labor agency or cabinet.

Section 8(a)(1) of the LMRA makes it an unfair labor practice for an employer to "interfere with, restrain, or coerce employees in the exercise of the rights guaranteed under Section 7 (of the Act) . . . ." This section prohibits not only unfavorable treatment and conduct by the employer but also favorable actions, such as increased pay or benefits, in order to interfere with the individual's right to join, form or otherwise participate in a union. Fire and emergency service organizations should be aware that the motives of the organization are not an essential element in proof of a violation and this type of charge can be refuted if appropriate business justification can be established for the conduct in question. For example, the act does not prohibit a fire service organization from implementing and administering rules of conduct and other personnel policies.[4]

Section 8(a)(2) of the Act prohibits the employer from dominating or interfering with the formation or administration of a union. Fire and emergency service organizations may commit an unfair labor practice by actively participating in union activities, providing financial support to union representation of firefighters, or even providing indirect support to the union. Indirect activities which may constitute a violation include the fire service organization payment of union dues,[5] payment of profits from vending machines to the union, and even fire service organization gifts to union agents.[6]

Section 8(a)(3) of the Act prohibits discrimination "in regard to hire or tenure of employment or any term or condition of employment to encourage or discourage membership in any labor organization." This section also includes coverage for strike discrimination, discriminatory discharge from employment, layoffs/recalls, and relocations, lockouts, and shutdowns.

[3]Section 8(a) (3) provides an exception whereby employees may forfeit their rights to refrain from union activity but may be subject to any union security clause.
[4]*NLRB v. Mylan-Sparta Co.,* 166 F.2d 485 (CA-6, 1948).
[5]*Dixie Bedding Mfg. Co v. NLRB,* 268 F.2d 901 (CA-5, 1959).
[6]*Superior Engraving Co. V. NLRB,* 183 F.2d 783 (CA-1, 1957).

Conversely, Section 8(b) of the Act established provisions that prohibit unions from interfering with employee rights to join and form a union. A union commits an 8(b) violation when it coerces or restrains an individual firefighter from exercising his/her Section 7 rights. Examples of violations of Section 8(b) include threats of violence, threats of loss of benefits, coercing strike participation, and interference with employee's statement to the NLRB.

An often-asked question is: What is the difference between a closed shop and a union shop? A **closed shop** usually results from an agreement between the union and employer that requires prospective employees to become members of the union prior to being hired. This is usually a violation of Section 8(b)(2) because the agreement does not allow for a thirty-day grace period as required by Section 9(a)(3). A **union shop** usually results from an agreement between the employer and union that requires membership in the union within a specified period of time following hiring. This type of shop is usually lawful. Another type of shop is an **agency shop**, where union membership is not required, however payment of union fees is mandatory. This type of shop is also usually lawful.

There are many similarities in the labor law area between the public and the private sector, however, there are many differences. The main areas of distinction between the public sector and the private sector include the specific labor laws focused on the fire and emergency service "industry," the source of funding, the structure of employment, and the other laws, including constitutional and statutory protection, which are afforded to public sector personnel. Most states have enacted specific laws for public sector employees, which is most fire and EMS personnel, and in states that have not enacted specific statutes[7] or have not prohibited public sector collective bargaining,[8] the right to join, form, or assist in collective bargaining is often an accepted practice. Particular requirements vary from state to state, however, most are modeled after the LMRA and other federal labor laws. Prudent fire and emergency service organizations should become familiar with the laws and regulations in their particular jurisdiction or seek appropriate legal assistance.

## OFTEN-ASKED LABOR QUESTIONS

*Can my fire department prohibit a union from distributing union membership materials at the fire house?*

As a general rule, an employer may validly post its property against employee distribution and solicitation, including the distribution and solicitation of employees on the employer's property by outside union organizers.[9] However, this prohibition must be balanced with the individual's Section 7 rights. Before

**closed shop**
a place of employment where prospective employees are required to become union members prior to being hired as a result of an agreement between the union and employer

**union shop**
a place of employment where employees are required to become union members within a specified time period following hiring as a result of an agreement the union and employer

**agency shop**
a place of employment where union membership is not required, but payment of union fees in mandatory

---

[7]Developments in the law, Public Employment, 97 Harv. L. Rev. 1661 (1984). States lacking statutes authorizing public sector bargaining include Arizona, Arkansas, Colorado, Louisiana, Mississippi, and Utah.
[8]For example, North Carolina General Statute § 95-98 (1981).

implementing policies and SOPs regarding the prohibition of solicitation, careful analysis of the current NLRB rules regarding solicitation and individual state laws is advisable as well as review by legal counsel.

*Do inquiries of my firefighters regarding union activities constitute interrogation?*

Interrogation of employees as to their union sympathies is generally not viewed by the NLRB as constitutionally protected free speech. Questioning employees by management as to their union affiliation or activities normally is a violation of Section 8(a)(1) "because of its natural tendency to instill in the minds of the employees, fear of discrimination on the basis of the information the employee has obtained."[10] However, casual or isolated inquiries normally are not a violation so long as the inquiry is free of coercion, threats, or promises and the NLRB has provided an exception to the general rule which permits questioning of employees within very specific parameters in preparation for an NLRB hearing.

*Can I deny a fire fighter's request to have his/or her union representative present during an investigatory interview?*

In most cases the answer is "no". When an employer denies an employee's request to have his or her union representative present at an investigatory interview where the employee reasonably believes that disciplinary action may result, this denial is often an 8(a)(1) violation. Additionally, the decision in *Weingarten*[11] extended this protection to nonunion employees and provided specific responsibilities for the employee, employer, and representative.

*Can my employer "spy" on my union activities?*

Surveillance by employers on employee's union activities normally constitutes a violation of Section 8(a)(1). As a general rule, the employer is permitted to observe normal on-site activities that are performed during work time, however, the employer or its agents are prohibited from other activities such as photographing employees in union-organizing campaigns.

*My boss promised we would get a big increase in pay if we voted against the union. Is this a violation?*

Section 8(a)(1) generally prohibits such promises as financial inducements for the purpose of preventing unionization. Generally, when such a promise is tied to preventing the union from organizing a particular employer, the employer is

---

[9]*NLRB v. Babcock & Wilcox Co.,* 351 U.S. 105 (1956).

[10]*NLRB v. West Casket Co.,* 205 F.2d 902 (CA-9, 1953).

[11]*NLRB v. Weingarten, Inc.,* 420 U.S. 251 (1975).

committing an unfair labor practice prohibited by Section 8(a)(1) and is also violating the employee's Section 7 rights. Again, this is a very difficult area and legal counsel should be retained whenever a fire or emergency service organization intends to provide any benefits, as well as implement any actions detrimental to its employees, in order to avoid the legal entanglements in this very complicated area of the law.

### ROCKLAND PROFESSIONAL FIRE FIGHTERS ASSOCIATION
### v. CITY OF ROCKLAND
261 A.2d 418 (Me. 1970)

WEATHERBEE, JUSTICE, In 1957 our Legislature enacted "An Act Relating to Arbitration Pursuant to Collective Bargaining Contracts." This legislation as amended is now 26 M.R.S.A. § 951, et seq. Section 951 reads:

> "A written provision in any collective bargaining contract to settle by arbitration a controversy thereafter arising out of such contract or out of the refusal to perform the whole or any part thereof, or an agreement in writing to submit to arbitration an existing controversy arising out of such a contract, or such refusal, herein designated in this subchapter as 'a written submission agreement,' shall be valid, irrevocable and enforceable, save upon such grounds, independent of the provisions for arbitration, as exist at law or in equity for the revocation of any contract."

Section 954 provides for the appointment of arbitrators by the Court in the event that the collective bargaining contract fails to establish a method for their appointment or if the method so provided is ignored by one party. Section 956 details methods for obtaining testimony of witnesses at the arbitration hearing, and for the fees of witnesses. Section 957 provides for the enforcement by the Superior Court of the award of arbitration upon application of a party that judgment shall be entered for the party.

Section 958 enumerates grounds which would justify vacating an award.

In 1965 the Legislature enacted a statute which is referred to as the Fire Fighters Arbitration Law which is now 26 M.R.S.A. § 980, et seq. In doing so the Legislature declared it to be the public policy of the State that this particular class of municipal employees in their position of higher responsibility should be given the right to organize and bargain collectively with the municipalities in arriving at a contract of employment:

Section 981 reads:

> "The protection of the public health, safety and welfare demands that the permanent uniformed members of any paid fire department in any municipality not be accorded the right to strike or engage in any work stoppage or slowdown. This necessary prohibition does not, however, require the denial to such municipal employees of other recognized rights of labor such as the right to organize, to be represented by a labor organization of their choice, and the right to bargain collectively concerning wages, rates of pay and other terms and conditions of employment."

Following the broad statement of policy of section 981 the Fire Fighters Arbitration Law proceeds to empower the fire fighters to compel the municipality to meet with the fire fighters' bargaining agent to bargain collectively in the formation of a written contract of employment. Section 986 provides that if the parties cannot reach a contract the unresolved issues shall be submitted to arbitration at the option of the association. Section 987 sets out the method of selecting the arbitrators. Section 988 describes the rules under which hearings shall be conducted and specifies that a less-than-unanimous decision of the arbitrators shall not be binding on either party.

Section 991 includes the words:

"Any collective bargaining agreement negotiated under this chapter shall specifically provide that the fire fighters who are subjected to its terms shall have no right to engage in any work stoppage, slowdown or strike, the consideration for such provision being the right to a resolution of disputed questions."

Under the authority of the Fire Fighters Arbitration Law the firemen of Rockland organized and entered into a contract of employment for the calender year of 1967 . . . [The contract established a grievance procedure] consisting of four steps, the third of which provided for a hearing before an Appeal Grievance Board . . . .

The contract then provided that:

"Arbitration shall be in accordance with Title 26, Sect. 987 of the Maine Revised Statutes Annotated."

[H]owever, section 987 only sets up machinery for the organization and operation of an arbitration board for the purpose of resolving disputes concerning the negotiation of a labor contract and makes no provision for arbitration of disputes arising later under the contract.

On April 27, 1967 Walter R. Dyer, a Rockland fireman, was suspended for six days allegedly for insubordinate conduct. Dyer and the Plaintiff exhausted the grievance processes provided in the contract and the matter went on to arbitration. The parties followed the provisions of section 987 in selecting the members of the arbitration board but on June 12, before the Board could hear the matter, Dyer was discharged for a similar reason.

Plaintiff and Defendant then combined the two grievances and submitted to the arbitration board the single issue of whether the discharge of Dyer was for sufficient cause. After hearing, a majority of the board returned an award which read:

"1. There was not sufficient cause for the discharge of Walter R. Dyer on June 12, 1967.

2. If Dyer resigns from office in Local 1584 of the International Association of Firefighters, the City of Rockland shall offer him reinstatement to his former position without back pay."

Dyer did resign from office in the union but the city refused to reinstate him. Plaintiff then brought an application for judgment upon the arbitration award in the Superior Court in Knox County. Defendant moved to dismiss the application for judgment on the grounds that:

1) The award was not unanimous and therefore invalid.

2) The board abused its discretion and exceeded the powers given it in going beyond the single question submitted to them.

3) The city has the right to discharge any employee when, in his best judgment, the City Manager considers that the city's interest requires it.

The matter was heard before a Justice of the Superior Court who granted Defendant's motion to dismiss the application for judgment on the award. The ruling of the Justice was based on his conclusion that

" . . . [T]he Legislature intended that whatever benefits it was conferring upon the Fire Fighters, the right to organize, bargain collectively, arbitrate and all other rights being granted, were to be viewed entirely in the light of the language of the Fire Fighters Arbitration Law . . . ."

As the Fire Fighters Arbitration Law provided only for arbitration of disputes surrounding the *making* of a contract, the Justice held the Fire Fighters had been given no right to arbitration of disputes arising *subsequent* to the contract and therefore had no right to enforcement of an award. . . .

We find that the Fire Fighters Arbitration Law, in spite of the scope suggested by its title, provides for arbitration only as to the negotiation of a labor contract. The parties here did negotiate a contract as the Fire Fighters Arbitration Law empowered them to do. The contract bound the parties to submit unresolved grievances to arbitration at the option of the union. The Legislature intended the Fire Fighters to have all the rights of labor organizations except those specifically withheld and in section 951, years earlier, the Legislature had given all labor organizations the right to provide in their contracts for arbitration of grievances arising out of such contracts. The fire fighters' authority to agree with the city to submit their unresolved grievances to arbitration is found in section 951. The provision in Plaintiff and Defendant's contract that arbitration of their labor grievances should be in accordance with section 987 appears to us to be best explained by the fact that the 1957 general statute authorizing arbitration makes no provision for the method in which the arbitrators shall be chosen (except that if the contract fails to provide a method or if a party refuses to proceed under the method agreed upon, resort may be had to the Court for appointment of the arbitrator). It leaves the parties to state in their contracts the method they prefer for selecting arbitrators.

We view the reference to section 987 as the parties' agreement that the arbitrators should be chosen by the method provided in section 987 . . . .

Applying the *in pari materia* rule of construction and reading in connection with the Fire Fighters Arbitration Law the 1957 general arbitration statute having the same general purpose, it appears clear that the grant to the fire fighters of "all the rights of labor other than the right to strike, or engage in any work stoppage or slowdown" found in section 981 included the right to submit grievances to binding arbitration. We hold that the Plaintiff was entitled to submit to arbitration grievances arising under the labor contract

and was not limited to the specific issue of formation of a contract.

The Justice in the Superior Court did not reach the issues concerning the validity of the award itself . . . . They remain to be determined in the Superior Court. . . .

### Notes

1. It has frequently been held that state statutes pertaining to employer-employee relations "must be construed to apply only to private industry, at least until such time as the legislature shows a definite intent to include political subdivisions." *Wichita Pub. Schools Employees Union, Local 513 v. Smith*, 194 Kan. 2, 397 P.2d 357 (1964). See also *Miami Water Works Local 654 v. City of Miami*, 157 Fla. 445, 26 So.2d 194 (1946). The 1957 grievance arbitration statute in the principal case was not made specifically applicable to the public sector. How, then, did the Supreme Court of Maine determine that this statute was applicable to the collective bargaining agreement between the city and the firemen's union? Did the court rule in effect that the Fire Fighters Arbitration Law was itself sufficient to compel the enforcement of the arbitration award? See also *Providence Teachers Union Local 958 v. School Comm.*, 108 R.I. 444, 276 A.2d 762 (1971), wherein a nearly identical result was reached.

2. Concerning the general subject of grievance arbitration, see D. Rothschild, L. Merrifield & H. Edwards, Collective Bargaining and Labor Arbitration (1979); F. Elkouri & E. Elkouri, How Arbitration Works (1973); R. Fleming, The Labor Arbitration Process (1965). Regarding grievance arbitration in the public sector, see Frazier, *Labor Arbitration in the Federal Service*, 45 Geo. Wash. L. Rev. 712 (1977); Rock, *The Role of the Neutral in Grievance Arbitration in Public Employment*, in Collective Bargaining in Government 141 (J. Loewenberg & M. Moskow eds. 1972); U.S. Bureau of Labor Statistics, Bull. No. 1661, Negotiation Impasse, Grievance, and Arbitration in Federal Agreements (1970); Note, *Legality and Propriety of Agreements to Arbitrate Major and Minor Disputes in Public Employment*, 54 Cornell L. Rev. 129 (1968); Krislov & Schmulowitz, *Grievance Arbitration in State and Local Government Units*, 18 Arb J. 171 (1963); Killingsworth, *Grievance Adjudication in Public Employment*, 13 Arb. J. 3 (1958). See also Granof & Moe, *Grievance Arbitration in the U.S. Postal Service: The Postal Service View*, 29 Arb. J. 1 (1974); Cohen, *Grievance Arbitration in the United States Postal Service*, 28 Arb. J. 258 (1973).

# APPENDIXES

Appendix

# A

# Additional Cases

## INTERNATIONAL UNION, UAW v. JOHNSON CONTROLS, INC.

### 111 S.Ct. 1196 (1991).

JUSTICE BLACKMUN delivered the opinion of the Court.

In this case we are concerned with an employer's gender-based fetal-protection policy. May an employer exclude a fertile female employee from certain jobs because of its concern for the health of the fetus the woman might conceive?

### I

Respondent Johnson Controls, Inc., manufactures batteries. In the manufacturing process, the element lead is a primary ingredient. Occupational exposure to lead entails health risks, including the risk of harm to any fetus carried by a female employee.

Before the Civil Rights Act of 1964 became law, Johnson Controls did not employ any woman in a battery-manufacturing job. In June 1977, however, it announced its first official policy concerning its employment of women in lead-exposure work:

> "[P]rotection of the health of the unborn child is the immediate and direct responsibility of the prospective parents. While the medical profession and the company can support them in the exercise of this responsibility, it cannot assume it for them without simultaneously infringing their rights as persons.

> * * *

> ". . . . Since not all women who can become mothers wish to become mothers (or will become mothers), it would appear to be illegal discrimination to treat all who are capable of pregnancy as though they will become pregnant."

Consistent with that view, Johnson Controls "stopped short of excluding women capable of bearing children from lead exposure," but emphasized that a woman who expected to have a child should not choose a job in which she would have such exposure. The company also required a woman who wished to be considered for employment to sign a statement that she had been advised of the risk of having a child while she was exposed to lead. The statement informed the woman that although there was evidence "that women exposed to lead have a higher rate of abortion," this evidence was "not as clear . . . as the relationship between cigarette smoking and cancer," but that it was, "medically speaking, just good sense not to run that risk if you want children and do not want to expose the unborn child to risk, however small. . . ."

Five years later, in 1982, Johnson Controls shifted from a policy of warning to a policy of exclusion. Between 1979 and 1983, eight employees became pregnant while maintaining blood lead levels in excess of 30 micrograms per deciliter. This appeared to be the critical level noted by the Occupational Health and Safety Administration (OSHA) for a worker who was planning to have a family. The company responded by announcing a broad exclusion of women from jobs that exposed them to lead:

> ". . . [I]t is [Johnson Controls'] policy that women who are pregnant or who are capable of bearing children will not be placed into jobs involving lead exposure or which could expose them to lead through the exercise of job bidding, bumping, transfer or promotion rights."

The policy defined "women . . . capable of bearing children" as "[a]ll women except those whose inability to bear children is medically documented." It further stated that an unacceptable work station was one where, "over the past year," an employee had recorded a blood lead level of more than 30 micrograms per deciliter or the work site had yielded an air sample containing a lead level in excess of 30 micrograms per cubic meter.

### II

In April 1984, petitioners filed in the United States District Court for the Eastern District of Wisconsin a class action challenging Johnson Controls' fetal-protection policy as sex discrimination that violated Title VII of the Civil Rights Act of 1964, as amended. Among the individual plaintiffs were petitioners Mary Craig, who had chosen to be sterilized in order to avoid los-ing her job, Elsie Nason, a 50-year-old divorcee, who had suffered a loss in compensation when she was transferred out of a job where she was exposed to lead, and Donald Penney, who had been denied a request for a leave of absence for the purpose of lowering his lead level because he intended to become a father. Upon stipulation of the parties, the District Court certified a class consisting of "all past, present and future production and maintenance employees" in United Auto Workers bargaining units at nine of Johnson Controls' plants "who have been and continue to be affected by [the employer's] Fetal Protection Policy implemented in 1982."

The District Court granted summary judgment for defendant-respondent Johnson Controls. Applying a three-part business necessity defense derived from fetal-protection cases in the Courts of Appeals for the Fourth and Eleventh Circuits, the District Court concluded that while "there is a disagreement among the experts regarding the effect of lead on the fetus," the hazard to the fetus through exposure to lead was established by "a considerable body of opinion"; that although "[e]xpert opinion has been provided which holds that lead also affects the reproductive abilities of men and women . . . [and] that these effects are as great as the effects of exposure of the fetus . . . a great body of experts are of the opinion that the fetus is more vulnerable to levels of lead that would not affect adults"; and that petitioners had "failed to establish that there is an acceptable alternative policy which would protect the fetus." The court stated that, in view of this disposition of the business necessity defense, it did not "have to undertake a bona fide occupational qualification's (BFOQ) analysis."

The Court of Appeals for the Seventh Circuit, sitting en banc, affirmed the summary judgment by a 7-to-4 vote. The majority held that the proper standard for evaluating the fetal-protection policy was the defense of business necessity; that Johnson Controls was entitled to summary judgment under that defense; and that even if the proper standard was a BFOQ, Johnson Controls still was entitled to summary judgment.

The Court of Appeals first reviewed fetal-protection opinions from the Eleventh and Fourth Circuits.

See *Hayes v. Shelby Memorial Hospital*, 726 F.2d 1543 (CA11 1984), and *Wright v. Olin Corp.*, 697 F.2d 1172 (CA4 1982). Those opinions established the three-step business necessity inquiry: whether there is a substantial health risk to the fetus; whether transmission of the hazard to the fetus occurs only through women; and whether there is a less discriminatory alternative equally capable of preventing the health hazard to the fetus. The Court of Appeals agreed with the Eleventh and Fourth Circuits that "the components of the business necessity defense the courts of appeals and the EEOC have utilized in fetal protection cases balance the interests of the employer, the employee and the unborn child in a manner consistent with Title VII."

\* \* \*

### III

The bias in Johnson Controls' policy is obvious. Fertile men, but not fertile women, are given a choice as to whether they wish to risk their reproductive health for a particular job. Section 703(a) of the Civil Rights Act of 1964 prohibits sex-based classifications in terms and conditions of employment, in hiring and discharging decisions, and in other employment decisions that adversely affect an employee's status. Respondent's fetal-protection policy explicitly discriminates against women on the basis of their sex. The policy excludes women with childbearing capacity from lead-exposed jobs and so creates a facial classification based on gender. Respondent assumes as much in its brief before this Court.

Nevertheless, the Court of Appeals assumed, as did the two appellate courts who already had confronted the issue, that sex-specific fetal-protection policies do not involve facial discrimination. These courts analyzed the policies as though they were facially neutral, and had only a discriminatory effect upon the employment opportunities of women. Consequently, the courts looked to see if each employer in question had established that its policy was justified as a business necessity. The business necessity standard is more lenient for the employer than the statutory BFOQ defense. The Court of Appeals here went one step further and invoked the burden-shifting framework set forth in *Wards Cove Packing Co. v. Atonio*, 490 U.S. 642 (1989), thus requiring petitioners to bear the burden of persuasion on all questions. The court assumed that because the asserted reason for the sex-based exclusion (protecting women's unconceived offspring) was ostensibly benign, the policy was not sex-based discrimination. That assumption, however, was incorrect.

First, Johnson Controls' policy classifies on the basis of gender and childbearing capacity, rather than fertility alone. Respondent does not seek to protect the unconceived children of all its employees. Despite evidence in the record about the debilitating effect of lead exposure on the male reproductive system, Johnson Controls is concerned only with the harms that may befall the unborn offspring of its female employees. Accordingly, it appears that Johnson Controls would have lost in the Eleventh Circuit under Hayes because its policy does not "effectively and equally protec[t] the offspring of all employees." This Court faced a conceptually similar situation in *Phillips v. Martin Marietta Corp.*, 400 U.S. 542 (1971), and found sex discrimination because the policy established "one hiring policy for women and another for men—each having pre-school-age children." Johnson Controls' policy is facially discriminatory because it requires only a female employee to produce proof that she is not capable of reproducing.

Our conclusion is bolstered by the Pregnancy Discrimination Act of 1978 (PDA), 92 Stat. 2076, 42 U.S.C. § 2000e(k), in which Congress explicitly provided that, for purposes of Title VII, discrimination "on the basis of sex" includes discrimination "because of or on the basis of pregnancy, childbirth, or related medical conditions."[3] "The Pregnancy Discrimination Act has now made clear that, for all Title VII purposes, discrimination based on a woman's pregnancy is,

on its face, discrimination because of her sex." In its use of the words "capable of bearing children" in the 1982 policy statement as the criterion for exclusion, Johnson Controls explicitly classifies on the basis of potential for pregnancy. Under the PDA, such a classification must be regarded, for Title VII purposes, in the same light as explicit sex discrimination. Respondent has chosen to treat all its female employees as potentially pregnant; that choice evinces discrimination on the basis of sex.

We concluded above that Johnson Controls' policy is not neutral because it does not apply to the reproductive capacity of the company's male employees in the same way as it applies to that of the females. Moreover, the absence of a malevolent motive does not convert a facially discriminatory policy into a neutral policy with a discriminatory effect. Whether an employment practice involves disparate treatment through explicit facial discrimination does not depend on why the employer discriminates but rather on the explicit terms of the discrimination.

\* \* \*

In sum, Johnson Controls' policy "does not pass the simple test of whether the evidence shows 'treatment of a person in a manner which but for that person's sex would be different.' " We hold that Johnson Controls' fetal-protection policy is sex discrimination forbidden under Title VII unless respondent can establish that sex is a "bona fide occupational qualification."

### IV

Under § 703(e)(1) of Title VII, an employer may discriminate on the basis of "religion, sex, or national origin in those certain instances where religion, sex, or national origin is a bona fide occupational qualification reasonably necessary to the normal operation of that particular business or enterprise." We therefore turn to the question whether Johnson Controls' fetal-protection policy is one of those "certain instances" that come within the BFOQ exception.

The BFOQ defense is written narrowly, and this Court has read it narrowly. Our emphasis on the restrictive scope of the BFOQ defense is grounded on both the language and the legislative history of § 703.

---

[3]The Act added subsection (k) to § 701 of the Civil Rights Act of 1964 and reads in pertinent part:

"The terms 'because of sex' or 'on the basis of sex' (in Title VII) include, but are not limited, because of or on the basis of pregnancy, childbirth, or related medical conditions; and women affected by pregnancy, childbirth, or related medical conditions shall be treated the same for all employment-related purposes as other persons not so affected but similar in their ability or inability to work. . . ."

* * *

Johnson Controls argues that its fetal-protection policy falls within the so-called safety exception to the BFOQ. Our cases have stressed that discrimination on the basis of sex because of safety concerns is allowed only in narrow circumstances. In *Dothard v. Rawlinson*, 433 U.S. 321 (1977), this Court indicated that danger to a woman herself does not justify discrimination. We there allowed the employer to hire only male guards in contact areas of maximum-security male penitentiaries only because more was at stake than the "individual woman's decision to weigh and accept the risks of employment." We found sex to be a BFOQ inasmuch as the employment of a female guard would create real risks of safety to others if violence broke out because the guard was a woman. Sex discrimination was tolerated because sex was related to the guard's ability to do the job—maintaining prison security. We also required in *Dothard* a high correlation between sex and ability to perform job functions and refused to allow employers to use sex as a proxy for strength although it might be a fairly accurate one.

Similarly, some courts have approved airlines' layoffs of pregnant flight attendants at different points during the first five months of pregnancy on the ground that the employer's policy was necessary to ensure the safety of passengers. In two of these cases, the courts pointedly indicated that fetal, as opposed to passenger, safety was best left to the mother.

We considered safety to third parties in *Western Airlines, Inc. v. Criswell*, 472 U.S. 400 (1985), in the context of the ADEA. We focused upon "the nature of the flight engineer's tasks," and the "actual capabilities of persons over age 60" in relation to those tasks. Our safety concerns were not independent of the individual's ability to perform the assigned tasks, but rather involved the possibility that, because of age-connected debility, a flight engineer might not properly assist the pilot, and might thereby cause a safety emergency. Furthermore, although we considered the safety of third parties in *Dothard* and *Criswell*, those third parties were indispensable to the particular business at issue. In *Dothard*, the third parties were the inmates; in *Criswell*, the third parties were the passengers on the plane. We stressed that in

order to qualify as a BFOQ, a job qualification must relate to the "essence," or to the "central mission of the employer's business."

The concurrence ignores the "essence of the business" test and so concludes that "the safety to fetuses in carrying out the duties of battery manufacturing is as much a legitimate concern as is safety to third parties in guarding prisons (*Dothard*) or flying airplanes (*Criswell*)." By limiting its discussion to cost and safety concerns and rejecting the "essence of the business" test that our case law has established, the concurrence seeks to expand what is now the narrow BFOQ defense. Third-party safety considerations properly entered into the BFOQ analysis in *Dothard* and *Criswell* because they went to the core of the employee's job performance. Moreover, that performance involved the central purpose of the enterprise. *Dothard* ("the essence of a correctional counselor's job is to maintain prison security"); *Criswell* (the central mission of the airline's business was the safe transportation of its passengers). The concurrence attempts to transform this case into one of customer safety. The unconceived fetuses of Johnson Controls' female employees, however, are neither customers nor third parties whose safety is essential to the business of battery manufacturing. No one can disregard the possibility of injury to future children; the BFOQ, however, is not so broad that it transforms this deep social concern into an essential aspect of batterymaking.

Our case law, therefore, makes clear that the safety exception is limited to instances in which sex or pregnancy actually interferes with the employee's ability to perform the job. This approach is consistent with the language of the BFOQ provision itself, for it suggests that permissible distinctions based on sex must relate to ability to perform the duties of the job. Johnson Controls suggests, however, that we expand the exception to allow fetal-protection policies that mandate particular standards for pregnant or fertile women. We decline to do so. Such an expansion contradicts not only the language of the BFOQ and the narrowness of its exception but the plan language and history of the Pregnancy Discrimination Act.

The PDA's amendment to Title VII contains a BFOQ standard of its own: unless pregnant employees differ from others "in their ability or inability to work," they must be "treated the same" as other employees "for all employment-related purposes." This language clearly sets forth Congress' remedy for discrimination on the basis of pregnancy and potential pregnancy. Women who are either pregnant or potentially pregnant must be treated like others "similar in their ability . . . to work." In other words, women as capable of doing their jobs as their male counterparts may not be forced to choose between having a child and having a job.

The concurrence asserts that the PDA did not alter the BFOQ defense. The concurrence arrives at this conclusion by ignoring the second clause of the Act which states that "women affected by pregnancy, childbirth, or related medical conditions shall be treated the same for all employment-related purposes . . . as other persons not so affected but similar in their ability or inability to work." Until this day, every Member of this Court had acknowledged that "[t]he second clause [of the PDA] could not be clearer: it mandates that pregnant employees 'shall be treated the same for all employment-related purposes' as nonpregnant employees similarly situated with respect to their ability or inability to work." The concurrence now seeks to read the second clause out of the Act.

The legislative history confirms what the language of the Pregnancy Discrimination Act compels. Both the House and Senate Reports accompanying the legislation indicate that this statutory standard was chosen to protect female workers from being treated differently from other employees simply because of their capacity to bear children.

* * *

This history counsels against expanding the BFOQ to allow fetal-protection policies. The Senate Report quoted above states that employers may not require a pregnant woman to stop working at any time during her pregnancy unless she is unable to do her work. Employment late in pregnancy often imposes risks on the unborn child, but Congress indicated that the employer may take into account only the woman's ability to get her job done. With the PDA, Congress made clear that the decision to become

pregnant or to work while being either pregnant or capable of becoming pregnant was reserved for each individual woman to make for herself.

We conclude that the language of both the BFOQ provision and the PDA which amended it, as well as the legislative history and the case law, prohibit an employer from discriminating against a woman because of her capacity to become pregnant unless her reproductive potential prevents her from performing the duties of her job. We reiterate our holdings in *Criswell* and *Dothard* that an employer must direct its concerns about a woman's ability to perform her job safely and efficiently to those aspects of the woman's job-related activities that fall within the "essence" of the particular business.

**V**

We have no difficulty concluding that Johnson Controls cannot establish a BFOQ. Fertile women, as far as appears in the record, participate in the manufacture of batteries as efficiently as anyone else. Johnson Controls' professed moral and ethical concerns about the welfare of the next generation do not suffice to establish a BFOQ of female sterility. Decisions about the welfare of future children must be left to the parents who conceive, bear, support, and raise them rather than to the employers who hire those parents. Congress has mandated this choice through Title VII, as amended by the Pregnancy Discrimination Act. Johnson Controls has attempted to exclude women because of their reproductive capacity. Title VII and the PDA simply do not allow a woman's dismissal because of her failure to submit to sterilization.

Nor can concerns about the welfare of the next generation be considered a part of the "essence" of Johnson Controls' business. Judge Easterbrook in this case pertinently observed: "It is word play to say that 'the job' at Johnson [Controls] is to make batteries without risk to fetuses in the same way 'the job' at Western Air Lines is to fly planes without crashing."

Johnson Controls argues that it must exclude all fertile women because it is impossible to tell which women will become pregnant while working with lead. This argument is somewhat academic in light of our conclusion that the company may not exclude fertile women at all; it perhaps is worth noting, however, that Johnson Controls has shown no "factual basis for believing that all or substantially all women would be unable to perform safely and efficiently the duties of the job involved." Even on this sparse record, it is apparent that Johnson Controls is concerned about only a small minority of women. Of the eight pregnancies reported among the female employees, it has not been shown that any of the babies have birth defects or other abnormalities. The record does not reveal the birth rate for Johnson Controls' female workers but national statistics show that approximately nine percent of all fertile women become pregnant each year. The birthrate drops to two percent for blue collar workers over age 30. Johnson Controls' fear of prenatal injury, no matter how sincere, does not begin to show that substantially all of its fertile women employees are incapable of doing their jobs.

## VI

A word about tort liability and the increased cost of fertile women in the workplace is perhaps necessary. One of the dissenting judges in this case expressed concern about an employer's tort liability and concluded that liability for a potential injury to a fetus is a social cost that Title VII does not require a company to ignore. It is correct to say that Title VII does not prevent the employer from having a conscience. The statute, however, does prevent sex-specific fetal-protection policies. These two aspects of Title VII do not conflict.

More than 40 States currently recognize a right to recover for a prenatal injury based either on negligence or on wrongful death. According to Johnson Controls, however, the company complies with the lead standard developed by OSHA and warns its female employees about the damaging effects of lead. It is worth noting that OSHA gave the problem of lead lengthy consideration and concluded that "there is no basis whatsoever for the claim that women of childbearing age should be excluded from the workplace in order to protect the fetus or the course of pregnancy." Instead, OSHA established a series of mandatory protections which, taken together, "should effectively minimize any risk to the fetus and newborn child." Without negligence, it

would be difficult for a court to find liability on the part of the employer. If, under general tort principles, Title VII bans sex-specific fetal-protection policies, the employer fully informs the woman of the risk, and the employer has not acted negligently, the basis for holding an employer liable seems remote at best.

Although the issue is not before us, the concurrence observes that "it is far from clear that compliance with Title VII will preempt state tort liability."

\* \* \*

If state tort law furthers discrimination in the workplace and prevents employers from hiring women who are capable of manufacturing the product as efficiently as men, then it will impede the accomplishment of Congress' goals in enacting Title VII. Because Johnson Controls has not argued that it faces any costs from tort liability, not to mention crippling ones, the preemption question is not before us. We therefore say no more than that the concurrence's speculation appears unfounded as well as premature.

The tort-liability argument reduces to two equally unpersuasive propositions. First, Johnson Controls attempts to solve the problem of reproductive health hazards by resorting to an exclusionary policy. Title VII plainly forbids illegal sex discrimination as a method of diverting attention from an employer's obligation to police the workplace. Second, the spectre of an award of damages reflects a fear that hiring fertile women will cost more. The extra cost of employing members of one sex, however, does not provide an affirmative Title VII defense for a discriminatory refusal to hire members of that gender. Indeed, in passing the PDA, Congress considered at length the considerable cost of providing equal treatment of pregnancy and related conditions, but made the "decision to forbid special treatment of pregnancy despite the social costs associated therewith."

We, of course, are not presented with, nor do we decide, a case in which costs would be so prohibitive as to threaten the survival of the employer's business. We merely reiterate our prior holdings that the incremental cost of hiring women cannot justify discriminating against them.

## VII

Our holding today that Title VII, as so amended, forbids sex-specific fetal-protection policies is neither remarkable nor unprecedented. Concern for a woman's existing or potential offspring historically has been the excuse for denying women equal employment opportunities. Congress in the PDA prohibited discrimination on the basis of a woman's ability to become pregnant. We do no more than hold that the Pregnancy Discrimination Act means what it says.

It is no more appropriate for the courts than it is for individual employers to decide whether a woman's reproductive role is more important to herself and her family than her economic role. Congress has left this choice to the woman as hers to make.

The judgment of the Court of Appeals is reversed and the case is remanded for further proceedings consistent with this opinion.

It is so ordered.

JUSTICE WHITE, with whom THE CHIEF JUSTICE and JUSTICE KENNEDY join, concurring in part and concurring in the judgment.

The Court properly holds that Johnson Controls' fetal protection policy overtly discriminates against women, and thus is prohibited by Title VII unless it falls within the bona fide occupational qualification (BFOQ) exception, set forth at 42 U.S.C. § 2000e–2(e). The Court erroneously holds, however, that the BFOQ defense is so narrow that it could never justify a sex-specific fetal protection policy. I nevertheless concur in the judgment of reversal because on the record before us summary judgment in favor of Johnson Controls was improperly entered by the District Court and affirmed by the Court of Appeals.

* * *

The Court dismisses the possibility of tort liability by no more than speculating that if "Title VII bans sex-specific fetal-protection policies, the employer fully informs the woman of the risk, and the employer has not acted negligently, the basis for holding an employer liable seems remote at best." Such speculation will be small comfort to employers. First, it is far from clear that compliance with Title VII will preempt state tort liability, and the Court offers no support for

that proposition. Second, although warnings may preclude claims by injured *employees,* they will not preclude claims by injured children because the general rule is that parents cannot waive causes of action on behalf of their children, and the parents' negligence will not be imputed to the children. Finally, although state tort liability for prenatal injuries generally requires negligence, it will be difficult for employers to determine in advance what will constitute negligence. Compliance with OSHA standards, for example, has been held not to be a defense to state tort or criminal liability. Moreover, it is possible that employers will be held strictly liable, if, for example, their manufacturing process is considered "abnormally dangerous."

Relying on *Los Angeles Dept. of Water and Power v. Manhart,* 435 U.S. 702 (1978), the Court contends that tort liability cannot justify a fetal protection policy because the extra costs of hiring women is not a defense under Title VII. This contention misrepresents our decision in *Manhart.* There, we held that a requirement that female employees contribute more than male employees to a pension fund, in order to reflect the greater longevity of women, constituted discrimination against women under Title VII because it treated them as a class rather than as individuals. We did not in that case address in any detail the nature of the BFOQ defense, and we certainly did not hold that cost was irrelevant to the BFOQ analysis. Rather, we merely stated in a footnote that "there has been no showing that sex distinctions are reasonably necessary to the normal operation of the Department's retirement plan." We further noted that although Title VII does not contain a "cost-justification defense comparable to the affirmative defense available in a price discrimination suit," "no defense based on the *total* cost of employing men and women was attempted in this case."

* * *

JUSTICE SCALIA, concurring in the judgment.

I generally agree with the Court's analysis, but have some reservations, several of which bear mention.

First, I think it irrelevant that there was "evidence in the record about the debilitating effect of lead exposure on the male reproductive system." Even without such evidence, treating women differently "on the basis of pregnancy" constitutes discrimination "on the basis of sex," because Congress has unequivocally said so.

Second, the Court points out that "Johnson Controls has shown no factual basis for believing that all or substantially all women would be unable to perform safely . . . the duties of the job involved," (internal quotations omitted). In my view, this is not only "somewhat academic in light of our conclusion that the company may not exclude fertile women at all," *ibid.;* it is entirely irrelevant. By reason of the Pregnancy Discrimination Act, it would not matter if all pregnant women placed their children at risk in taking these jobs, just as it does not matter if no men do so. As Judge Easterbrook put it in his dissent below, "Title VII gives parents the power to make occupational decisions affecting their families. A legislative forum is available to those who believe that such decisions should be made elsewhere."

Third, I am willing to assume, as the Court intimates, that any action required by Title VII cannot give rise to liability under state tort law. That assumption, however, does not answer the question whether an action *is* required by Title VII (including the BFOQ provision) even if it is subject to liability under state tort law. It is perfectly reasonable to believe that Title VII has *accommodated* state tort law through the BFOQ exception. However, all that need be said in the present case is that Johnson has not demonstrated a substantial risk of tort liability—which is alone enough to defeat a tort-based assertion of the BFOQ exception.

## Notes and Questions

1.   Based on the majority opinion, would it violate Title VII for a hospital to require that already-pregnant x-ray technicians be reassigned to other jobs in the hospital for the duration of their pregnancy?

2.   Justice Blackmun's opinion relies on the principle of autonomy.

The bias in Johnson Controls' policy is obvious. Fertile men, but not fertile women, are given a choice as to whether they wish to risk their reproductive health for a particular job.

To what extent should decision about risk acceptability be left to employees? Does OSHA rely on autonomy or paternalism? What about the Americans with Disabilities Act?

3.   How reassured are employers likely to be that their potential tort liability is "remote at best"? How would you address the issue of possible tort liability?

## DALHEIM v. KDFW–TV

918 F.2d 1220 (5th Cir.1990).

ALVIN B. RUBIN, Circuit Judge:

\* \* \*

Plaintiffs are nineteen present and former general-assignment reporters, producers, directors, and assignment editors employed in the news and programming departments of television station KDFW–TV (KDFW). As its call letters imply, KDFW serves the Dallas–Fort Worth area which, with approximately 3.5 million viewers, is the eighth largest television market in the nation. The news and programming departments are responsible for producing KDFW's local news broadcasts and its public affairs programming.

KDFW's general-assignment reporters usually receive a new coverage assignment each day. The assignment manager or an assignment editor tells the reporter the story to be covered, what she is expected to "shoot," and the intended angle or focus of the story. After the reporter interviews the persons that she or another KDFW employee has arranged to interview, she obtains pertinent video footage, and then writes and records the text of the story, subject to review by the producer. Some reporters help assemble the video and text narration; others rely on a video editor to put the final package together. General-assignment reporters are only infrequently assigned to do a series of reports focusing on a single topic or related topics. Successful reporters usually have a pleasant physical appearance and a strong and appealing voice, and are able to present themselves as credible and knowledgeable.

Producers are responsible for determining the content of the ten-to-twelve minute news portion of KDFW's thirty-minute newscast. They participate in meetings to decide which stories and story angles will be covered; they also decide the amount of time to be given a particular story, the sequence in which stories will be aired, and when to take commercial breaks. Producers have the authority to revise reporters' stories. All of the producers' actions are subject to approval by the executive producer.

Directors review the script for the newscast in order to prepare technical instructions for "calling" the show. The director decides which camera to use and on which machine to run videotaped segments or preproduction graphics. During the broadcast, the director cues the various technical personnel, telling them precisely when to perform their assigned tasks. The overall appearance of KDFW's newscasts, however, is prescribed by station management. The director therefore has no discretion concerning lighting, camera-shot blocking, closing-shot style, or the sequence of opening and closing graphics. KDFW's directors also direct some public affairs programming, which have no prescribed format but involve only simple camera work and a basic set. In addition, KDFW's directors screen commercials to be aired by the station to ensure that they meet the standards set by KDFW's parent, Times Mirror Corporation.

Assignment editors are primarily responsible for pairing reporters with both photographers and videotape editors. They also monitor the wire services, police and fire department scanners, newspapers, and press releases for story ideas that conform to KDFW's general guidelines. Assignment editors have no authority to decide the stories to be covered, but they may reassign reporters if they learn of a story requiring immediate action. Assignment editors operate under the supervision of the assignment manager.

Plaintiffs brought this suit in May, 1985, alleging that KDFW's reporters, producers, directors, and assignment editors were required to work more than forty hours per week without overtime pay, in willful violation of § 7 of the FLSA, and seeking to recover back wages from May, 1982 to the present. After an eight-day bench trial, the district court concluded that none of the plaintiffs was exempt from § 7 as a bona fide executive, administrative, or professional employee under § 13(a)(1) and that KDFW had violated the FLSA by failing to pay overtime. The court further concluded, however, that KDFW's violation was not willful, and that KDFW therefore was not liable for damages outside the FLSA's two-year statute of limitations for nonwillful violations.

Section 7 of the FLSA requires employers to pay overtime to employees who work more than forty hours per week. Section 13(a)(1) exempts from the maximum hour provision employees occupying "bona fide executive, administrative, or professional" positions. That same section empowers the Secretary of Labor to define by regulation the terms "executive," "administrative," and "professional." She has done so at 29 C.F.R. § 541.0 et seq., setting out "long" tests for employees earning more than $155 per week but less than $250 per week, which include specific criteria, and "short" tests, described in less detail, for employees earning more than $250 per week. In addition, the Secretary has issued interpretations of those regulations, which are codified at 29 C.F.R. § 541.100 et seq. The § 13(a)(1) exemptions are "construed narrowly against the employer seeking to assert them," and the employer bears the burden of proving that employees are exempt.

The short test for the executive exemption requires that an employee's "primary duty" consist of the "management of the enterprise" in which she is employed "or a customarily recognized subdivision thereof." In addition, the executive employee's work must include "the customary and regular direction of the work" of two or more employees. The regulations define an exempt administrative employee as one whose "primary duty" consists of "office or non-manual work directly related to management policies or general business operations" that "includes work requiring the exercise of discretion and independent judgment." The exemption for creative professionals requires that the employee's "primary duty" consist of work that is "original and creative in character in a recognized field or artistic endeavor," the result of which depends "primarily on the invention, imagination, or talent of the employee."

KDFW challenges the holding of the district court on four distinct grounds, claiming that the district court (1) erroneously construed the term "primary duty" to mean duties occupying more than half of an employee's time; (2) erroneously concluded that reporters' work is not "original and creative" as those terms are used in the regulations; (3) misconstrued the requirement that administrative work be "directly related to management policies and general business operations," in that it (a) erroneously applied the concept of "production," as that term is used in the Secretary's interpretations, to the work of white-col-

lar employees like producers, directors, and assignment editors, and (b) erroneously concluded that the work of producers, directors, and assignment editors should not be deemed "directly related" to business operations because they "carr[y] out major assignments in conducting the operations of the business," within the meaning of the interpretations; and (4) erroneously failed to "tack" exemptions for producers, directors, and assignment editors as provided for in the regulations.

* * *

## IV

### A.    KDFW's General–Assignment Reporters

KDFW argues that its general-assignment reporters are exempt artistic professionals. Under the regulations, KDFW must prove that the reporter's "primary duty" consists of work that is "original and creative in character in a recognized field of artistic endeavor," the result of which depends "primarily on the invention, imagination, or talent of the employee."

The regulations and interpretations at issue here, §§ 541.3(a)(2) and 541.303(e) and (f), have not changed in any material respect since 1949, long before broadcast journalism evolved into its modern form. To apply the Secretary's interpretation literally to the plaintiffs would be to assume that those occupations exist today as they did forty years ago. No one disputes that the technological revolution that has swept this society into the so-called Information Age has rendered that assumption untenable. The question is what role, if any, § 541.303(e) and (f) may have in determining the exempt status of modern broadcast journalists.

KDFW argues that the district court gave the interpretation undue weight, thus blinding itself to the realities of modern broadcast journalism. Rather than focusing on the "essential nature" of reporters' duties, KDFW contends, the district court "pigeonholed" reporters according to standards that are decades out of date. Amicus National Association of Broadcasters (NAB) goes even further, contending that the Secretary's interpretation is based on "erroneous, outmoded assumptions about journalism and journalists," and that "[i]nsofar as the District Court took these 1940 assumptions about print journalists and applied them to the present-day duties of KDFW television reporters, [it] erred as a matter of law."

\* \* \*

The Secretary's interpretations make it abundantly clear that § 541.3(a)(2) was intended to distinguish between those persons whose work is creative in nature from those who work in a medium capable of bearing creative expression, but whose duties are nevertheless functional in nature. The factual inquiry in this case was directed precisely at determining on which side of that line KDFW's reporters stand. The district court found that, at KDFW, the emphasis was on "good reporting, in the aggregate," and not on individual reporters with the "presence" to draw on audience. The district court found that the process by which reporters meld sound and pictures relies not upon the reporter's creativity, but upon her skill, diligence, and intelligence. More importantly, the district court found that "[r]eporters are told the story that the station intends they cover, what they are expected to shoot, and the intended angle or focus of the story."

In essence, the district court found that KDFW failed to prove that the work constituting its reporters' primary duty is original or creative in character. The district court recognized, and we think correctly, that general-assignment reporters may be exempt creative professionals, and that KDFW's reporters did, from time to time, do original and creative work. Nevertheless, at KDFW, the approach reporters take to their day-to-day work is in large part dictated by management, and the stories they daily produce are neither analytic nor interpretive nor original. In neither form nor substance does a reporter's work "depend primarily on [her] invention, imagination, or talent." Those inferences, while not compelled by the evidence, are certainly supported by it. Based on those inferences and the underlying historical facts, which we review only for clear error, we think the legal conclusion that reporters are nonexempt follows as a matter of course. We therefore conclude that the district court did not err in holding that KDFW's general-assignment reporters are not exempt professionals.

B.   KDFW's News Producers

KDFW argues that its news producers are exempt either as creative professionals, administrators, executives, or a combination thereof. We address each argument in turn.

1.   *Producers as Creative Professionals.*—KDFW does not press this argument much, for good reason. The district court found that KDFW failed to prove that the work producers do in rewriting reporters' copy and in formatting the newscast are products of their "invention, imagination, and talent." Rather, producers perform their work within a well-defined framework of management policies and editorial convention. To the extent that they exercise discretion, it is governed more by skill and experience than by originality and creativity. Because the district court's findings are supported by the record, we find no error.

2.   *Producers as Administrators.*—The argument KDFW pursues most vigorously with respect to producers is that they are exempt administrative employees. Section 541.2 of the regulations requires that an exempt administrator perform (1) office or nonmanual work (2) that is directly related to the employer's management policies or general business operations and (3) involves the exercise of discretion and independent judgment. The Secretary's interpretation, § 541.205(a), defines the "directly related" prong by distinguishing between what it calls "the administrative operations of a business" and "production." Administrative operations include such duties as "advising the management, planning, negotiating, representing the company, purchasing, promoting sales, and business research and control." Work may also be "directly related" if it is of "substantial importance" to the business operations of the enterprise in that it involves "major assignments in conducting the operations of the business, or . . . affects business operations to a substantial degree." KDFW argues that the district court erred in finding that producers' work failed the "directly related" requirement because it is neither related to the administrative operations of KDFW, nor is it of "substantial importance" to the enterprise. Whether an employee's work is or should be deemed "directly

related" to business operations is an inference drawn from the historical facts; we review such inferences for clear error.

* * *

That is not the case with KDFW's news producers. Their responsibilities begin and end with the ten-to-twelve minute portion of the newscast they are working on. They are not responsible for setting business policy, planning the long- or short-term objectives of the news department, promoting the newscast, negotiating salary or benefits with other department personnel, or any of the other types of "administrative" tasks noted in § 541.205(b). The district court determined, based on the facts before it, that "[t]he duties of a producer clearly relate to the production of a KDFW news department product and not to defendant's administrative operations." That determination was not erroneous.

KDFW next asserts that the district court erred in holding that producers' work does not consist of carrying out "major assignments" of "substantial importance" to KDFW's business. Again, KDFW disputes the district court's factual conclusions and inferences, and again, its contention is without merit. KDFW was charged with proving that its producers are exempt employees. The only record evidence KDFW points to in support of its contention that producers' work is of "substantial importance," other than the evidence of what producers do, is that KDFW operates in the nation's eighth largest television market, and that local news is an important source of revenue for the station. The "importance" of producers' work we are left to infer is that, if a producer performs poorly, KDFW's bottom line might suffer.

As a matter of law, that is insufficient to establish the direct relationship required by § 541.2 by virtue of the "substantial importance" contemplated by § 541.205(c). The Secretary's interpretations specifically recognize that the fact that a worker's poor performance may have a significant profit-and-loss impact is not enough to make that worker an exempt administrator. "An employee's job can even be 'indispensable' and still not be of the necessary 'substantial importance' to meet the 'directly related' element." In assessing whether an employee's work is of sub-

stantial importance, it is necessary yet again to look to "the *nature* of the work, not its ultimate consequence." The nature of producers' work, the district court found, is the application of "techniques, procedures, repetitious experience, and specific standards" to the formatting of a newscast. KDFW was obliged to demonstrate how work of that nature is so important to KDFW that it should be deemed "directly related" to business operations. It did not do so. Indeed, the evidence shows that the work one would think of as being "substantially important"—such as setting news department policy and designing the uniform "look" of the newscast—is done by employees who seem clearly to be exempt administrators: the executive producer and the news director, for example. We therefore conclude that the district court did not err in holding that producers are not exempt administrators.

3.    Producers as Executives.—

* * *

To qualify for an executive exemption under the short test, an employee's primary duty must consist of the "management of the enterprise" in which she is employed "or a customarily recognized subdivision thereof." In addition, the employee must customarily and regularly direct the work of two or more employees. The district court found that management was not the producers' primary duty, and that producers do not customarily direct the work of two or more employees.

We agree with the district court. The evidence establishes that, while the producer plays an important role in coordinating and formatting a portion of the newscast, the other members of the ensemble—that is, the reporters, technicians, assignment editors, and so on—are actually supervised by other management personnel. Producers perform none of the executive duties contemplated by the regulations, such as training, supervising, disciplining, and evaluating employees. Indeed, this court previously upheld a determination by the National Labor Relations Board that neither producers nor directors nor assignment editors are "supervisors" within § 2(11) of the National Labor Relations Act. Producers, therefore, do not "manage," and are not exempt executives.

C.   KDFW's Directors and Assignment Editors

For the same reasons it asserts with respect to its producers, KDFW claims that the district court erred in concluding that its directors and assignment editors are not exempt either as executives or administrators or a combination thereof. KDFW's arguments with respect to directors and assignment editors thus fail for the reasons set out above. First, the evidence wholly fails to establish that the work of either directors or assignment editors is "directly related" to management policies or business operations, as required by § 541.2. Second, the evidence does not demonstrate that either directors or assignment editors "manage" anything, as required by § 541.1. KDFW's directors are, as the district court found, highly skilled coordinators, but they are not managers. Assignment editors have no real authority, and participate in no decisions of consequence. Finally, because neither directors nor assignment editors do any exempt work, the district court did not err in failing to consider a combination exemption under § 541.600.

For the reasons stated above, the judgment of the district court is AFFIRMED.

## NLRB v. GISSEL PACKING CO.

395 U.S. 575, 89 S.Ct. 1918, 23 L.Ed.2d 547 (1969).

[The Court considered four cases, all raising common issues of employer coercion during organizing campaigns and the authority of the Board to order the employer, as a remedy for such coercion, to bargain with a union which had demonstrated majority support through means other than a Board-supervised representation election. In one of the cases, the petitioner Sinclair Company operated two plants at which the Teamsters in 1965 had begun an organizing campaign, had obtained from 11 of 14 employees engaged in wire weaving signed cards authorizing the Teamsters to act as bargaining representative, and had made a request for recognition by the company, which the company refused. The Teamsters then petitioned for an election. In company communications to the employees, several references were made to a three-month strike in 1952, when the employees were represented by a different union; the strike had led to the union's loss of support among company employees. (Other portions of the Court's opinion are set forth at pages 328–41 infra.)]

When petitioner's president first learned of the Union's drive in July, he talked with all of his employees in an effort to dissuade them from joining a union. He particularly emphasized the results of the long 1952 strike, which he claimed "almost put our company out of business," and expressed worry that the employees were forgetting the "lessons of the past." He emphasized, secondly, that the Company was still on "thin ice" financially, that the Union's "only weapon is to strike," and that a strike "could lead to the closing of the plant," since the parent company had ample manufacturing facilities elsewhere. He noted, thirdly, that because of their age and the limited usefulness of their skills outside their craft, the employees might not be able to find reemployment if they lost their jobs as a result of a strike. Finally, he warned those who did not believe that the plant could go out of business to "look around Holyoke and see a lot of them out of business." The president sent letters to the same effect to the employees in early November, emphasizing that the parent company had no reason to stay in Massachusetts if profits went down.

During the two or three weeks immediately prior to the election on December 9, the president sent the employees a pamphlet captioned: "Do you want another 13-week strike?" stating, *inter alia*, that: "We have no doubt that the Teamsters Union can again close the Wire Weaving Department and the entire plant by a strike. We have no hopes that the Teamsters Union Bosses will not call a strike. . . . The Teamsters Union is a strike happy outfit." Similar communications followed in late November, including one stressing the Teamsters' "hoodlum control." Two days before the election, the Company sent out another pamphlet that was entitled: "Let's Look at the Record," and that purported to be an obituary of companies in the Holyoke-Springfield, Massachusetts, area that had allegedly gone out of business because of union demands, eliminating some 3,500 jobs; the first page carried a large cartoon showing the preparation of a grave for the Sinclair Company and other headstones containing the names of other plants allegedly victimized by the unions. Finally, on the day before the election, the president made another personal appeal to his employees to reject the Union. He repeated that the Company's financial condition was precarious; that a possible strike would jeopardize the continued operation of the plant; and that age and lack of education would make reemployment difficult. The Union lost the election 7–6, and then filed both objections to the election and unfair labor practice charges which were consolidated for hearing before the trial examiner.

\* \* \*

[The Board, finding that the president's communications were reasonably read by the employees in the circumstances to threaten loss of jobs if the union were to win the election, held that the employer had violated Section 8(a)(1) and that the election should be set aside. It also held that the employer violated Section 8(a)(5) by refusing to bargain in good faith with the Teamsters Union which at the time of its request for recognition demonstrated majority support through its authorization cards. The court of appeals sustained the conclusions of the Board, as well as its order that the employer bargain with the union on request. The Supreme Court, affirming the judgment of the court of appeals, considered Sinclair's claim that the Board and court had acted in violation of the First Amendment to the federal Constitution and Section 8(c) of the Labor Act.]

Any assessment of the precise scope of employ-

er expression, of course, must be made in the context of its labor relations setting. Thus, an employer's rights cannot outweigh the equal rights of the employees to associate freely, as those rights are embodied in § 7 and protected by § 8(a)(1) and the proviso to § 8(c). And any balancing of those rights must take into account the economic dependence of the employees on their employers, and the necessary tendency of the former, because of that relationship, to pick up intended implications of the latter that might be more readily dismissed by a more disinterested ear. Stating these obvious principles is but another way of recognizing that what is basically at stake is the establishment of a nonpermanent, limited relationship between the employer, his economically dependent employee and his union agent, not the election of legislators or the enactment of legislation whereby that relationship is ultimately defined and where the independent voter may be freer to listen more objectively and employers as a class freer to talk. Cf. *New York Times Co. v. Sullivan*, 376 U.S. 254, 84 S.Ct. 710, 11 L.Ed.2d 686 (1964).

Within this framework, we must reject the Company's challenge to the decision below and the findings of the Board on which it was based. The standards used below for evaluating the impact of an employer's statements are not seriously questioned by petitioner and we see no need to tamper with them here. Thus, an employer is free to communicate to his employees any of his general views about unionism or any of his specific views about a particular union, so long as the communications do not contain a "threat of reprisal or force or promise of benefit." He may even make a prediction as to the precise effects he believes unionization will have on his company. In such a case, however, the prediction must be carefully phrased on the basis of objective fact to convey an employer's belief as to demonstrably probable consequences beyond his control or to convey a management decision already arrived at to close the plant in case of unionization. See *Textile Workers v. Darlington Mfg. Co.*, 380 U.S. 263, 274, n. 20, 85 S.Ct. 994, 13 L.Ed.2d 827 (1965). If there is any implication that an employer may or may not take action solely on his own initiative for reasons unrelated to economic necessities and known only to him, the statement is no longer a reasonable prediction based on available facts but a threat of retaliation based on misrepresentation and coercion, and as such without the protection of the First Amendment. We therefore agree with the court below that "[c]onveyance of the employer's belief, even though sincere, that unionization will or may result in the closing of the plant is not a statement of fact unless, which is most improbable, the eventuality of closing is capable of proof." 397 F.2d 157, 160. As stated elsewhere, an employer is free only to tell "what he reasonably believes will be the likely economic consequences of unionization that are outside his control," and not "threats of economic reprisal to be taken solely on his own volition." *NLRB v. River Togs, Inc.*, 382 F.2d 198, 202 (C.A.2d Cir.1967).

Equally valid was the finding by the court and the Board that petitioner's statements and communications were not cast as a prediction of "demonstrable 'economic consequences,' " 397 F.2d, at 160, but rather as a threat of retaliatory action. The Board found that petitioner's speeches, pamphlets, leaflets, and letters conveyed the following message: that the company was in a precarious financial condition; that the "strike-happy" union would in all likelihood have to obtain its potentially unreasonable demands by striking, the probable result of which would be a plant shutdown, as the past history of labor relations in the area indicated; and that the employees in such a case would have great difficulty finding employment elsewhere. In carrying out its duty to focus on the question: "[W]hat did the speaker intend and the listener understand?" (A. Cox. Law and the National Labor Policy 44 (1960)), the Board could reasonably conclude that the intended and understood import of that message was not to predict that unionization would inevitably cause the plant to close but to threaten to throw employees out of work regardless of the economic realities. In this connection, we need go no further than to point out (1) that petitioner had no support for its basic assumption that the union, which had not yet even presented any demands, would have to strike to be heard, and that it admitted at the hearing that it had no basis for attributing other plant closings in the area to unionism; and (2) that the Board has often found that employees, who are particularly sensitive to rumors of plant closings, take such hints as coercive threats rather than honest forecasts.

Petitioner argues that the line between so-called permitted predictions and proscribed threats is too vague to stand up under traditional First Amendment analysis and that the Board's discretion to curtail free speech rights is correspondingly too uncontrolled. It is true that a reviewing court must recognize the Board's competence in the first instance to judge the impact of utterances made in the context of the employer-employee relationship, see *NLRB v. Virginia Electric & Power Co.*, 314 U.S. 469, 479, 62 S.Ct. 344, 349, 86 L.Ed. 348 (1941). But an employer, who has control over that relationship and therefore knows it best, cannot be heard to complain that he is without an adequate guide for his behavior. He can easily make his views known without engaging in " 'brinkmanship' " when it becomes all too easy to "overstep and tumble [over] the brink." *Wausau Steel Corp. v. NLRB*, 377 F.2d 369, 372 (C.A.7th Cir.1967). At the least he can avoid coercive speech simply by avoiding conscious overstatements he has reason to believe will mislead his employees.

\* \* \*

## DURKIN v. BOARD OF POLICE AND FIRE COMMISSIONERS

48 Wis. 112, 180 N.W.2d 1 (1970)

The Board of Police and Fire Commissioners for the City of Madison, Wisconsin, appellant, entered an order determining that Edward D. Durkin, respondent, had violated sec. 111.70 (4) (l), Stats., SLL 60:244, and certain rules of the fire department of the city of Madison, and thereupon suspended the respondent from the fire department. Respondent petitioned the circuit court for Dane County for review. The trial court entered judgment reversing the order of the Board and the Board has appealed from the judgment of the circuit court.

This action was commenced before the Board of Police and Fire Commissioners for the City of Madison, Wisconsin, upon a complaint of an elector of the city of Madison against Edward D. Durkin, a captain of the fire department for the city of Madison and president of the City Fire Fighters Union Local No. 311.

On March 27, 28 and 29, 1969, there occurred a strike by the firemen for the city of Madison. During this period negotiations were conducted between the city of Madison (City) and the Fire Fighters Union Local No. 311 (Union), which resulted in the signing of an agreement by the City and the Union and an end to the strike. The agreement included a clause whereby the Union and members of the fire department were granted amnesty for their activity in connection with the strike.

Following the settlement, an investigation of the strike was conducted by the appellant, but no charges were filed against any member of the fire department. Thereafter, on June 27, 1969, an elector of the city of Madison filed a complaint with the appellant charging respondent with having counseled, abetted and led a strike by the Union, and with having participated in that strike by absenting himself from duty during his assigned hours. The complaint alleged that respondent was guilty of violating sec. 111.70 (4) (l), Stats., which prohibits strikes by municipal employees, and of "contumacious conduct endangering the public safety." A hearing on the complaint pursuant to sec. 62.13 (5), was requested.

A hearing on the complaint was conducted by

appellant, and on August 24, 1969, an order was entered finding that respondent had participated in the strike and was guilty of violating sec. 111.70 (4) (l), Stats., "contumacious conduct endangering the public safety," and of violating Rules 34, 89 and 100 of the Rules of the Fire Department of Madison, Wisconsin. Respondent was suspended from the fire department for a period of one hundred eighty days.

On petition of the respondent, the order was reviewed by the circuit court for Dane County. The court entered judgment reversing appellant's order on a finding that the order was affected by an error of law in that appellant did not consider itself bound by the terms of an amnesty clause in the collective bargaining agreement signed by the City and the Union. The circuit court held that the amnesty clause prevented appellant from suspending respondent and, therefore, ordered the respondent reinstated, with back pay . . . .

The collective bargaining agreement between the City and the Union included a clause by which the City agreed to dismiss all legal proceedings commenced by it and pending against the Union and its members, to waive all other causes of action arising out of the negotiations or the strike, and to refrain from directly or indirectly commencing an action that would in any way discipline any employee for participation in the strike.

The narrow issue presented by this case is whether the amnesty clause above referred to and contained in the collective bargaining agreement abrogates the statutory right of an elector to file a complaint with the appellant contained in sec. 62.13 (5) (b), Stats. We are of the opinion that it does not.

The first paragraph of the agreement specifically refers to proceedings commenced by the City and to causes of action by the City. The filing of a complaint by an elector with the Board constitutes neither.

The second paragraph of the agreement recites, "Consistent with appropriate Wisconsin statutes, it is the express policy of the City that *it* will not directly or indirectly commence an action that will in any way discipline. . . ." (Emphasis added.)

It is the contention of the appellant that the processing of the elector's complaint by the appellant constitutes the City indirectly commencing an action to discipline the respondent. However, the elector has a statutory right to file charges and if the city council could somehow foreclose the right of the Board to process charges filed by the elector, it follows that the lawful right of an elector to file charges as provided in sec. 62.13 (5) (b), Stats., would be rendered meaningless. The Board is required to process charges filed with it by an elector in accordance with the statutes of the State of Wisconsin and such rules and regulations as it may adopt which are not inconsistent therewith. The ultimate disposition by the Board of the charges so filed by an elector will be considered later in this opinion.

We find no authority which is particularly helpful on this issue and have considered all authorities advanced by both parties. Among other authorities, our attention has been directed to *Muskego-Norway C.S.J. S.D. No. 9 v. W.E.R.B.* (1967), 35 Wis. 2d 540, 151 N.W.2d 617, and *Joint School Dist. No. 8 v. W.E.R.B.* (1967), 37 Wis. 2d 483, 155 N.W.2d 78. Respondent advances an argument on the same principle as that adopted in *Joint School Dist. No. 8 v. W.E.R.B.*, supra, that since sec. 111.70, Stats., was enacted subsequent to sec. 62.13 (5), the latter must yield to sec. 111.70 (4) (i), SLL 60:244, and thus a city may agree to a provision granting amnesty when entering into a binding, collective bargaining agreement pursuant to sec. 111.70 (4) (i) even though it would defeat the lawful right of an elector to file charges. The argument, however, assumes that sec. 111.70 (4) (i), does authorize a municipality to agree to an amnesty clause in a collective bargaining agreement which would abrogate the right of an elector to file charges. We find no authority to support such a position. . . .

We conclude that the case must be remanded to the Board for further proceedings for the reason that the respondent was not afforded due process.

Respondent had notice of and an opportunity to defend against charges that he was guilty of violating sec. 111.70 (4) (l), Stats., and of "contumacious conduct endangering the public safety." The charges filed by the elector against the respondent did not allege that he had violated any rules of the fire department. It appears the respondent had no notice of these alleged rule violations until it was pronounced by the

Board that he had so violated the three rules in its written decision filed after the hearing. Due process of law requires that an individual have notice of and an opportunity to defend against charges proffered against him. In *General Electric Co. v. Wisconsin E. R. Board* (1958), 3 Wis. 2d 227, 241, 88 N.W.2d 691, 700, 42 LRRM 2187, 2192, this court held:

> "The principle of fair play is an important factor in a consideration of due process of law. Parties in a legal proceeding have a right to be apprised of the issues involved, and to be heard on such issues. A finding or order made in a proceeding in which there has not been a 'full hearing' is a denial of due process and is void. . . ."

No court of review has the means of determining whether the Board would have imposed the same penalty had it found the respondent in violation of only two of the violations charged in the complaint.

On this appeal, the respondent contends that based upon the evidence before it, the order of the Board was arbitrary and discriminatory and also unreasonable. However, this issue was not passed upon by the circuit court; and also in view of our remand to the Board for further proceedings because of the lack of due process, it is not properly raised in this court. Nevertheless, we would observe that the Board does have the authority to dismiss the complaint after it has been processed if, in its judgment it should determine such was a proper disposition of the charges filed by the elector. Also, should the Board decide further proceedings are necessary, on the basis of the record now before us, various factors should be taken into consideration by the Board in its ultimate decision. Among these are: (1) The amnesty clause in the agreement which unequivocally sets forth the position of the city council in its relation with the Union and its members; (2) the decisions of the Board, as such, and its individual members, not to file charges against any fireman; and (3) the fact that the Board had knowledge of the fact that over 270 firemen participated in the strike and that no charges were filed against anyone except the respondent.

We reach our conclusions as to the disposition of this case upon different grounds than those considered by the circuit court. However, the effect of our decision is that the judgment of the circuit court which reverses the order of the Board of Police and Fire Commissioners is affirmed. That part of the judgment ordering the Madison fire department to forthwith reinstate the respondent, and that he be paid as though he had been in continuous service, is re-versed, and the cause is remanded to the Board of Police and Fire Commissioners for further proceedings consistent with this opinion.

*By the Court* — AFFIRMED in part; REVERSED in part . . . .

## CITY OF CHARLOTTE v. LOCAL 660, INT'L ASS'N OF FIREFIGHTERS

426 U.S. 283, 96 S. Ct. 2036, 48 L. Ed. 2d 636 (1976)

JUSTICE MARSHALL delivered the opinion of the Court.

The city of Charlotte, N.C., refuses to withhold from the paychecks of its firefighters dues owing to their union, Local 660, International Association of Firefighters. We must decide whether this refusal violates the Equal Protection Clause of the Fourteenth Amendment.

### I

Local 660 represents some 351 of the 543 uniformed members of the Charlotte Fire Department. Since 1969 the union and individual members have repeatedly requested the city to withhold dues owing to the union from the paychecks of those union members who agree to a checkoff. The city has refused each request. After the union learned that it could obtain a private group life insurance policy for its membership only if it had a dues checkoff agreement with the city, the union and its officers filed suit in federal court alleging, *inter alia*, that the city's refusal to withhold the dues of union members violated the Equal Protection Clause of the Fourteenth Amendment.[1] The complaint asserted that since the city withheld amounts from its employees' paychecks for payment to various other organizations, it could

not arbitrarily refuse to withhold amounts for payment to the union.

On cross-motions for summary judgment, the District Court for the Western District of North Carolina ruled against the city. The court determined that, although the city had no written guidelines, its "practice has been to allow checkoffs from employees' pay to organizations or programs as required by law or where the checkoff option is available to all City employees or where the checkoff option is available to all employees within a single employee unit such as the Fire Department." 381 F. Supp. 500, 502 (1974). The court further found that the city has "not allowed checkoff options serving only single employees or programs which are not available either to all City employees or to all employees engaged in a particular section of City employment." *Ibid.* Finding, however, that withholding union dues from the paychecks of union members would be no more difficult than processing any other deduction allowed by the city, the District Court concluded that the city had not offered a rational explanation for its refusal to withhold for the union. Accordingly, the District Court held that the city's refusal to withhold moneys when requested to do so by the respondents for the benefit of Local 660 "constitutes a violation of the individual [respondents'] rights to equal protection of laws under the Fourteenth Amendment." *Id.,* at 502-503. The court ordered that so long as the city continued "without clearly stated and fair standards, to withhold moneys from the paychecks of City employees for other purposes," it was enjoined from refusing to withhold union dues from the paychecks of the re-

[1]Respondents brought suit under 42 U.S.C. § 1983, grounding jurisdiction in 28 U.S.C. §§ 1331 and 1343. . . .

spondents. *Id.,* at 503. The Court of Appeals for the Fourth Circuit affirmed, 518 F.2d 83 (1975), and we granted certiorari. 423 U.S. 890 (1975). We reverse.

## II

Since it is not here asserted—and this Court would reject such a contention if it were made—that respondents' status as union members or their interest in obtaining a dues checkoff is such as to entitle them to special treatment under the Equal Protection Clause, the city's practice must meet only a relatively relaxed standard of reasonableness in order to survive constitutional scrutiny.

The city presents three justifications for its refusal to allow the dues checkoff requested by respondents. First, it argues, North Carolina law makes it illegal for the city to enter into a contract with a municipal union, N.C. Gen. Stat. § 95-98 (1975), and an agreement with union members to provide a dues checkoff, with the union as a third-party beneficiary, would in effect be such a contract. See 40 N.C. Op. Atty. Gen. 591 (1968-1970). Thus, compliance with the state law, and with the public policy it represents of discouraging dealing with municipal unions, is said to provide a sufficient basis for refusing respondents' request. Second, it claims, a dues checkoff is a proper subject of collective bargaining, which the city asserts Congress may shortly require of state and local governments. Under this theory, the desire to preserve the checkoff as a bargaining chip in any future collective-bargaining process is in itself an adequate basis for the refusal. Lastly, the city contends, allowing withholding only when it benefits all city or departmental employees is a legitimate method for avoiding the burden of withholding money for all persons or organizations that request a checkoff. Because we find that this explanation provides a sufficient justification for the challenged practice, we have no occasion to address the first two reasons proffered.

The city submitted affidavits to show that it would be unduly burdensome and expensive for it to withhold money for every organization or person that requested it, App. 17, 45, 55, and respondents did not contest this showing. As respondents concede, it was therefore reasonable, and permissible under the Equal Protection Clause, for the city to develop standards or restrictions to determine who would be eligible for withholding. *Mathews v. Diaz,* [426 U.S. 67,] at 82-83. See Brief for Respondents 9. Within the limitations of the Equal Protection Clause, of course, the choice of those standards is for the city and not for the courts. Thus, our inquiry is not whether standards might be drawn that would include the union but whether the standards that were drawn were reasonable ones with "some basis in practical experience." *South Carolina v. Katzenbach,* 383 U.S. 301, 331 (1966). Of course, the fact that the standards were drawn and applied in practice rather than pursuant to articulated guidelines is of no import for equal protection purposes.

The city allows withholding for taxes, retirement-insurance programs, savings programs, and certain charitable organizations. These categories, the District Court found, are those in which the checkoff option can, or must, be availed of by all city employees, or those in an entire department. Although the District Court found that this classification did not present a rational basis for rejecting respondents' requests, 381 F. Supp., at 502, we disagree. The city has determined that it will provide withholding only for programs of general interest in which all city or departmental employees can, without more, participate. Employees can participate in the union checkoff only if they join an outside organization—the union. Thus, Local 660 does not fit the category of groups for which the city will withhold. We cannot say that denying withholding to associational or special interest groups that claim only some departmental employees as members and that employees must first join before being eligible to participate in the checkoff marks an arbitrary line so devoid of reason as to violate the Equal Protection Clause. Rather, this division seems a reasonable method for providing the benefit of withholding to employees in their status as employees, while limiting the number of instances of withholding and the financial and administrative burdens attendant thereon.

Given the permissibility of creating standards and the reasonableness of the standards created, the District Court's conclusion that it would be no more difficult for the city to withhold dues for the union than to process other deductions is of no import. We may accept, *arguendo,* that the difficulty involved in processing any individual deduction is neither great nor different in kind from that involved in processing any other deduction. However, the city has not drawn its lines in order to exclude individual deductions, but in order to avoid the cumulative burden of processing deductions every time a request is made; and inherent in such a line-drawing process are difficult choices and "some harsh and apparently arbitrary consequences. . . ." *Mathews v. Diaz,* [426 U.S.,] at 83. See [*id.*] at 82-84; *Dandridge v. Williams,* 397 U.S. 471, 485 (1970). Cf. *Schilb v. Kuebel,* 404 U.S. 357, 364 (1971); *Williamson v. Lee Optical Co.,* 348 U.S. 483, 489 (1955).

Respondents recognize the legitimacy of such a process and concede that the city "is free to develop fair and reasonable standards to meet any possible cost problem." Brief for Respondents 9. Respondents have wholly failed, however, to present any reasons why the present standards are not fair and reasonable—other than the fact that the standards exclude them. This fact, of course, is insufficient to transform the city policy into a constitutional violation. Since we find a reasonable basis for the challenged classification, the judgment of the Court of Appeals for the Fourth Circuit must be reversed, and the case remanded for further proceedings consistent with this opinion.

It is so ordered.

MR. JUSTICE STEWART concurs in the judgment upon the ground that the classification challenged in this case is not invidiously discriminatory and does not, therefore, violate the Equal Protection Clause of the Fourteenth Amendment.

---

**CITY OF LOUISVILLE, MOVANT**

vs.

**LOUISVILLE PROFESSIONAL FIREFIGHTERS ASSOCIATION, LOCAL UNION NO. 345, IAFF, AFL-CIO,** by and through its president, **RONALD E. GNAGIE,** and **RONALD E. GNAGIE,** Individually, and **STATE LABOR RELATIONS BOARD, RESPONDENT**

N. 89-SC-441-D

SUPREME COURT OF KENTUCKY

813 S.W.2d 804

May 9, 1991

Appeal From Court of Appeals; 88-CA-1208; Jefferson Circuit Court; Hon. Martin E. Johnstone, Judge; 87-CI-4911.

**COUNSEL**

ATTORNEYS FOR MOVANT: Hon. Frank X. Quickert, Jr., Director of Law, City of Louisville, Hon. David Leightty, Hon. Cecil A. Blye, Jr., Assistant Directors of Law, Louisville, Kentucky.

ATTORNEYS FOR RESPONDENT: Hon. Herbert L. Segal, Hon. Irwin H. Cutler, Jr., Segal, Isenberg, Sales, Stewart, and Cutler, Louisville, Kentucky.

ATTORNEYS FOR STATE LABOR RELATIONS BOARD: Hon. Rex Hunt, General Counsel, Commonwealth of Kentucky Labor Cabinet, Frankfort, Kentucky.

**JUDGES**

Justice Spain. All concur.
AUTHOR: SPAIN

**OPINION**

This action arises out of a grievance filed by the Louisville Professional Firefighters Association, Local Union No. 345, IAFF, AFL-CIO (union), charging the

City of Louisville, Kentucky (city), with an unfair labor practice.

In 1986, the city and the union entered into a two-year collective bargaining agreement (agreement) which established the work schedule of the city's firefighters. However, the agreement omitted the schedule of the arson investigation unit.

On August 4, 1986, the city unilaterally implemented a third shift into the two-shift schedule of the arson unit. The city did not consult nor bargain with the union concerning the schedule change because it believed the agreement did not prevent such a change in hours. The union formally objected to this change and filed a grievance with the State Labor Relations Board (Board). The union claimed that the city did not have the right to make such a change in the schedule of the arson squad. A hearing was held before an arbitrator on August 7, 1986. The arbitrator, in a non-binding decision, found that the new shift was a change in the employees' working conditions and ordered the city to negotiate with the union regarding the change in hours. The city agreed to negotiate but refused to change the hours of the arson squad back to the *status quo ante*. *See NLRB v. Truckdrivers Union Local 164*, 753 F.2d 53 (6th Cir. 1985).

The union filed suit in the Jefferson Circuit Court (Division Seven), seeking injunctive relief from the changed work schedule. The union claimed that the city was in violation of the agreement, and of the arbitration award. The city continued to argue that the contract permitted it to change the hours without notice to the union. The parties stipulated in this first action that the union "is the exclusive representative under KRS Chapter 345 for purposes of collective bargaining for a unit of the Defendant's employees in its Division of Fire." The city also argued vehemently before Division Seven that KRS Chapter 345 governed the labor relations between the city and the union and that the dispute between the parties should be decided by the Board. The city stated in its brief to Division Seven:

As public employees of the City of Louisville, this Union, the employees in question, and the City of Louisville are not covered by the Federal labor laws which regulate labor relations in the private sector. In-

deed, public employers are expressly excluded from the definition of employer in the National Labor Relations Act. 29 U.S.C. Sec. 152(2).

However, there is a state statutory enactment which governs the labor relations between the City of Louisville and the Firefighters Union. The Statute in question is KRS Chapter 345, under which the Union was recognized as collective bargaining representative for Firefighters of the City of Louisville . . .

Further, the city, in supplemental briefs requested by the trial court stated:

In Kentucky, the Firefighter's Collective Bargaining Act, embodied in KRS Chapter 345, creates a State Labor Relations Board, which has exclusive jurisdiction over the unfair labor practices which are established in that chapter. It is KRS Chapter 345 which creates the obligation to bargain, and makes failure to bargain an unfair labor practice. The procedures prescribed in Chapter 345 are therefore the exclusive remedies for an alleged unfair labor practice.

The Division Seven trial court found in its Findings of Fact, Conclusions of Law, and Order, that the city had committed an unfair labor practice when it unilaterally added the third shift to the schedule of the arson squad. Further, the trial court in this first action held that "the City had a [statutory] duty to bargain before adding a third shift unless the Union had given up the right to bargain on this issue, and the contract contains no such provision." Finally, the trial court agreed with the city's argument that the union was required to seek redress before the Board to remedy an unfair labor practice. The case was dismissed without prejudice and remanded to the Board to determine whether the city had committed an unfair labor practice. Neither party appealed from the first decision of the trial court.

The union then compliantly filed an administrative complaint before the Board claiming that the city had committed an unfair labor practice when it failed to bargain collectively concerning the new shift. This administrative complaint was expressly filed under the provisions of KRS Chapter 345. The Board, without deciding the issue of whether the city had committed an unfair labor practice, treated the dispute as a matter of contract interpretation, and dismissed the complaint.

The union appealed the decision of the Board to the Jefferson Circuit Court (Division Three). The union argued that the Memorandum and Order of the Board should be reversed and remanded because its decision was arbitrary and capricious without an adequate evidentiary basis for its findings of fact and conclusions of law. The union also requested that Division Three find that the unilateral change of hours was an issue for collective bargaining and within the jurisdiction granted the Board by KRS 345.070. The city for the first time now argued in the second trial court action that KRS Chapter 345 was unconstitutional because the statute was "special" or "local" legislation under Sections 59 and 60 of the Kentucky Constitution. The city did a further "about-face" and argued that KRS Chapter 345 was inapplicable because the population of Louisville, Kentucky, had dropped below 300,000 people; the threshold population requirement for the application of KRS Chapter 345.

The Division Three trial court in this second action held that KRS Chapter 345 was constitutional and ruled that the city was barred under the doctrines of *res judicata* and waiver from asserting the argument that KRS Chapter 345 was inapplicable. The trial court also set aside the Board's Orders of February 17, 1987, and May 9, 1987, and held that its decision was "undeniably" arbitrary and capricious. The trial court remanded the dispute to the Board for further adjudication on the issue of whether the city had committed an unfair labor practice when it unilaterally added the third shift, and on the issue of whether the city had committed a statutory violation under KRS 345.070 by failing to bargain collectively with the union.

The city appealed the decision of Division Three to the Court of Appeals, which affirmed on all issues. We granted discretionary review.

The issues which we review are: 1) whether KRS Chapter 345 is constitutional; 2) whether the city is barred under the doctrines of *res judicata* and waiver from raising the issue of the applicability of KRS Chapter 345; and 3) whether the State Labor Relations Board acted arbitrarily and capriciously. We will address the *res judicata* and waiver issue first.

Under the doctrine of *res judicata* or "claim preclusion," a judgment on the merits in a prior suit in-

volving the same parties or their privies bars a subsequent suit based upon the same cause of action. *Lawlor v. National Screen Service Corporation*, 349 U.S. 322, 75 S.Ct. 865, 99 L.Ed. 1122 (1955); *Vaughn's Adm'r v. Louisville & N.R. Co.*, Ky., 297 Ky. 309, 179 S.W.2d 441 (1944). In *Newman v. Newman*, Ky., 451 S.W.2d 417 (1970), we set forth the elements of *res judicata:*

The general rule for determining the question of *res judicata* as between parties in actions embraces several conditions. First, there must be identity of the parties. Second, there must be identity of the two causes of action. Third, the action must be decided on its merits. In short, the rule of *res judicata* does not act as a bar if there are different issues or the questions of law presented are different.

*Id.* at 419.

The term "estoppel" has frequently been used in connection with the doctrine of *res judicata*, not only with respect to the relitigation of particular issues in a subsequent action on a different cause of action, but also with respect to the relitigation of the same cause of action. 46 Am Jur 2d, Judgments, Section 397. (Footnotes omitted.) The more recent trend is to describe the latter aspect of the doctrine of *res judicata* as a "collateral estoppel" or a "collateral estoppel by judgment," as distinguished from the "direct estoppel by judgment" or "merger" where the earlier and later causes of action are identical. *Id.* (Footnotes omitted.) The use of collateral estoppel regarding previously adjudicated issues has been endorsed by the United States Supreme Court and adopted by this Court. *Parklane Hosiery Company v. Shore*, 439 U.S. 322, 99 S.Ct. 645, 58 L.Ed.2d 552 (1979); *Sedley v. City of West Buechel*, Ky., 461 S.W.2d 556 (1971).

In *Lawlor, supra* at 326, the U.S. Supreme Court distinguished the doctrines of *res judicata* and collateral estoppel as follows:

The basic distinction between the doctrines of *res judicata* and collateral estoppel, as those terms are used in this case, has frequently been emphasized. Thus, under the doctrine of *res judicata*, a judgment 'on the merits' in a prior suit involving the same parties or their privies bars a second suit on the same cause of action. Under the doctrine of collateral estoppel, on the other hand, such a judgment precludes relitigation of issues actually litigated and determined in the prior suit, regardless of whether it was based on the same cause of action as the second suit.

And in *Cream Top Creamery v. Dean Milk Company, Inc.*, 383 F.2d 358, 362 (6th Cir. 1967), the United States Court of Appeals determined that the related doctrine of collateral estoppel, also known as the "issue preclusion" doctrine, is applicable in the following situation:

Where the second action between the same parties is upon a different claim or demand, the judgment in the prior action operates as an estoppel only as to those matters in issue or points controverted, upon the determination of which the finding or verdict was rendered. In all cases, therefore, where it is sought to apply to estoppel of a judgment rendered upon one cause of action to matters arising in a suit upon a different cause of action, the inquiry must always be as to the point or question actually litigated and determined in the original action, not what might have been thus litigated and determined. Only upon such matters is the judgment conclusive in another action. (Citation omitted.)

We agree with the Court of Appeals that the parties and the subject matter in the first and second circuit court actions are the same. The subject matter in both actions from the viewpoint of the union was whether the city had committed an unfair labor practice by its unilateral implementation of a third shift in the schedule of the arson squad. The only change in the two cases was the position of the city. The city argued in the first action that the Board had exclusive jurisdiction under KRS Chapter 345 to resolve the parties' dispute. The city then reversed its position in the second action and claimed that KRS Chapter 345 *did not* govern the dispute and, therefore, the Board did not have jurisdiction to hear the case. By invoking the statute and the Board's jurisdiction in the first action, the city cannot later deny jurisdiction and the applicability of KRS Chapter 345 in the second action.

The city argues that the earlier Division Seven decision was a nullity because in it the trial court ultimately decided that it did not have jurisdiction to hear the case. The city also argues that, accordingly, the language in the trial court's opinion is dicta and non-binding in the subsequent action because the decision was dismissed "without prejudice."

The general rule is that a former adjudication is regarded as not being on the merits, within the scope of the doctrine of *res judicata*, where it was based upon the fact that the court lacked jurisdiction. 46 Am Jur 2d Judgments, Section 500.

(Footnotes omitted.) We also note the rule which states that the inclusion of the term "without prejudice" in a judgment of dismissal ordinarily indicates the absence of a decision on the merits, and leaves the parties free to litigate the matter in a subsequent action, as though the dismissed action had not been commenced. *Overstreet v. Greenwell*, Ky., 441 S.W.2d 443, 447 (1979).

However, where a judgment disposes of an action without a determination on the merits, it is nevertheless conclusive as to the issues or technical points actually decided therein, and this rule has been applied to a judgment based on a lack of jurisdiction, so as to render conclusive the prior court's determination of its lack of jurisdiction, as well as questions material to the issue of jurisdiction and actually decided by the judgment. 46 Am Jur 2d, *supra* at section 500. (Footnotes omitted.)

The first trial court action decided the ultimate issue of whether the circuit court or the Board had subject matter jurisdiction to hear the case. The decision made final and conclusive the issue of whether KRS Chapter 345 applied to the parties' dispute, even though the trial court dismissed the case "without prejudice." To reach this result, the trial court was required to make findings of fact and conclusions of law on its subject matter jurisdiction. The trial court then determined that the agreement was governed by KRS Chapter 345 and that the Board had exclusive jurisdiction under the statute to hear the dispute. Therefore, we find that to this extent, the merits of the case were reached by the trial court. We agree with the trial court and the Court of Appeals that the city is barred under the doctrine of *res judicata* and waiver from asserting the issue of the applicability of the KRS Chapter 345 in the subsequent trial court case. We also find that the city is collaterally estopped from relitigating the issue. *Sedley v. City of West Buechel, supra*.

Since we agree that the decision of the first action was *res judicata*, we do not reach the merits of whether KRS Chapter 345 was unconstitutional because we find that issue to be moot.

We also agree with the Court of Appeals and the trial court that the State Board's decision was both arbitrary and capricious and without an adequate evidentiary basis. The trial court specifically decided that the jurisdiction to resolve the parties' dispute was vested under KRS Chapter 345 in the Board. The case should be remanded to the Board on the issues of whether the city had committed an unfair labor practice when it unilaterally added the third shift, and whether the city had failed to collectively bargain with the union under KRS Chapter 345.

The decisions of the Court of Appeals and the Jefferson Circuit Court are affirmed. The case is remanded to the State Labor Relations Board with instructions that this administrative body consider the unilateral change in hours of the arson squad as an issue for collective bargaining within its exclusive jurisdiction, as mandated by KRS Chapter 345.

*AFFIRMING.*

**DISPOSITION**

AFFIRMING.

# Sample Collective Bargaining Agreement

## Between Local 1, United Firefighters Association and the XYZ Fire Department

THIS AGREEMENT made and entered on this 1st day of January, 1996 between the XYZ Fire Department, City of ZZZ, and Local 1 of the United Firefighters Union (hereinafter referred to as the "union").

This contract supersedes any and all agreements, addendums or memos of understanding predating the effective date of this Agreement.

WITNESSETH:

WHEREAS, it is the intent and purpose of the parties hereto that this Agreement shall foster, promote, and improve the industrial relationship between the Fire Department and its employees and to set forth herein a basic agreement covering wages, hours, and working conditions, and other conditions of employment, to be carried out, observed and performed by the parties hereto.

Now, THEREFORE, in consideration of the covenants, agreements, understandings, terms and conditions herein contained and in consideration of other goods and valuable considerations, it is hereby mutually agreed between the parties hereto as follows:

## ARTICLE I-SUCCESSORS

Section 1 - *Binding on Successors.* This agreement shall be binding upon the parties hereto, their successors, administrators, executors, and assigns. In the event the entire operation or any part thereof is sold, leased, transferred, or taken over by sale, transfer, lease, assignment, receivership, or bankruptcy proceedings, such operation shall continue to be subject to the terms and conditions of this Agreement for the life thereof. It is understood by this provision that the parties hereto shall not use any leasing device to a third party to evade the contract.

## ARTICLE II - SCOPE OF AGREEMENT

Section 1 - *Bargaining Unit Employees.* This Agreement shall apply to all fire department employees of XYZ Fire Department and at all other departments located within a one hundred (100) mile radius of the XYZ Fire Department, excluding all office clerical employees, professional employees, and supervisors as defined in the Act.

## ARTICLE III - UNION RECOGNITION

Section 1 - *Bargaining Agent.* The Fire Department hereby and herewith recognizes the Union as the sole and exclusive bargaining agent on behalf of all employees described in Article I herein, pursuant to the Certification by the National Labor Relations Board for the Third Region dated August 13, 1962.

Section 2 - *Agreement to Negotiate.* The Fire Department agrees to negotiate at all times necessary, in the manner provided herein, with the chosen accredited representative of the Local Union and representatives of the Union for the purpose of settling grievance disputes which may arise in connection with hours, rates, working conditions, and other conditions of employment.

Section 3 - *Fire Department Not to Interfere with Union Activities.* The Fire Department will not interfere with, restrain, or coerce an employee or employees for the purpose of discouraging membership in the Union or for the purpose of discouraging Union activities.

Section 4 - *Conferences.* Conferences shall take place between the Union and the Fire Department, for the discussions of any questions or grievances which may arise, at such times and places as may be agreed upon between them. Minutes or a record shall be kept by the Fire Department of any conference between the Fire Department and the Union if requested by either party. Initialed copies shall be given to both parties, if requested.

Section 5 - *Calling of Meetings.* Meetings may be called either by the Fire Department or the Union upon no less than two (2) days notice for the consideration of matters pertaining to the contract.

## ARTICLE IV - REPRESENTATION

Section 1 - *Fire Department Entrance and Payment of Shop Committee.* The Union shall be represented by (a) International Representatives, (b) Local Union Representatives, and (c) a shop committee of two (2) plus an alternate to function in the absence of a committeeman. In the event of a second and third shift there shall be one (1) committeeman for each additional twenty-five (25) employees or a major fraction thereof for the second or third shift.

Section 2 - *List of Fire Department and Union Representatives.* The Fire Department shall give to the Union a list of supervisors, including their names and titles, and the Union likewise shall give the Company a list of its representatives.

## ARTICLE V - UNION SHOP

Section 1 - *Union Membership as Condition of Employment.* Employees covered by this Agreement at the time it becomes effective, and newly hired employees, who are covered by this Agreement, shall be required as a condition of continued employment to become members of the Union on or before the fifth (5) day following the thirtieth (30) calendar day of such employment.

Section 2 - *Termination of Membership.* Employees to whom membership in the Union is terminated by reason of the failure of the employee to tender the periodic dues and initiation fees uniformly levied against all Union members in conformity with the Constitution and By-Laws of the Union shall not be retained in the employ of the Fire Department.

## ARTICLE VI - DUES COLLECTION

Section 1 - *Authorization Card.* The Fire Department agrees, during the term of this Agreement, upon receipt of an individual, separate authorization and request in writing, duly executed by a member of the Union pursuant to the provisions of Section 302(c) of the Labor Management Relations Act of 1947, to deduct monthly dues and such initiation fees as may be fixed by the Union pursuant to the terms of said authorization, and remit the same, by check payable to Local Union # 55, UAW to the Financial Secretary of said Local Union No. 55. All such deductions shall be made from the employee's first pay for each month during the term of this contract and transmitted to the Union the full amount collected for initiation fees and dues not later than the twentieth (20) day of the month in which the collection was made.

Section 2 - *Deduction of Initiation Fees and Dues.* Deductions of initiation fees and dues of new employees shall be made in the first check-off period after joining the Union. The Fire Department will notify all employees of these conditions at the time of hire.

Section 3 - *Monthly Dues Deduction Record.* The Fire Department and the Union shall work out a mutually satisfactory arrangement by which the Fire Department will furnish the Financial Secretary of the Local Union monthly a record of those for whom deductions have been made, together with the amount of such deductions.

## ARTICLE VII - GRIEVANCE PROCEDURE

Section 1 - *Supervisor's Disposition - Step 1.* Any employee having a grievance shall present it to his committeeman who, together with the aggrieved employee, will attempt to negotiate the matter with the designated Supervisor. If the grievance is not satisfactorily settled, such grievance shall be reduced to writing on triplicate forms provided by the Fire Department. The grievance shall be dated and signed by the Committeeman. The Supervisor shall state his disposition of the grievance and reasons therefore in writing one (1) working day after the grievance had been presented to him.

Section 2 - *Grievance Becomes Property of Union.* After the grievance has been signed by the aggrieved employee, the grievance shall become the property of the Union and representative of the Fire Department shall not contact the aggrieved employee relative to the grievance, except in the presence of a Union representative.

Section 3 - *Disposition of Officers of the Fire Department - Step 2.* (a) If the above steps shall fail to secure satisfactory settlement, the grievance may be appealed to the designated representatives of the Fire Department. The grievance shall be presented to the said designated representatives of the Fire Department and be processed by them and the accredited representatives of the Union. If at this meeting the grievance is not satisfactorily settled, the Fire Department shall state its disposition in writing and reasons therefor, within three (3) work days from the date of the meeting. If the Fire Department fails to give a written answer within the aforementioned three (3) work days, unless it is mutually agreed to extend the time limit in writing, the adjustment sought by the aggrieved employee or the shop committee as the case may be, shall be considered as the final adjustment to be effectuated.

(b) In the event the Fire Department representative to function in steps 1, 2, and/or 3 of the grievance procedure is the same person, the grievance shall be appealed directly to the highest step of the grievance procedure in which the Fire Department Representative functions.

Section 5 - *Step 3.* (a) If the Union shall fail to secure satisfactory settlement, the Fire Department and the Union may, by mutual agreement, request the office of the New York State Board of Mediation to submit the name of a Staff Arbitrator who will arbitrate the grievance or grievances pending. In the event no agreement is reached, as heretofore stated, the Union may make a request to the American Arbitration Association or other agreed upon party for a panel of nine (9) names from which an Arbitrator shall be selected, either by mutual agreement or by each party alternately striking off a name from the panel. The remaining name shall be the arbitrator who shall arbitrate the grievances or grievances pending.

(b) The arbitrator shall fix and notify the parties of the time and place for arbitration of the grievance.

(c) Any issue involving the interpretation or application of any term of this Agreement may be initiated by the Union directly with the Fire Department at Step 2 above. Upon failure of the parties to agree, the Union may then appeal the issue directly to arbitration for a decision.

(d) The decision of the arbitrator shall be final and binding upon both parties, but he shall have no power either to add to, subtract from, or modify any of the terms, conditions, or limitations of this Agreement or any agreement made supplementary hereto.

(e) All of the costs and expenses of the arbitrator shall be divided equally between the Fire Department and the Union.

## ARTICLE VIII - SENIORITY

Section 1 - *Seniority.* Seniority shall be established on a Fire Department basis with each employee's seniority determined as of his date of hire.

Section 2 - *Twenty (20) Work Days Probationary Period.* (a) All new employees and those rehired after a break in the continuity of service shall be probationary employees for the first twenty (20) days worked after their employment or rehiring. If such employees are detained after the probationary period, they shall become eligible for seniority rights from the date of the new hiring or rehiring. During that probationary period, the Fire Department may lay off or discharge such employees as it may determine in its sole judgment. This probationary period may be extended by mutual agreement between the Fire Department and the Union.

(b) When an employee is laid off prior to completion of their probationary period and the employee is recalled to work within ninety (90) days from the date of their lay off, all days worked for the Fire Department prior to the lay off date shall be counted toward completion of their probationary period.

## ARTICLE IX - JOB POSTING

Section 1 - *Posting of Job Vacancy.* It is agreed that, in cases of vacancies in old or new nonsupervisory jobs, first consideration in filling the jobs will be given to employees in order of their seniority, provided they are capable of filling the jobs.

In filling vacancies occurring in old or new nonsupervisory jobs, there shall be posted on the bulletin board for a period of not less than two (2) working days, a notice setting forth the general work classification and the hourly rate of the opening. This notice shall also state the time limit for filing applications. Employees are permitted during such stated period to make application in writing and on forms obtained from the Fire Chief.

## ARTICLE X - LEAVE OF ABSENCE

Section 1 - *Union Leave.* The Fire Department shall grant a leave of absence for a period up to one (1) year, but with the privilege of yearly renewals, to any employee elected or selected to a Union office, and his seniority shall accumulate during such leave, and he shall be reinstated to his job or a similar job at the current rate of pay at the time of his reemployment provided he is then able to do and is qualified for the job and to which his seniority entitles him. Such employee's wages will be terminated on leaving active duty until such time as he returns from such leave of absence.

Section 2 - *Personal Leave.* Leave of absence may be granted for good cause to any employee. If any employee desires a leave of absence, he shall make his request in writing, listing the reasons therefore, and present that request to the committee. If the committee approves that request, then it shall present it, with the committee's approval noted thereon, to the Fire Department for the Fire Department's final and binding decision. The Fire Department will note its approval or disapproval in writing on the request and return it to the committee. Such employee's wages shall be terminated on leaving active duty until such time as he shall return to active duty from his leave of absence.

Section 3 - *Sick Leave.* Employees who are sick shall automatically be on sick leave or absence and, during such leave, they shall accumulate seniority.

Section 4 - *Veteran's Leave.* This section concerns all employees who:

(a) Enter the Armed Forces of the United States of America, the United States Merchant Marines, the Coast Guard or the Peace Corps;

(b) Are called for duty under the Selective Service Act of the United States of America;

(c) Shall, by order or directive of the Government or any of its agencies, be required to work elsewhere; shall be granted leave during which period of leave their seniority shall continue to accumulate.

Section 5 - *Military Leave - Temporary.* An employee, except part-time and/or probationary, who is a qualified member of a Federally sponsored military unit and who is called to temporary active service (maximum 1/2 month) for training requirements, will be granted a leave of absence under the following conditions:

(a) Before service, the employee shall submit official notification.

(b) If he had less than one (1) year of seniority, he shall be paid his regular rate less the gross amount on income he received while on military leave. In order to obtain this differential, he must submit documents to substantiate the compensation he received for military duty.

Section 6 - *Military Service - Long Term.* (a) When an employee is inducted into the United States Armed Services he will be eligible for additional compensation as follows:

One (1) year or more seniority - One (1) month's wages less his earned gross income in the service. This payment to be made on the regular pay days and only after such employee's reporting to his first military station and upon receipt of employee's address, rank and serial number from his Commanding Officer.

(b) During a period of peace time, while there is no Selective Service Act in effect any employee enlisting in the armed services shall be considered a quit. This Section shall be null and void if in violation of any law.

Section 7 - *Bereavement Leave with Pay.* Employees shall be paid for time lost, not exceeding three (3) work days, at their guaranteed hourly rate when a death occurs in the employee's immediate family, namely, father, mother, brother, sister, husband, wife, son, daughter, mother-in-law, father-in-law, grandmother, grandfather, or grandchild of employee. Father, mother, etc. shall mean step-father, step-mother, etc.

Section 8 - *Jury Duty Leave with Pay.* When an employee is called for jury duty, the Fire Department will make up the difference in pay between the employee's pay and the amount received for jury duty. An employee released from jury duty during the regular working hours is expected to return to work, being allowed ample time for travel, lunch and change of clothing.

Section 9 - *Time Paid at Compensation Hearing.* Employees who are injured on the job and sent home because of such injury shall be paid at their hourly rate for the balance of the shift in which the injury occurred. Any employee required to report to the State Compensation Board for a hearing relating to any injury at this Company, shall be paid at his hourly rate for all time lost from work in attending the hearing, as specified herein.

## ARTICLE XI - HEALTH AND SAFETY

Section 1 - *Promotion of Health and Safety.* The Fire Department shall maintain all facilities and equipment in such a manner as to adequately safeguard the health and safety of all employees. The Fire Department agrees to cooperate with the Union and all community, state and federal agencies interested in promoting safe, healthful working conditions for the purpose of preventing accidents and occupational diseases.

Section 2 - *Joint Health and Safety Committee.* A joint Safety Committee shall be composed of one representative to be chosen by the Union and one representative chosen by the Fire Department. The Fire Department and the Union shall investigate all accidents and work to maintain the proper safety standards. The Union Representative shall suffer no loss in pay for time necessarily spent during working hours in the pursuit of his duties.

Section 3 - *Protective Clothing Equipment.* The Fire Department agrees to furnish all necessary safety equipment, such as safety turnout gear, helmets, and SCBA's. The Fire Department will furnish safety prescription glasses. The employee must pay for the prescription. The Fire Department agrees to maintain clean, sanitary rest rooms, adequate hot and cold water, and adequate washing facilities for its employees.

Section 4 - *Medical Supplies.* The Fire Department shall keep adequate and reasonable medical supplies on hand at all times.

Section 5 - *Physical Examinations.* (a) New employees and those recalled after layoff will be subject to thorough pre-employment physical examination.

(b) In case of extended time off for accident or sickness, the Fire Department retains the right to have its own physician examine the employee if there is reasonable doubt as to his ability to perform his normal duties.

Section 7 - *Employment for Disabled.* Any employee suffering an injury or disease shall continue to be employed in line with his seniority and insofar as his injuries or disease will permit, with normal efficiency; but nothing shall require the Fire Department to create a job if there are no vacancies.

## ARTICLE XII - HOLIDAYS

Section 1 - *Paid Holidays.* (a) Employees are not required to work but shall receive the straight time hourly rate of eight (8) hours for the following holidays:

1. Thanksgiving Day
2. Day after Thanksgiving Day
3. Last Working Day before Christmas
4. Christmas Day
5. The Day after Christmas
6. The Second Day after Christmas
7. The Third Day after Christmas
8. Last Working Day before New Years Day
9. New Years Day
10. Good Friday
11. Memorial Day
12. Independence Day
13. Labor Day

## ARTICLE XIII

Section 1 - *Vacation Schedule.* Employees shall receive vacation time and vacation pay based on the following schedule:

(a) Employees with seniority as of June 1, from the date of hire, and every June 1 thereafter, shall be paid as per the schedule below at their regular hourly rate.

| *Seniority* | *Vacation Time Off With Pay* |
| --- | --- |
| less than one (1) year | 1 week |
| 1–4 years | 2 weeks |
| 5–9 years | 3 weeks |
| 10–12 years | 4 weeks |
| 13–15 years | 5 weeks |
| 16 years | 5 weeks plus 1 day |
| 17 years | 5 weeks plus 2 days |
| 18 years | 5 weeks plus 3 days |
| 19 years | 5 weeks plus 4 days |
| 20 years | 6 weeks |

## ARTICLE XIV - HOURS OF WORK, WAGES, OVERTIME

Section 1 - *Regular Work Day and Regular Work Week.* The normal work week starts at 8:00 A.M. on Monday and ends at 4:00 P.M. on Friday. The normal work day includes the start of work at 8:00 A.M., a paid lunch period from 12:00 noon to 12:30 P.M. and a quitting time at 4:00 P.M.

Section 2 - *Reduced Working Day, Break Periods and Wash-Up Time.* (a) The Fire Department may expect that each employee will be on his job ready for work and will stay on his job until the lunch hour or quitting time bell has rung. It is further understood that employees' personal time spent getting coffee will be continued as in past practice and that there will be no flagrant misuse of these allowances.

(b) It is mutually agreed that wash-up time is to be on the employees' own and not the Fire Department's time.

Section 3 - *Additional Shifts.* (a) The Fire Department, at its discretion, may add additional shifts which may result in a change in the starting and quitting time of the existing shifts.

(b) Employees desiring a transfer in the same classification to another shift may make application for such transfer and will be transferred according to seniority as openings occur.

(c) Employees working on a regularly assigned second or third shift shall receive an increase in their regular wage rate of six percent (6%).

Section 4 - *Temporary Transfers.* (a) Temporary transfers of employees to other classifications shall be by mutual agreement by the Fire Department and the Union.

(b) Any employee temporarily transferred to a higher rated job will receive the higher rate upon transfer. Employees temporarily transferred to a lower rated job shall suffer no reduction in rate, unless such transfer was made upon their request or was made in accordance with their exercise of seniority at the time of a reduction in force.

Section 5 - *Travel Expenses.* The policy with regard to expenses incurred by an employee who is required to travel for Fire Department purposes is as follows.

A. If an employee uses his own car, mileage will be paid at the rate of twenty-nine cents (29) per mile.

B. All meals, incidental expenses (such as parking, tolls, etc.) and hotel expenses will be paid for at cost.

C. Normally, an employee sent on a trip out of the metropolitan area will draw and advance sufficient to meet the expenses to be incurred, and the difference between this advance and the total expense involved will be returned to the Fire Department after the trip is completed. An expense account covering expenses incurred must be submitted to substantiate the reimbursement to the employee.

Section 6 - *Overtime Payments.* (a) All time worked in excess of forty (40) hours per week or eight (8) hours per day will be paid for at the overtime rates specified in the contract.

(b) For the purpose of computing overtime, all time off with pay including the lunch period will be considered as hours worked for computing either daily or weekly overtime.

(c) Time and one-half shall be paid for the first eight (8) hours of work performed on Saturday.

(d) Double time shall be paid for all work performed in excess of eight (8) hours on Saturday and for all work performed on Sunday.

(e) An employee not on layoff, who is called back to work in an emergency after his regular working day and who had left the plant property, shall be guaranteed a minimum of four (4) hours work at his regular wage rate times the applicable overtime rate.

(f) There shall be no pyramiding of overtime under the provisions of this Article.

(g) Overtime assignment of less than one total hour shall be paid for at the rate of one full hour.

(h) Employees are not required to work beyond the regular work day or work as specified herein except as otherwise mutually agreed.

Section 7 - *Subcontracting of Work.* The Fire Department may subcontract work out provided that:

1. It does not result in a layoff of any employee who is capable of doing the work.

2. It does not result in reduced earnings or deprive employees from overtime opportunities from time to time.

3. It does not deprive a laid off employee from being recalled to work who is capable of doing the work.

Nothing in this section is intended to restrict the Fire Department from contracting outwork which has normally been contracted out in the past, such as office cleaning, window washing, gardening and such other work as may arise requiring specific licensing or other skills such as plumbing, electrical work, heating and ventilation, rigging and building construction.

Section 8 - *Non-Bargaining Unit Employees Cannot Perform Work.* Foreman, supervisory employees, or other employees outside the bargaining unit shall not perform work normally performed by employees in the bargaining unit, except for the purpose of instruction or in cases of emergency when no other employee is available.

Section 9 - *Pay Day.* Pay Day shall be on the Friday following the end of one week work period. Payment will be in cash if requested by the Union.

## ARTICLE XV - COST OF LIVING

A cost of living allowance is provided for herein and shall be determined as follows:

(a) The cost-of-living allowance shall be added to employees straight time hourly earnings and will be adjusted up or down each three (3) months, as provided herein.

(b) The cost-of-living allowance will be determined in accordance with changes in the Consumer Price Index for Urban Wage Earners and Clerical Workers, U.S. Department of Labor (1967 = 100) and hereinafter referred to as the B.L.S. Consumers Price Index.

(c) The amount of cost-of-living allowance at the time shall be included in computing overtime premium, vacation payments, holiday payments and call-in pay, etc.

(d) In the event that the Bureau of Labor Statistics does not issue the Consumers Price Index on or before the beginning of the pay period referred to in paragraph (c), the adjustment regained will be made at the beginning of the first pay period after receipt of the Index.

(e) No adjustments, retroactive or otherwise, shall be made due to any revisions which later may be made in the published figures for the B.L.S. Consumers Price Index for any base month.

## ARTICLE XVI - INSURANCE BENEFITS

Section 1 - *Insurance Program.* Effective January 1, 1996, and each month thereafter, for the duration of this collective Bargaining Agreement, the Company shall provide to each employee, in accordance with the terms of this Article XV, the insurance program listed below in Section 2.

Section 2 - *Insurance Benefits.* The Company will pay the entire cost for and insurance program which will provide the following benefits to each employee and the employees dependents where applicable.

a. Life Insurance - $7,000 (employee only)

b. AD&D Principal Sum - $7,000 (employee only)

c. HMO Insurance Level AAA - full payment (employee only)

Section 3 - *Coverage for Inactive and Terminated Employees.* (a) Employees who are discharged, quit, or are on a personal leave of absence will be continued under the Plan for the remainder of the month in which the separation or leave occurs.

## ARTICLE XVII - RETIREMENT FUND

Section 1 - *Local 1 Retirement Income Fund.* (a) Effective January 1, 1996, the Fire Department shall continue to be a "Contributing Employer" and shall be bound by the provisions of the Local 1 established pursuant to an Agreement and Declaration of Trust dated January 1, 1995.

## ARTICLE XVIII - STRIKES AND LOCKOUTS

In accordance with State Law, members of Local 1 are prohibited from striking and the Fire Department is prohibited from locking out employees.

## ARTICLE XIX - MISCELLANEOUS

Section 1 - *Bulletin Boards.* The Fire Department shall furnish bulletin boards for the exclusive use of the Union for the posting of officially signed notices. The Union agrees that bulletin boards will be limited to the display of notices of Union meetings and functions.

Section 2 - *No Divulging of Information.* Except where required by law, a copy of any report, written or oral, given to any person, Fire Department or outside party, will be given to the employee involved at the time it is given out or mailed to the last known address of a former employee. If the employee believes that all or part of the contents of that report is without foundation, that employee or former employee has the right within 30 calendar days upon receipt of such report to seek redress, financial or otherwise, through the grievance machinery presented herein; such report shall contain information only as to that employment record. No information with respect to the employee's Union activity, financial status, garnishees or personal problems not related to his work will be contained in the report, unless such information is requested in writing by the employee.

Section 3 - *Warning Notices.* Warning notices issued to employees shall be expunged from their record twelve (12) months from the date of their issuance.

Section 4 - *Telephones.* The telephone will be available to all employees on an unrestricted basis, during non-working hours. All incoming calls will continue to be made through the regular Fire Department telephone and will be restricted to calls of an emergency nature, and Union business. Outgoing telephone calls of an emergency nature only, may be made during working hours with the permission of the employee's supervisor.

Section 5 - *No Discrimination.* The Fire Department agrees that there shall be no discrimination with respect to hire, wages, or any other term or condition of employment because of sex, creed, color, religion, or national, origin of an employee or prospective employee.

## ARTICLE XX - TERMINATION

This agreement shall continue in full force and effect until 12:01 a.m. on the 1st day of January, 1999. Thereafter it shall be considered automatically renewed for each following twelve (12) month period unless either party shall serve written notice on the other, sixty (60) days prior to each anniversary date of their intent to modify or amend the Agreement. If such notice is given, the parties will enter into negotiations in an attempt to reach an Agreement on the provisions of a modified or amended contract. Failing to agree, the contract will be effective only on a day-to-day basis commencing on the above termination date, until a new agreement is reached or until the Union serves the Company with a written notice to terminate the day-to-day Agreement.

IN WITNESS THEREOF, the parties have caused their names to be subscribed by their duly authorized officers and/or representatives, this _____ day of _____, 19___ .


_____

Signature - Fire Department


_____

Signature - Union

## SAMPLE AUTHORIZATION CARD

Date of signing _____

I,_____ , now employed by
*(print your name here)*

_____ (name of company)
have volumtarily accepted membership in the United Factory Workers
Union (AFL-CIO) and designate said Union as my collective bargaining
agent in all matters pertaining to wages, hours and other conditions of
employment. I hereby further subscribe to the dues deduction provisions
printed on the reverse side of this card.

Signed _____

_____     _____
                                          *(Signer's home address)*

                                       _____
                                                    *(Phone No.)*

*Appendix*

# C

# Sample Complaint Forms

JS 44
(Rev. 07/89)

# CIVIL COVER SHEET

The JS-44 civil cover sheet and the information contained herein neither replace nor supplement the filing and service of pleadings or other papers as required by law, except as provided by local rules of court. This form, approved by the Judicial Conference of the United States in September 1974, is required for the use of the Clerk of Court for the purpose of initiating the civil docket sheet. (SEE INSTRUCTIONS ON THE REVERSE OF THE FORM.)

## I (a) PLAINTIFFS

## DEFENDANTS

**(b)** COUNTY OF RESIDENCE OF FIRST LISTED PLAINTIFF _____
(EXCEPT IN U.S. PLAINTIFF CASES)

COUNTY OF RESIDENCE OF FIRST LISTED DEFENDANT _____
(IN U.S. PLAINTIFF CASES ONLY)
NOTE: IN LAND CONDEMNATION CASES, USE THE LOCATION OF THE
TRACT OF LAND INVOLVED

**(c)** ATTORNEYS (FIRM NAME, ADDRESS, AND TELEPHONE NUMBER)

ATTORNEYS (IF KNOWN)

## II. BASIS OF JURISDICTION (PLACE AN × IN ONE BOX ONLY)

☐ 1 U.S. Government
Plaintiff

☐ 2 U.S. Government
Defendant

☐ 3 Federal Question
(U.S. Government Not a Party)

☐ 4 Diversity
(Indicate Citizenship of
Parties in Item III)

## III. CITIZENSHIP OF PRINCIPAL PARTIES (PLACE AN × IN ONE BOX FOR PLAINTIFF AND ONE BOX FOR DEFENDANT)
(For Diversity Cases Only)

| | PTF | DEF | | PTF | DEF |
|---|---|---|---|---|---|
| Citizen of This State | ☐ 1 | ☐ 1 | Incorporated or Principal Place of Business in This State | ☐ 4 | ☐ 4 |
| Citizen of Another State | ☐ 2 | ☐ 2 | Incorporated and Principal Place of Business in Another State | ☐ 5 | ☐ 5 |
| Citizen or Subject of a Foreign Country | ☐ 3 | ☐ 3 | Foreign Nation | ☐ 6 | ☐ 6 |

## IV. CAUSE OF ACTION (CITE THE U.S. CIVIL STATUTE UNDER WHICH YOU ARE FILING AND WRITE A BRIEF STATEMENT OF CAUSE.)

DO NOT CITE JURISDICTIONAL STATUTES UNLESS DIVERSITY)

## V. NATURE OF SUIT (PLACE AN × IN ONE BOX ONLY)

| CONTRACT | TORTS | | FORFEITURE/PENALTY | BANKRUPTCY | OTHER STATUTES |
|---|---|---|---|---|---|
| ☐ 110 Insurance | **PERSONAL INJURY** | **PERSONAL INJURY** | ☐ 610 Agriculture | ☐ 422 Appeal 28 USC 158 | ☐ 400 State Reapportionment |
| ☐ 120 Marine | ☐ 310 Airplane | ☐ 362 Personal Injury— Med. Malpractice | ☐ 620 Other Food & Drug | ☐ 423 Withdrawal 28 USC 157 | ☐ 410 Antitrust |
| ☐ 130 Miller Act | ☐ 315 Airplane Product Liability | ☐ 365 Personal Injury— Product Liability | ☐ 625 Drug Related Seizure of Property 21 USC 881 | | ☐ 430 Banks and Banking |
| ☐ 140 Negotiable Instrument | ☐ 320 Assault, Libel & Slander | ☐ 368 Asbestos Personal Injury Product Liability | ☐ 630 Liquor Laws | **PROPERTY RIGHTS** | ☐ 450 Commerce/ICC Rates/etc. |
| ☐ 150 Recovery of Overpayment & Enforcement of Judgment | ☐ 330 Federal Employers' Liability | | ☐ 640 R.R. & Truck | ☐ 820 Copyrights | ☐ 460 Deportation |
| ☐ 151 Medicare Act | ☐ 340 Marine | **PERSONAL PROPERTY** | ☐ 650 Airline Regs | ☐ 830 Patent | ☐ 470 Racketeer Influenced and Corrupt Organizations |
| ☐ 152 Recovery of Defaulted Student Loans (Excl. Veterans) | ☐ 345 Marine Product Liability | ☐ 370 Other Fraud | ☐ 660 Occupational Safety/Health | ☐ 840 Trademark | ☐ 810 Selective Service |
| ☐ 153 Recovery of Overpayment of Veteran's Benefits | ☐ 350 Motor Vehicle | ☐ 371 Truth in Lending | ☐ 690 Other | **SOCIAL SECURITY** | ☐ 850 Securities/Commodities/ Exchange |
| ☐ 160 Stockholders' Suits | ☐ 355 Motor Vehicle Product Liability | ☐ 380 Other Personal Property Damage | **LABOR** | ☐ 861 HIA (1395ff) | ☐ 875 Customer Challenge 12 USC 3410 |
| ☐ 190 Other Contract | ☐ 360 Other Personal Injury | ☐ 385 Property Damage Product Liability | ☐ 710 Fair Labor Standards Act | ☐ 862 Black Lung (923) | ☐ 891 Agricultural Acts |
| ☐ 195 Contract Product Liability | | | ☐ 720 Labor/Mgmt. Relations | ☐ 863 DIWC/DIWW (405(g)) | ☐ 892 Economic Stabilization Act |
| **REAL PROPERTY** | **CIVIL RIGHTS** | **PRISONER PETITIONS** | ☐ 730 Labor/Mgmt. Reporting & Disclosure Act | ☐ 864 SSID Title XVI | ☐ 893 Environmental Matters |
| ☐ 210 Land Condemnation | ☐ 441 Voting | ☐ 510 Motions to Vacate Sentence Habeas Corpus: | ☐ 740 Railway Labor Act | ☐ 865 RSI (405(g)) | ☐ 894 Energy Allocation Act |
| ☐ 220 Foreclosure | ☐ 442 Employment | | | **FEDERAL TAX SUITS** | ☐ 895 Freedom of Information Act |
| ☐ 230 Rent Lease & Ejectment | ☐ 443 Housing/ Accommodations | ☐ 530 General | ☐ 790 Other Labor Litigation | ☐ 870 Taxes (U.S. Plaintiff or Defendant) | ☐ 900 Appeal of Fee Determination Under Equal Access to Justice |
| ☐ 240 Torts to Land | ☐ 444 Welfare | ☐ 535 Death Penalty | ☐ 791 Empl. Ret. Inc. Security Act | ☐ 871 IRS—Third Party 26 USC 7609 | ☐ 950 Constitutionality of State Statutes |
| ☐ 245 Tort Product Liability | ☐ 440 Other Civil Rights | ☐ 540 Mandamus & Other | | | ☐ 890 Other Statutory Actions |
| ☐ 290 All Other Real Property | | ☐ 550 Other | | | |

## VI. ORIGIN (PLACE AN × IN ONE BOX ONLY)

☐ 1 Original
Proceeding

☐ 2 Removed from
State Court

☐ 3 Remanded from
Appellate Court

☐ 4 Reinstated or
Reopened

☐ 5 Transferred from
another district
(specify)

☐ 6 Multidistrict
Litigation

☐ 7 Appeal to District
Judge from
Magistrate
Judgment

## VII. REQUESTED IN COMPLAINT:

CHECK IF THIS IS A **CLASS ACTION**
☐ UNDER F.R.C.P. 23

**DEMAND $**

Check YES only if demanded in complaint:
**JURY DEMAND:** ☐ YES ☐ NO

## VIII. RELATED CASE(S) IF ANY (See instructions):

JUDGE _____ DOCKET NUMBER _____

DATE

SIGNATURE OF ATTORNEY OF RECORD

UNITED STATES DISTRICT COURT

S.F. 1 (REV. FEBRUARY, 1991)
EMPLOYER'S FIRST REPORT
OF INJURY OR ILLNESS AND
SUPPLEMENTARY RECORD UNDER
THE OCCUPATIONAL SAFETY
AND HEALTH ACT

DEPARTMENT OF WORKERS' CLAIMS
**WORKERS' COMPENSATION BOARD**
1270 Louisville Road
Perimeter Park West, Building C
Frankfort, Kentucky 40601

IF THIS CASE WAS OSHA RECORDABLE, INDICATE REASON
FOR RECORDING AND GIVE OSHA CASE OR FILE NUMBER

*Reason for recording (e.g. "loss of consciousness")*

KRS 342.990 AUTHORIZES A FINE FOR EMPLOYER'S FAILURE TO SUBMIT THIS ORIGINAL REPORT
WITHIN ONE WEEK OF KNOWLEDGE OF INJURY TO THE WORKERS' COMPENSATION BOARD. TO
COMPLY WITH THIS LAW, EACH QUESTION SHALL BE ANSWERED COMPLETELY, ACCURATELY AND
LEGIBLY. IMPROPERLY PREPARED REPORTS WILL BE REFUSED AND RETURNED. PLEASE USE
TYPEWRITER OR PRINT IN INK. COMPLETE ALL QUESTIONS!

*OSHA Case or File Number (from your OSHA Form 200)*

**EMPLOYER**

1. EMPLOYER'S NAME                    EMPLOYER NUMBER

2. STREET OR ROAD          LOCATION AT WHICH EMPLOYEE WORKED

DO NOT WRITE IN THIS COLUMN

File No.

3. IF INDIVIDUAL OR PARTNERSHIP, NAME OF BUSINESS

4. CITY     COUNTY     STATE     ZIP

Employer No.

5. MAILING ADDRESS

6. AREA CODE TELEPHONE     7. UNEMPLOYMENT INSURANCE I.D. No.

U.I. No.

8. CITY     COUNTY     STATE     ZIP

9. NATURE OF BUSINESS (e.g., tree trimming, boot mfg.)

Industry

10. WORKERS'S COMPENSATION INSURANCE CARRIER (IF SELF-INSURED, CHECK HERE □)     POLICY NUMBER

11. SPECIFY PRODUCT OR SERVICE COMPRISING MAJORITY OF SALES (e.g., ski boots)

Soc. Sec. No.

**EMPLOYEE**

12. EMPLOYEE'S NAME     FIRST     MIDDLE     LAST

13. AREA CODE TELEPHONE (HOME)     14. SOCIAL SECURITY NO.

Age

15. EMPLOYEE'S HOME ADDRESS

16. SINGLE □   MALE □
    MARRIED □   FEMALE □     17. DATE OF BIRTH

Sex

Marital Status

18. CITY     STATE     ZIP

19. DEPARTMENT IN WHICH REGULARLY EMPLOYED

Occupation

20. REGULAR OCCUPATION (JOB TITLE)

21. DEPARTMENT WHERE WORKING WHEN INJURY OCCURRED

Department

22. HOW LONG EMPLOYED BY YOU?     23. HOW LONG IN PRESENT JOB?

24. NUMBER OF HOURS WORKED  PER DAY   PER WK.     25. NUMBER OF DAYS WORKED  PER WK.

Months on Job

Shift

26. EMPLOYEE'S WAGE RATE $     HR.
    or $     /DAY, or $     /WK.

27. COMMISSION OR PIECE WORK EARNINGS     $     IN     HRS. IN PAST 12 MO.

28. WEEKLY DOLLAR VALUE OF PAY IN KIND (LODGING, FOOD, ETC.)$

Weekly Wage

29. NO. OF DEPENDENTS (Please complete back of form)

30. PLACE OF ACCIDENT OR EXPOSURE (LOCATION, INCLUDING COUNTY)

31. DATE EMPLOYER NOTIFIED

County of Injury

**THE ACCIDENT OR EXPOSURE**

32. ON EMPLOYER'S PREMISES?
    YES □     NO □

33. DATE OF OCCURENCE

34. TIME OF DAY

35. TIME WORKDAY BEGAN AND WOULD NORMALLY (A.M.) (A.M.)
    END FROM (P.M.) (P.M.)

Nature of Injury

Body Part

36. HOW DID THE ACCIDENT OR EXPOSURE OCCUR? (Begin by telling what the employee was doing just before the accident or exposure? Be specific. If employee was using tools or equipment, or handling material, name them and tell what employee was doing with them.)

Accident Type

37. (Now describe fully the events which resulted in injury or illness. Tell what happened and how it happened. Specify how objects or substances were involved. Give full details of all factors which led or contributed to the accident or exposure.)

Source of Injury

38. WHAT THING DIRECTLY PRODUCED THIS INJURY OR ILLNESS? (Name objects struck against or struck by, vapor, poison, chemical, or radiation. If strain or hernia, the thing being lifted, pulled, pushed, etc. If injury resulted solely from bodily motion, the stretching, twisting, etc. which resulted in injury.)

39. DESCRIBE THE INJURY OR ILLNESS IN DETAIL AND INDICATE THE PART OF BODY AFFECTED. (e.g. amputation of right index finger at second joint, fracture of 2 ribs, lead poisoning, dermatitis of left hand, etc.)
    FATAL?  YES □  NO □

Date Returned

Time Present Job

40. NAME AND ADDRESS OF TREATING PHYSICIAN

41. NAME AND ADDRESS OF HOSPITAL
    IN PATIENT □
    OUT PATIENT □

Extent of Disability

42. MEDICAL TREATMENT GIVEN (DESCRIBE)     IF RESTRICTIONS OF DUTY OR PERMANENT TRANSFER TO ANOTHER JOB, CHECK □

Lost Workdays

Injury Date

43. DATE STOPPED WORK BECAUSE OF THIS INJURY OR ILLNESS

44. DATE RETURNED TO WORK

45. NUMBER OF SCHEDULED WORK DAYS LOST TO DATE

46. WAS EMPLOYEE PAID FOR FULL DAY ON DATE OF INJURY?  YES □  NO □

Injury Hour

47. IF DEATH, GIVE NAME AND ADDRESS OF NEXT OF KIN

48. DATE OF DEATH

Date of Disability

49. REPORT PREPARED BY

50. TITLE

51. DATE OF THIS REPORT

Date of Report

*EVERY QUESTION MUST BE ANSWERED AND FORM SIGNED*

| PERSONS ACTUALLY DEPENDENT ON INJURED EMPLOYEE, LIST YOUNGEST FIRST | | |
|---|---|---|
| NAME | DATE OF BIRTH | RELATIONSHIP |
| | | |
| | | |
| | | |
| | | |

## INSTRUCTIONS

This form is designed for completion with a typewriter. Vertical spacing matches carriage advance of most typewriters. Horizontal spacing (4 steps) can be set up on tabulator.

**PLEASE USE TYPEWRITER OR COMPLETE LEGIBLY IN INK!**

### EMPLOYER

1., 3., 5., 8. — Give the name and address exactly as it appears on your certificate of workers' compensation insurance. If you are an individual or a partner in business enter your name, or names of partners on line 1, and the name of your business enterprise on line 3. If a corporation, enter name of corporation on line 1 and leave line 3 blank.

2., 4. — Enter location of the establishment at which the employee was regularly employed at the time of the injury or illness.

6. — Enter telephone number at which person in charge of injury records can be reached.

7. — The employer number under which you pay unemployment insurance

9. — Classification of industry or business

10. — Name of company (not agent) carrying your workers' compensation insurance in Kentucky.

11. — The product or service which is responsible for the largest percentage of your gross sales.

### EMPLOYEE

19., 21. — Use descriptive word or phrase which identifies the kind of work performed in the department.

23. — In present department and with present job title.

24., 25. — On the average over the most recent quarter.

27. — Earnings in dollars and hours worked (if known) in past 12 months.

28. — Include value of all materials or services (auto, utilities, etc.) furnished for private use of employee or his family.

### THE ACCIDENT OR EXPOSURE

29. — Enter the number of dependents in space 29., then turn to back of form and fill in the ages and relationships of each person principally dependent on the employee at the time of injury.

31. — Date that employer first knew of the injury or illness.

33. — Date of injury if known, or date injury or illness was diagnosed.

35. — Employee's work shift on the day of the injury.

36.-39. — Follow instructions on front of form with care. Forms which are incompletely filled out will be returned for completion, and submission of a completed form will be required. The information from these questions is used to compile statistical information which is essential to the study of accidents and occupational hazards.

41. — Complete only if employee was taken to a hospital. Check "in patient" if employee was admitted to the hospital. Check "out patient" if he was treated in the emergency room, for example, and released without being admitted. In either case, give the name and address of the hospital.

42. — Indicate treatment given both at scene and at medical facility (if any).

45. — Use the OSHA criteria for counting lost work days.

| CHARGE OF DISCRIMINATION | AGENCY | CHARGE NUMBER |
|---|---|---|
| This form is affected by the Privacy Act of 1974; See Privacy Act Statement before completing this form. | ☐ FEPA<br>☒ EEOC | |

_____ and EEOC
*State or local Agency, if any*

| NAME *(Indicate Mr., Ms., Mrs.)* | | HOME TELEPHONE *(Include Area Code)* |
|---|---|---|
| STREET ADDRESS | CITY, STATE AND ZIP CODE | DATE OF BIRTH / / |

NAMED IS THE EMPLOYER, LABOR ORGANIZATION, EMPLOYMENT AGENCY APPRENTICESHIP COMMITTEE, STATE OR LOCAL GOVERNMENT AGENCY WHO DISCRIMINATED AGAINST ME *(If more than one list below.)*

| NAME | NUMBER OF EMPLOYEES, MEMBERS | TELEPHONE *(Include Area Code)* |
|---|---|---|
| STREET ADDRESS | CITY, STATE AND ZIP CODE | COUNTY |
| NAME | | TELEPHONE NUMBER *(Include Area Code)* |
| STREET ADDRESS | CITY, STATE AND ZIP CODE | COUNTY |

CAUSE OF DISCRIMINATION BASED ON *(Check appropriate box(es))*

☐ RACE   ☐ COLOR   ☐ SEX   ☐ RELIGION   ☐ NATIONAL ORIGIN
☐ RETALIATION   ☐ AGE   ☐ DISABILITY   ☐ OTHER *(Specify)*

DATE DISCRIMINATION TOOK PLACE
*EARLIEST*        *LATEST*
/  /          /  /
☐ CONTINUING ACTION

THE PARTICULARS ARE   *(If additional space is needed, attach extra sheet(s)):*

☐ I want this charge filed with both the EEOC and the State or local Agency, if any. I will advise the agencies if I change my address or telephone number and cooperate fully with them in the processing of my charge in accordance with their procedures.

I declare under penalty of perjury that the foregoing is true and correct.

Date _____ Charging Party *(Signature)*

NOTARY - (When necessary for State and Local Requirements)

I swear or affirm that I have read the above charge and that it is true to the best of my knowledge, information and belief.

SIGNATURE OF COMPLAINANT

SUBSCRIBED AND SWORN TO BEFORE ME THIS DATE
(Day, month, and year)

# OSHA COMPLAINT

U.S. DEPARTMENT OF LABOR
OCCUPATIONAL SAFETY AND HEALTH ADMINISTRATION

OMB No. 044R1449

| For Official Use Only | | |
|---|---|---|
| Area | Date Received | Time |
| Region | Received By | |

## COMPLAINT

This form is provided for the assistance of any complainant and is not intended to constitute the exclusive means by which a complaint may be registered with the U.S. Department of Labor.

The undersigned (check one)
☐ Employee   ☐ Representative of employees   ☐ Other (specify) _____
believes that a violation at the following place of employment of an occupational safety or health standard exists which is a job safety or health hazard.

Does this hazard(s) immediately threaten death or serious physical harm?   ☐ Yes   ☐ No

Employer's Name _____
Address _____
(Street
(City _____ State _____ Telephone _____ Zip Code _____

1. Kind of business _____

2. Specify the particular building or worksite where the alleged violation is located, including address. _____

3. Specify the name and phone number of employer's agent(s) in charge. _____

4. Describe briefly the hazard which exists there including the approximate number of employees exposed to or threatened by such hazard. _____

_(Continue on reverse side if necessary)_

Sec. 8(f)(1) of the Williams-Steiger Occupational Safety and Health Act, 29 U.S.C. 651, provides as follows: Any employees or representative of employees who believe that a violation of a safety or health standard exists that threatens physical harm, or that an imminent danger exists, may request an inspection by giving notice to the Secretary or his authorized representative of such violation or danger. Any such notice shall be reduced to writing, shall set forth with reasonable particularity the grounds for the notice, and shall be signed by the employees or representative of employees, and a copy shall be provided the employer or his agent no later than at the time of inspection, except that, upon request of the person giving such notice, his name and the names of individual employees referred to therein shall not appear in such copy or on any record published, released, or made available pursuant to subsection (g) of this section. If upon receipt of such notification the Secretary determines there are reasonable grounds to believe that such violation or danger exists, he shall make a special inspection in accordance with the provisions of this section as soon as practicable, to determine if such violation or danger exists. If the Secretary determines there are no reasonable grounds to believe that a violation or danger exists he shall notify the employees or representative of the employees in writing of such determination.

[E5938]

Form OSHA-7
Rev. Jan. 1972

---

5. List by number and/or name the particular standard (or standards) issued by the Department of Labor which you claim has been violated, if known. _____

6. (a) To your knowledge has this violation been considered previously by any Government agency? _____

   (b) If so, please state the name of the agency _____

   (c) and, the approximate date it was so considered. _____

7. (a) Is this complaint, or a complaint alleging a similar violation, being filed with any other Government agency? _____

   (b) If so, give the name and address of each. _____

8. (a) To your knowledge, has this violation been the subject of any union/management grievance or have you (or anyone you know) otherwise called it to the attention of, or discussed it with, the employer or any representative thereof? _____

   (b) If so, please give the results thereof, including any efforts by management to correct the violation. _____

9. Please indicate your desire:
   ☐ I do not want my name revealed to the employer.
   ☐ My name may be revealed to the employer.

_Continue Item 4 here, if additional space is needed._

## COMPLAINANT'S NAME

Signature _____ Date _____

Typed or Printed Name _____ Telephone _____

Address _____
(Street
(City _____ State _____ Zip Code _____

If you are a representative of employees, state the name of your organization _____

[E5939]

# Appendix D

# Civil Rights Acts

## CIVIL RIGHTS ACT OF 1964

78 Stat. 253 (1964), as amended; 42 U.S.C. §§ 2000e et seq. (1982).

### TITLE VII—EQUAL EMPLOYMENT OPPORTUNITY

### DEFINITIONS

**Sec. 701.**   (§ 2000e) For the purposes of this title—

(a) The term "person" includes one or more individuals, governments, governmental agencies, political subdivisions, labor unions, partnerships, associations, corporations, legal representatives, mutual companies, joint-stock companies, trusts, unincorporated organizations, trustees, trustees in bankruptcy, or receivers.

(b) The term "employer" means a person engaged in an industry affecting commerce who has fifteen or more employees for each working day in each of twenty or more calendar weeks in the current or preceding calendar year, and any agent of such a person, but such term does not include (1) the United States, a corporation wholly owned by the Government of the United States, an Indian tribe, or any department or agency of the District of Columbia subject by statute to procedures of the competitive service (as defined in section 2102 of Title 5), or (2) a bona fide private membership club (other than a labor organization) which is exempt from taxation under section 501(c) of Title 26, except that during the first year after March 24, 1972, persons having fewer than twenty-five employees (and their agents) shall not be considered employers.

(c) The term "employment agency" means any person regularly undertaking with or without compensation to procure employees for an employer or to procure for employees opportunities to work for an employer and includes an agent of such a person.

(d) The term "labor organization" means a labor organization engaged in an industry affecting commerce, and any agent of such an organization, and includes any organization of any kind, any agency, or employee representation committee, group, association, or plan so engaged in which employees participate and which exists for the purpose, in whole or in part, of dealing with employers concerning grievances, labor disputes, wages, rates of pay, hours, or other terms or conditions of employment, and any conference, general committee, or joint or system board, or joint council so engaged which is subordinate to a national or international labor organization.

(e) A labor organization shall be deemed to be engaged in an industry affecting commerce if (1) it maintains or operates a hiring hall or hiring office which procures employees for an employer or procures for employees opportunities to work for an employer, or (2) the number of its members (or, where it is a labor organization composed of other labor organizations or their representatives, if the aggregate number of the members of such other labor organization) is (A) twenty-five or more during the first year after March 24, 1972, or (B) fifteen or more thereafter, and such labor organization—

(1) is the certified representative of employees under the provisions of the National Labor Relations Act, as amended, or the Railway Labor Act, as amended;

(2) although not certified, is a national or international labor organization or a local labor organization recognized or acting as the representative of employees of an employer or employers engaged in an industry affecting commerce; or

(3) has chartered a local labor organization or subsidiary body which is representing or actively seeking to represent employees of employers within the meaning of paragraph (1) or (2); or

(4) has been chartered by a labor organization representing or actively seeking to represent employees within the meaning of paragraph (1) or (2) as the local or subordinate body through which such employees may enjoy membership or become affiliated with such labor organization; or

(5) is a conference, general committee, joint or system board, or joint council subordinate to a national or international labor organization, which includes a labor organization engaged in an industry affecting commerce within the meaning of any of the preceding paragraphs of this subsection.

(f) The term "employee" means an individual employed by an employer, except that the term "employee" shall not include any person elected to public office in any State or political subdivision of any State by the qualified voters thereof, or any person chosen by such officer to be on such officer's personal staff, or an appointee on the policy making level or an immediate adviser with respect to the exercise of the constitutional or legal powers of the office. The exemption set forth in the preceding sentence shall not include employees subject to the civil service laws of a State government, governmental agency or political subdivision.

(g) The term "commerce" means trade, traffic, commerce, transportation, transmission, or communication among the several States; or between a State and any place outside thereof; or within the District of Columbia, or a possession of the United States; or between points in the same State but through a point outside thereof.

(h) The term "industry affecting commerce" means any activity, business, or industry in commerce or in which a labor dispute would hinder or obstruct commerce or the free flow of commerce and includes any activity or industry "affecting commerce" within the meaning of the Labor-Management Reporting and Disclosure Act of 1959, and further includes any governmental industry, business, or activity.

(i) The term "State" includes a State of the United States, the District of Columbia, Puerto Rico, the Virgin Islands, American Samoa, Guam, Wake Island, the Canal Zone, and Outer Continental Shelf lands defined in the Outer Continental Shelf Lands Act.

(j) The term "religion" includes all aspects of religious observance and practice, as well as belief, unless an employer demonstrates that he is unable to reasonably accommodate to an employee's or prospective employee's religious observance or practice without undue hardship on the conduct of the employer's business.

(k) The terms "because of sex" or "on the basis of sex" include, but are not limited to, because of or on the basis of pregnancy, childbirth, or related medical conditions; and women affected by pregnancy, childbirth, or related medical conditions shall be treated the same for all employment-related purposes, including receipt of benefits under fringe benefit programs, as other persons not so affected but similar in their ability or inability to work, and nothing in section 2(h) of this Act shall be interpreted to permit otherwise. This subsection shall not require an employer to pay for health insurance benefits for abortion, except where the life of the mother would be endangered if the fetus were carried to term, or except where medical complications have arisen from an abortion: *Provided,* That nothing herein shall preclude an employer from providing abortion benefits or otherwise affect bargaining agreements in regard to abortion.

## EXEMPTION

**Sec. 702.**   (§ 2000e–1) This Subchapter shall not apply to an employer with respect to the employment of aliens outside any State, or to a religious corporation, association, educational institution, or society with respect to the employment of individuals of a particular religion to perform work connected with the carrying on by such corporation, association, educational institution, or society of its activities.

## DISCRIMINATION BECAUSE OF RACE, COLOR, RELIGION, SEX, OR NATIONAL ORIGIN

**Sec. 703.**   (§ 2000(e)–2) (a) It shall be an unlawful employment practice for an employer—

(1) to fail or refuse to hire or to discharge any individual, or otherwise to discriminate against any individual with respect to his compensation, terms, conditions, or privileges of employment, because of such individual's race, color, religion, sex, or national origin; or

(2) to limit, segregate, or classify his employees or applicants for employment in any way which would deprive or tend to deprive any individual of employment opportunities or otherwise adversely affect his status as an employee, because of such individual's race, color, religion, sex, or national origin.

(b) It shall be an unlawful employment practice for an employment agency to fail or refuse to refer for employment, or otherwise to discriminate against, any individual because of his race, color, religion, sex, or national origin, or to classify or refer for employment any individual on the basis of his race, color, religion, sex, or national origin.

(c) It shall be an unlawful employment practice for a labor organization—

(1) to exclude or to expel from its membership, or otherwise to discriminate against, any individual because of his race, color, religion, sex, or national origin;

(2) to limit, segregate, or classify its membership or applicants for membership, or to classify or fail or refuse to refer for employment any individual, in any way which would deprive or tend to deprive any individual of employment opportunities, or would limit such employment opportunities or otherwise adversely affect his status as an employee or as an applicant for employment, because of such individual's race, color, religion, sex, or national origin; or

(3) to cause or attempt to cause an employer to discriminate against an individual in violation of this section.

(d) It shall be an unlawful employment practice for any employer, labor organization, or joint labor-management committee controlling apprenticeship or other training or retraining, including on-the-job training programs to discriminate against any individual because of his race, color, religion, sex, or national origin in admission to, or employment in, any program established to provide apprenticeship or other training.

(e) Notwithstanding any other provision of this Subchapter (1) it shall not be an unlawful employment practice for an employer to hire and employ employees, for an employment agency to classify, or refer for employment any individual, for a labor organization to classify its membership or to classify or refer for employment any individual, or for an employer, labor organization, or joint labor-management committee controlling apprenticeship or other training or retraining programs to admit or employ any individual in any such program, on the basis of his religion, sex, or national origin in those certain instances where religion, sex, or national origin is a bona fide occupational qualification reasonably necessary to the normal operation of that particular business or enterprise, and (2) it shall not be an unlawful employment practice for a school, college, university, or other educational institution or institution of learning to hire and employ employees of a particular religion if such school, college, university, or other educational institution or institution of learning is, in whole or in substantial part, owned, supported, controlled, or managed by a particular religion or by a particular religious corporation, association, or society, or if the curriculum of such school, college, university, or other educational institution or institution of learning is directed toward the propagation of a particular religion.

(f) As used in this Subchapter, the phrase "unlawful employment practice" shall not be deemed to include any action or measure taken by an employer, labor organization, joint labor-management committee, or employment agency with respect to an individual who is a member of the Communist Party of the United States or of any other organization required to register as a Communist-action or Communist-front organization by final order of the Subversive Activities Control Board pursuant to the Subversive Activities Control Act of 1950.

(g) Notwithstanding any other provision of this Subchapter, it shall not be an unlawful employment practice for an employer to fail or refuse to hire and employ any individual for any position, for an employer to discharge any individual from any position, or for an employment agency to fail or refuse to refer any individual for employment in any position, or for a labor organization to fail or refuse to refer any individual for employment in any position, if—

(1) the occupancy of such position, or access to the premises in or upon which any part of the duties of such position is performed or is to be performed, is subject to any requirement imposed in the interest of the national security of the United States under any security program in effect pursuant to or administered under any statute of the United States or any Executive order of the President; and

(2) such individual has not fulfilled or has ceased to fulfill that requirement.

(h) Notwithstanding any other provision of this Subchapter, it shall not be an unlawful employment practice for an employer to apply different standards of compensation, or different terms, conditions, or privileges of employment pursuant to a bona fide seniority or merit system, or a system which measures earnings by quantity or quality of production or to employees who work in different locations, provided that such differences are not the result of an intention to discriminate because of race, color, religion, sex, or national origin, nor shall it be an unlawful employment practice for an employer to give and to act upon the results of any professionally developed ability test provided that such test, its administration or action upon the results is not designed, intended or used to discriminate because of race, color, religion, sex or national origin. It shall not be an unlawful employment practice under this subchapter for any employer to differentiate upon the basis of sex in determining the amount of the wages or compensation paid or to be paid to employees of such employer if such differentiation is authorized by the provisions of section 206(d) of Title 29.

(i) Nothing contained in this Subchapter shall apply to any business or enterprise on or near an Indian reservation with respect to any publicly announced employment practice of such business or enterprise under which a preferential treatment is given to any individual because he is an Indian living on or near a reservation.

(j) Nothing contained in this Subchapter shall be interpreted to require any employer, employment agency, labor organization, or joint labor-management committee subject to this subchapter to grant preferential treatment to any individual or to any group because of the race, color, religion, sex, or national origin of such individual or group on account of an imbalance which may exist with respect to the total number of percentage of persons of any race, color, religion, sex, or national origin employed by any employer, referred or classified for employment by any employment agency or labor organization, admitted to membership or classified by any labor organization, or admitted to, or employed in, any apprenticeship or other training program, in comparison with the total number or percentage of persons of such race, color, religion, sex, or national origin in any community, State, section, or other area, or in the available work force in any community, State, section, or other area.

## OTHER UNLAWFUL EMPLOYMENT PRACTICES

**Sec. 704.**   (§ 2000e–3) (a) It shall be an unlawful employment practice for an employer to discriminate against any of his employees or applicants for employment, for an employment agency, or joint labor-management committee controlling apprenticeship or other training or retraining, including on-the-job training programs, to discriminate against any individual, or for a labor organization to discriminate against any member thereof or applicant for membership, because he has opposed any practice made an unlawful employment practice by this subchapter, or because he has made a charge, testified, assisted, or participated in any manner in an investigation, proceeding, or hearing under this subchapter.

(b) It shall be an unlawful employment practice for an employer, labor organization, employment agency, or joint labor-management committee controlling apprenticeship or other training or retraining; including on-the-job training programs, to print or publish or cause to be printed or published any notice or advertisement relating to employment by such an employer or membership in or any classification or referral for employment by such a labor organization, or relating to any classification or referral for employment by such an employment agency, or relating to admission to, or employment in, any program established to provide apprenticeship or other training by such a joint labor management committee, indicating any preference, limitation, specification, or discrimination, based on race, color, religion, sex, or national origin, except that such a notice or advertisement may indicate a preference, limitation, specification, or discrimination based on religion, sex, or national origin when religion, sex, or national origin is a bona fide occupational qualification for employment.

## EQUAL EMPLOYMENT OPPORTUNITY COMMISSION

**Sec. 705.** (§ 2000e–4) (a) There is hereby created a Commission to be known as the Equal Employment Opportunity Commission, which shall be composed of five members, not more than three of whom shall be members of the same political party. Members of the Commission shall be appointed by the President by and with the advice and consent of the Senate for a term of five years. Any individual chosen to fill a vacancy shall be appointed only for the unexpired term of the member whom he shall succeed, and all members of the Commission shall continue to serve (1) for more than sixty days when the Congress is in session unless a nomination to fill such vacancy shall have been submitted to the Senate, or (2) after the adjournment sine die of the session of the Senate in which such nomination was submitted. The President shall designate one member to serve as Chairman of the Commission, and one member to serve as Vice Chairman. The Chairman shall be responsible on behalf of the Commission for the administrative operations of the Commission, and, except as provided in subsection (b) of this section, shall appoint, in accordance with the provisions of Title 5, governing appointments in the competitive service, such officers, agents, attorneys, administrative law judges, and employees as he deems necessary to assist it in the performance of its functions and to fix their compensation in accordance with the provisions of chapter 51 and subchapter III of chapter 53 of Title 5, relating to classification and General Schedule pay rates: *Provided,* That assignment, removal, and compensation of administrative law judges, shall be in accordance with sections 3105, 3344, 5362, and 7521 of Title 5.

(b)(1) There shall be a General Counsel of the Commission appointed by the President, by and with the advice and consent of the Senate, for a term of four years. The General Counsel shall have responsibility for the conduct of litigation as provided in sections 706 and 707 of this subchapter. The General Counsel shall have such other duties as the Commission may prescribe or as may be provided by law and shall concur with the Chairman of the Commission on the appointment and supervision of regional attorneys. The General Counsel of the Commission on the effective date of this Act shall continue in such position and perform the functions specified in this subsection until a successor is appointed and qualified.

(2) Attorneys appointed under this section may, at the direction of the Commission, appear for and represent the Commission in any case in court, provided that the Attorney General shall conduct all litigation to which the Commission is a party in the Supreme Court pursuant to this subchapter.

(c) A vacancy in the Commission shall not impair the right of the remaining members to exercise all the powers of the Commission and three members thereof shall constitute a quorum.

(d) The Commission shall have an official seal which shall be judicially noticed.

(e) The Commission shall at the close of each fiscal year report to the Congress and to the President concerning the action it has taken; the names, salaries, and duties of all individuals in its employ and the moneys it has disbursed; and shall make such further reports on the cause of and means of eliminating discrimination and such recommendations for further legislation as may appear desirable.

(f) The principal office of the Commission shall be in or near the District of Columbia, but it may meet or exercise any or all its powers at any other place. The Commission may establish such regional or State offices as it deems necessary to accomplish the purpose of this subchapter.

(g) The Commission shall have power—

(1) to cooperate with an, with their consent, utilize regional, State, local, and other agencies, both public and private, and individuals;

(2) to pay to witnesses whose depositions are taken or who are summoned before the Commission or any of its agents the same witness and mileage fees as are paid to witnesses in the courts of the United States;

(3) to furnish to persons subject to this subchapter such technical assistance as they may request to further their compliance with this subchapter or an order issued thereunder;

(4) upon the request of (i) any employer, whose employees or some of them, or (ii) any labor organization, whose members or some of them, refuse or threaten to refuse to cooperate in effectuating the provisions of this subchapter, to assist in such effectuation by conciliation or such other remedial action as is provided by this subchapter;

(5) to make such technical studies as are appropriate to effectuate the purposes and policies of this subchapter and to make the results of such studies available to the public;

(6) to intervene in a civil action brought under section 706 of this title by an aggrieved party against a respondent other than a government, governmental agency or political subdivision.

(h) The Commission shall, in any of its educational or promotional activities, cooperate with other departments and agencies in the performance of such educational and promotional activities.

(i) All officers, agents, attorneys, and employees of the Commission shall be subject to the provisions of section 118i of Title 5, notwithstanding any exemption contained in such section.

## PREVENTION OF UNLAWFUL EMPLOYMENT PRACTICES

**Sec. 706.** (§ 2000e–5) (a) The Commission is empowered, as hereinafter provided, to prevent any person from engaging in any unlawful employment practice as set forth in section 703 or 704 of this title.

(b) Whenever a charge is filed by or on behalf of a person claiming to be aggrieved, or by a member of the Commission, alleging that an employer, employment agency, labor organization, or joint labor-management committee controlling apprenticeship or other training or retraining, including on-the-job training programs, has engaged in an unlawful employment practice, the Commission shall serve a notice of the charge (including the date, place and circumstances of the alleged unlawful employment practice) on such employer, employment agency, labor organization, or joint labor-management committee (hereinafter referred to as the "respondent") within ten days, and shall make an investigation thereof. Charges shall be in writing under oath or affirmation and shall contain such information and be in such form as the Commission requires. Charges shall not be made public by the Commission. If the Commission determines after such investigation that there is not reasonable cause to believe that the charge is true, it shall dismiss the charge and promptly notify the person claiming to be aggrieved and the respondent of its action. In determining whether reasonable cause exists, the Commission shall accord substantial weight to final findings and orders made by State or local authorities in proceedings commenced under State or local law pursuant to the requirements of subsections (c) and (d) of this section. If the Commission determines after such investigation that there is reasonable cause to believe that the charge is true, the Commission shall endeavor to eliminate any such alleged unlawful employment practice by informal methods of conference, conciliation, and persuasion. Nothing said or done during and as a part of such informal endeavors may be made public by the Commission, its officers or employees, or used as evidence in a subsequent proceeding without the written consent of the persons concerned. Any person who makes public information in violation of this subsection shall be fined not more than $1,000 or imprisoned for not more than one year, or both. The Commission shall make its determination on reasonable cause as promptly as possible and, so far as practicable, not later than one hundred and twenty days from the filing of the charge or, where applicable under subsection (c) or (d) of this section, from the date upon which the Commission is authorized to take action with respect to the charge.

(c) In the case of an alleged unlawful employment practice occurring in a State, or political subdivision of a State, which has a State or local law prohibiting the unlawful employment practice alleged and establishing or authorizing a State or local authority to grant or seek relief from such practice or to institute criminal proceedings with respect thereto upon receiving notice thereof, no charge may be filed under subsection (b) of this section by the person aggrieved before the expiration of sixty days after proceedings have been commenced under the State or local law, unless such proceedings have been earlier terminated, provided that such sixty-day period shall be extended to one hundred and twenty days during the first year after the effective date of such State or local law. If any requirement for the commencement of such proceedings is imposed by a State or local authority other than a requirement of the filing of a written and signed statement of the facts upon which the proceeding is based, the proceeding shall be deemed to have been commenced for the purposes of this subsection at the time such statement is sent by registered mail to the appropriate State or local authority.

(d) In the case of any charge filed by a member of the Commission alleging an unlawful employment practice occurring in a State or political subdivision of a State which has a State or local law prohibiting the practice alleged and establishing or authorizing a State or local authority to grant or seek relief from such practice or to institute criminal proceedings with respect thereto upon receiving notice thereof, the Commission shall, before taking any action with respect to such charge, notify the appropriate State or local officials and, upon request, afford them a reasonable time, but not less than sixty days (provided that such sixty-day period shall be extended to one hundred and twenty days during the first year after the effective day of such State or local law), unless a shorter period is requested, to act under such State or local law to remedy the practice alleged.

(e) A charge under this section shall be filed within one hundred and eighty days after the alleged unlawful employment practice occurred and notice of the charge (including the date, place and circumstances of the alleged unlawful employment practice) shall be served upon the person against whom such charge is made within ten days thereafter, except that in a case of an unlawful employment practice with respect to which the person aggrieved has initially instituted proceedings with a State or local agency with authority to grant or seek relief from such practice or to institute criminal proceedings with respect thereto upon receiving notice thereof, such charge shall be filed by or on behalf of the person aggrieved within three hundred days after the alleged unlawful employment practice occurred, or within thirty days after receiving notice that the State or local agency has terminated the proceedings under the State or local law, whichever is earlier, and a copy of such charge shall be filed by the Commission with the State or local agency.

(f)(1) If within thirty days after a charge is filed with the Commission or within thirty days after expiration of any period of reference under subsection (c) or (d) of this section, the Commission has been unable to secure from the respondent a conciliation agreement acceptable to the Commission, the Commission may bring a civil action against any respondent not a government, governmental agency, or political subdivision named in the charge. In the case of a respondent which is a government, governmental agency, or political subdivision, if the Commission has been unable to secure from the respondent a conciliation agreement acceptable to the Commission, the Commission shall take no further action and shall refer the case to the Attorney General who may bring a civil action against such respondent in the appropriate United States district court. The person or persons aggrieved shall have the right to intervene in a civil action brought by the Commission or the Attorney General in a case involving a government, governmental agency, or political subdivision. If a charge filed with the Commission pursuant to subsection (b) of this section is dismissed by the Commission, or if within one hundred and eighty days from the filing of such charge or the expiration of any period of reference under subsection (c) or (d) of this section, whichever is later, the Commission has not filed a civil action under this section or the Attorney General has not filed a civil action in a case involving a government, governmental agency, or political subdivision, or the Commission has not entered into a conciliation agreement to which the person aggrieved is a party, the Commission, or the Attorney General in a case involving a government, governmental agency, or political subdivision, shall so notify the person aggrieved and within ninety days after the giving of such notice a civil action may be brought against the respondent named in the charge (A) by the person claiming to be aggrieved or (B) if such charge was filed by a member of the Commission, by any person whom the charge alleges was aggrieved by the alleged unlawful employment practice. Upon application by the complainant and in such circumstances as the court may deem just, the court may appoint an attorney for such complainant and may authorize the commencement of the action without the payment of fees, costs, or security. Upon timely application, the court may, in its discretion, permit the Commission, or the Attorney General in a case involving a government, governmental agency, or political subdivision, to intervene in such civil action upon certification that the case is of general public importance. Upon request, the court may, in its discretion, stay further proceedings for not more than sixty days pending the termination of State or local proceedings described in subsection (c) or (d) of this section or further efforts of the Commission to obtain voluntary compliance.

(2) Whenever a charge is filed with the Commission and the Commission concludes on the basis of a preliminary investigation that prompt judicial action is necessary to carry out the purposes of this Act, the Commission, or the Attorney General in a case involving a government, governmental agency, or political subdivision, may bring an action for appropriate temporary or preliminary relief pending final disposition of such charge. Any temporary restraining order or other order granting preliminary or temporary relief shall be issued in accordance with rule 65 of the Federal Rules of Civil Procedure. It shall be the

duty of a court having jurisdiction over proceedings under this section to assign cases for hearing at the earliest practicable date and to cause such cases to be in every way expedited.

(3) Each United States district court and each United States court of a place subject to the jurisdiction of the United States shall have jurisdiction of actions brought under this subchapter. Such an action may be brought in any judicial district in the State in which the unlawful employment practice is alleged to have been committed, in the judicial district in which the employment records relevant to such practice are maintained and administered, or in the judicial district in which the aggrieved person would have worked but for the alleged unlawful employment practice, but if the respondent is not found within any such district, such an action may be brought within the judicial district in which the respondent has his principal office. For purposes of sections 1404 and 1406 of Title 28, the judicial district in which the respondent has his principal office shall in all cases be considered a district in which the action might have been brought.

(4) It shall be the duty of the chief judge of the district (or in his absence, the acting chief judge) in which the case is pending immediately to designate a judge in such district to hear and determine the case. In the event that no judge in the district is available to hear and determine the case, the chief judge of the district, or the acting chief judge, as the case may be, shall certify this fact to the chief judge of the circuit (or in his absence, the acting chief judge) who shall then designate a district or circuit judge of the circuit to hear and determine the case.

(5) It shall be the duty of the judge designated pursuant to this subsection to assign the case for hearing at the earliest practicable date and to cause the case to be in every way expedited. If such judge has not scheduled the case for trial within one hundred and twenty days after issue has been joined, that judge may appoint a master pursuant to rule 53 of the Federal Rules of Civil Procedure.

(g) If the court finds that the respondent has intentionally engaged in or is intentionally engaging in an unlawful employment practice charged in the complaint, the court may enjoin the respondent from engaging in such unlawful employment practice, and order such affirmative action as may be appropriate, which may include, but is not limited to, reinstatement or hiring of employees, with or without back pay (payable by the employer, employment agency, or labor organization, as the case may be, responsible for the unlawful employment practice), or any other equitable relief as the court deems appropriate. Back pay liability shall not accrue from a date more than two years prior to the filing of a charge with the Commission. Interim earnings or amounts earnable with reasonable diligence by the person or persons discriminated against shall operate to reduce the back pay otherwise allowable. No order of the court shall require the admission or reinstatement of an individual as a member of a union, or the hiring, reinstatement, or promotion of an individual as an employee, or the payment to him of any back pay, if such individual was refused admission, suspended, or expelled, or was refused employment or advancement or was suspended or discharged for any reason other than discrimination on account of race, color, religion, sex, or national origin or in violation of section 704 of this title.

(h) The provisions of sections 101 to 115 of Title 29 shall not apply with respect to civil actions brought under this section.

(i) In any case in which an employer, employment agency, or labor organization fails to comply with an order of a court issued in a civil action brought under this section, the Commission may commence proceedings to compel compliance with such order.

(j) Any civil action brought under this section and any proceedings brought under subsection (i) of this section shall be subject to appeal as provided in sections 1291 and 1292, Title 28.

(k) In any action or proceeding under this Title the court, in its discretion, may allow the prevailing party, other than the Commission or the United States, a reasonable attorney's fee as part of the costs, and the Commission and the United States shall be liable for costs the same as a private person.

**Sec. 707.** (§ 2000e–6) (a) Whenever the Attorney General has reasonable cause to believe that any person or group of persons is engaged in a pattern or practice of resistance to the full enjoyment of any of the rights secured by this subchapter, and that the pattern or practice is of such a nature and is intended to deny the full exercise of the rights herein described, the Attorney General may bring a civil action in the appropriate district court of the United States by filing with it a complaint (1) signed by him (or in his absence the Acting Attorney General), (2) setting forth facts pertaining to such pattern or practice, and (3) requesting such relief, including an application for a permanent or temporary injunction, restraining order or other order against the person or persons responsible for such pattern or practice, as he deems necessary to insure the full enjoyment of the rights herein described.

(b) The district courts of the United States shall have and shall exercise jurisdiction of proceedings instituted pursuant to this section, and in any such proceeding the Attorney General may file with the clerk of such court a request that a court of three judges be convened to hear and determine the case. Such request by the Attorney General shall be accompanied by a certificate that, in his opinion, the case is of general public importance. A copy of the certificate and request for a three-judge court shall be immediately furnished by such clerk to the chief judge of the circuit (or in his absence, the presiding circuit judge of the circuit) in which the case is pending. Upon receipt of such request it shall be the duty of the chief judge of the circuit or the presiding circuit judge, as the case may be, to designate immediately three judges in such circuit, of whom at least one shall be a circuit judge and another of whom shall be a district judge of the court in which the proceeding was instituted, to hear and determine such case, and it shall be the duty of the judges so designated to assign the case for hearing at the earliest practicable date, to participate in the hearing and determination thereof, and to cause the case to be in every way expedited. An appeal from the final judgment of such court will lie to the Supreme Court.

In the event the Attorney General fails to file such a request in any such proceeding, it shall be the duty of the chief judge of the district (or in his absence, the acting chief judge) in which the case is pending immediately to designate a judge in such district to hear and determine the case. In the event that no judge in the district is available to hear and determine the case, the chief judge of the district, or the acting chief judge, as the case may be, shall certify this fact to the chief judge of the circuit (or in his absence, the acting chief judge) who shall then designate a district or circuit judge of the circuit to hear and determine the case.

It shall be the duty of the judge designated pursuant to this section to assign the case for hearing at the earliest practicable date and to cause the case to be in every way expedited.

(c) Effective two years after March 24, 1972, the functions of the Attorney General under this section shall be transferred to the Commission, together with such personnel, property, records, and unexpended balances of appropriations, allocations, and other funds employed, used, held, available, or to be made available in connection with such functions unless the President submits, and neither House of Congress vetoes, a reorganization plan pursuant to chapter 9 of Title 5, inconsistent with the provisions of this subsection. The Commission shall carry out such functions in accordance with subsections (d) and (e) of this section.

(d) Upon the transfer of functions provided for in subsection (c) of this section, in all suits commenced pursuant to this section prior to the date of such transfer, proceedings shall continue without abatement, all court orders and decrees shall remain in effect, and the Commission shall be substituted as a party for the United States of America, the Attorney General, or the Acting Attorney General, as appropriate.

(e) Subsequent to March 24, 1972, the Commission shall have authority to investigate and act on a charge of a pattern or practice of discrimination, whether filed by or on behalf of a person claiming to be aggrieved or by a member of the Commission. All such actions shall be conducted in accordance with the procedures set forth in section 706 of this title.

## EFFECT ON STATE LAWS

**Sec. 708.** (§ 2000e–7) Nothing in this subchapter shall be deemed to exempt or relieve any person from any liability, duty, penalty, or punishment provided by any present or future law of any State or political subdivision of a State, other than any such law which purports to require or permit the doing of any act which would be an unlawful employment practice under this subchapter.

## INVESTIGATIONS, INSPECTIONS, RECORDS, STATE AGENCIES

**Sec. 709.** (§ 2000e–8) (a) In connection with any investigation of a charge filed under section 706 of this title, the Commission or its designated representative shall at all reasonable times have access to, for the purposes of examination, and the right to copy any evidence of any person being investigated or proceeded against that relates to unlawful employment practices covered by this subchapter and is relevant to the charge under investigation.

(b) The Commission may cooperate with State and local agencies charged with the administration of State fair employment practices laws and, with the consent of such agencies, may, for the purpose of carrying out its functions and duties under this subchapter and within the limitation of funds appropriated specifically for such purpose, engage in and contribute to the cost of research and other projects of mutual interest undertaken by such agencies, and utilize the services of such agencies and their employees, and, notwithstanding any other provision of law, pay by advance or reimbursement such agencies and their employees for services rendered to assist the Commission in carrying out this subchapter. In furtherance of such cooperative efforts, the Commission may enter into written agreements with such State or local agencies and such agreements may include provisions under which the Commission shall refrain from processing a charge in any cases or class of cases specified in such agreements or under which the Commission shall relieve any person or class of persons in such State or locality from requirements imposed under this section. The Commission shall rescind any such agreement whenever it determines that the agreement no longer serves the interest of effective enforcement of this subchapter.

(c) Every employer, employment agency, and labor organization subject to this Title shall (1) make and keep such records relevant to the determinations of whether unlawful employment practices have been or are being committed, (2) preserve such records for such periods, and (3) make such reports therefrom as the Commission shall prescribe by regulation or order, after public hearing, as reasonable, necessary, or appropriate for the enforcement of this subchapter or the regulations or orders thereunder. The Commission shall, by regulation, require each employer, labor organization, and joint labor-management committee subject to this subchapter which controls an apprenticeship or other training program to maintain such records as are reasonably necessary to carry out the purposes of this subchapter, including, but not limited to, a list of applicants who wish to participate in such program, including the chronological order in which applications were received, and to furnish to the Commission upon request, a detailed description of the manner in which persons are selected to participate in the apprenticeship or other training program. Any employer, employment agency, labor organization, or joint labor-management committee which believes that the application to it of any regulation or order issued under this section would result in undue hardship may apply to the Commission for an exemption from the application of such regulation or order, and, if such application for an exemption is denied, bring a civil action in the United States district court for the district where such records are kept. If the Commission or the court, as the case may be, finds that the application of the regulation or order to the employer, employment agency, or labor organization in question would impose an undue hardship, the Commission or the court, as the case may be, may grant appropriate relief. If any person required to comply with the provisions of this subsection fails or refuses to do so, the United States district court for the district in which such person is found, resides, or transacts business, shall, upon application of the Commission, or the Attorney General in a case involving a government, governmental agency or political subdivision, have jurisdiction to issue to such person an order requiring him to comply.

(d) In prescribing requirements pursuant to subsection (c) of this section, the Commission shall consult with other interested State and Federal agencies and shall endeavor to coordinate its requirements with those adopted by such agencies. The Commission shall furnish upon request and without cost to any State or local agency charged with the administration of a fair employment practice law information obtained pursuant to subsection (c) of this section from any employer, employment agency, labor organization, or joint labor-management committee subject to the jurisdiction of such agency. Such information shall be furnished on condition that it not be made public by the recipient agency prior to the institution of a proceeding under State or local law involving such information. If this condition is violated by a recipient agency, the Commission may decline to honor subsequent requests pursuant to this subsection.

(e) It shall be unlawful for any officer or employee of the Commission to make public in any manner whatever any information obtained by the Commission pursuant to its authority under this section prior to the institution of any proceeding under this subchapter involving such information. Any officer or employee of the Commission who shall make public in any manner whatever any information in violation of this subsection shall be guilty of a misdemeanor and upon conviction thereof, shall be fined not more than $1,000, or imprisoned not more than one year.

## INVESTIGATORY POWERS

**Sec. 710.**   (§ 2000e–9) For the purpose of all hearings and investigations conducted by the Commission or its duly authorized agents or agencies, section 161 of Title 29 shall apply.

## NOTICES TO BE POSTED

**Sec. 711.**   (§ 2000e–10) (a) Every employer, employment agency, and labor organization, as the case may be, shall post and keep posted in conspicuous places upon its premises where notices to employees, applicants for employment, and members are customarily posted a notice to be prepared or approved by the Commission setting forth excerpts from or, summaries of, the pertinent provisions of this subchapter and information pertinent to the filing of a complaint.

(b) A willful violation of this section shall be punishable by a fine of not more than $100 for each separate offense.

## VETERANS' RIGHTS OR PREFERENCE

**Sec. 712.**    (§ 2000e–11) Nothing contained in this subchapter shall be construed to repeal or modify any Federal, State, territorial, or local law creating special rights or preference for veterans.

## RULES AND REGULATIONS

**Sec. 713.**    (§ 2000e–12) (a) The Commission shall have authority from time to time to issue, amend, or rescind suitable procedural regulations to carry out the provisions of this subchapter. Regulations issued under this section shall be in conformity with the standards and limitations of the Administrative Procedure Act.

(b) In any action or proceeding based on any alleged unlawful employment practice, no person shall be subject to any liability or punishment for or on account of (1) the commission by such person of an unlawful employment practice if he pleads and proves that the act or omission complained of was in good faith, in conformity with, and in reliance on any written interpretation or opinion of the Commission, or (2) the failure of such person to publish and file any information required by any provision of this subchapter if he pleads and proves that he failed to publish and file such information in good faith, in conformity with the instructions of the Commission issued under this Title regarding the filing of such information. Such a defense, if established, shall be a bar to the action or proceeding, notwithstanding that (A) after such act or omission, such interpretation or opinion is modified or rescinded or is determined by judicial authority to be invalid or of no legal effect, or (B) after publishing or filing the description and annual reports, such publication or filing is determined by judicial authority not to be in conformity with the requirements of this Title.

## FORCIBLY RESISTING THE COMMISSION OR ITS REPRESENTATIVES

**Sec. 714.**    (§ 2000e–13) The provisions of sections 111 and 1114, Title 18, shall apply to officers, agents, and employees of the Commission in the performance of their official duties. Notwithstanding the provisions of sections 111 and 1114 of Title 18, whoever in violation of the provisions of section 114 of such title kills a person while engaged in or on account of the performance of his official functions under this Act shall be punished by imprisonment for any term of years or for life.

## COORDINATION OF EFFORTS

**Sec. 715.**    (§ 2000e–14) The Equal Employment Opportunity Commission shall have the responsibility for developing and implementing agreements, policies and practices designed to maximize effort, promote efficiency, and eliminate conflict, competition, duplication and inconsistency among the operations, functions and jurisdictions of the various departments, agencies and branches of the Federal Government responsible for the implementation and enforcement of equal employment opportunity legislation, orders, and policies. On or before October 1 of each year, the Equal Employment Opportunity Commission shall transmit to the President and to the Congress a report of its activities, together with such recommendations for legislative or administrative changes as it concludes are desirable to further promote the purposes of this section.

## PRESIDENTIAL CONFERENCES

**Sec. 716.** (§ 2000e–15) The President shall, as soon as feasible after July 2, 1964, convene one or more conferences for the purpose of enabling the leaders of groups whose members will be affected by this Title to become familiar with the rights afforded and obligations imposed by its provisions, and for the purpose of making plans which will result in the fair and effective administration of this Title when all of its provisions become effective. The President shall invite the participation in such conference or conferences of (1) the members of the President's Committee on Equal Employment Opportunity, (2) the members of the Commission on Civil Rights, (3) representatives of State and local agencies engaged in furthering equal employment opportunity, (4) representatives of private agencies engaged in furthering equal employment opportunity, and (5) representatives of employers, labor organizations, and employment agencies who will be subject to this Title.

## NON-DISCRIMINATION IN FEDERAL GOVERNMENT EMPLOYMENT

**Sec. 717.** (§ 2000e–16) (a) All personnel actions affecting employees or applicants for employment (except with regard to aliens employed outside the limits of the United States) in military departments as defined in section 102 of Title 5, in executive agencies (other than the General Accounting Office) as defined in section 105 of Title 5 (including employees and applicants for employment who are paid from nonappropriated funds), in the United States Postal Service and the Postal Rate Commission, in those units of the Government of the District of Columbia having positions in the competitive service, and in those units of the legislative and judicial branches of the Federal Government having positions in the competitive service, and in the Library of Congress shall be made free from any discrimination based on race, color, religion, sex, or national origin.

(b) Except as otherwise provided in this subsection, the Equal Employment Opportunity Commission shall have authority to enforce the provisions of subsection (a) of this section through appropriate remedies, including reinstatement or hiring of employees with or without back pay, as will effectuate the policies of this section, and shall issue such rules, regulations, orders and instructions as it deems necessary and appropriate to carry out its responsibilities under this section. The Equal Employment Opportunity Commission shall—

(1) be responsible for the annual review and approval of a national and regional equal employment opportunity plan which each department and agency and each appropriate unit referred to in subsection (a) of this section shall submit in order to maintain an affirmative program of equal employment opportunity for all such employees and applicants for employment;

(2) be responsible for the review and evaluation of the operation of all agency equal employment opportunity programs, periodically obtaining and publishing (on at least a semiannual basis) progress reports from each such department, agency, or unit; and

(3) consult with and solicit the recommendations of interested individuals, groups, and organizations relating to equal employment opportunity.

The head of each such department, agency, or unit shall comply with such rules, regulations, orders, and instructions which shall include a provision that an employee or applicant for employment shall be notified of any final action taken on any complaint of discrimination filed by him thereunder. The plan submitted by each department, agency, and unit shall include, but not be limited to—

(1) provision for the establishment of training and education programs designed to provide a maximum opportunity for employees to advance so as to perform at their highest potential; and

(2) a description of the qualifications in terms of training and experience relating to equal employment opportunity for the principal and operating officials of each such department, agency, or unit responsible for carrying out the equal employment opportunity program and of the allocation of personnel and resources proposed by such department, agency, or unit to carry out its equal employment opportunity program.

With respect to employment in the Library of Congress, authorities granted in this subsection to the Equal Employment Opportunity Commission shall be exercised by the Librarian of Congress.

(c) Within thirty days of receipt of notice of final action taken by a department, agency, or unit referred to in subsection (a) of this section, or by the Equal Employment Opportunity Commission upon an appeal from a decision or order of such department, agency, or unit on a complaint of discrimination based on race, color, religion, sex or national origin, brought pursuant to subsection (a) of this section, Executive Order 11478 or any succeeding Executive orders, or after one hundred and eighty days from the filing of the initial charge with the department, agency, or unit or with the Equal Employment Opportunity Commission on appeal from a decision or order of such department, agency, or unit until such time as final action may be taken by a department agency, or unit, an employee or applicant for employment, if aggrieved by the final disposition of his complaint, or by the failure to take final action on his complaint, may file a civil action as provided in section 2000e–5 of this title, in which civil action the head of the department, agency, or unit, as appropriate, shall be the defendant.

(d) The provisions of section 705(f) through (k) of this title, as applicable, shall govern civil actions brought hereunder.

(e) Nothing contained in this Act shall relieve any Government agency or official of its or his primary responsibility to assure nondiscrimination in employment as required by the Constitution and statutes or of its or his responsibilities under Executive Order 11478 relating to equal employment opportunity in the Federal Government.

## SPECIAL PROVISION WITH RESPECT TO DENIAL, TERMINATION AND SUPERVISION OF GOVERNMENT CONTRACTS

**Sec. 718.**    (§ 2000e–17) No Government contract, or portion thereof, with any employer, shall be denied, withheld, terminated, or suspended, by any agency or officer of the United States under any equal employment opportunity law or order, where such employer has an affirmative action plan which has previously been accepted by the Government for the same facility within the past twelve months without first according such employer full hearing and adjudication under the provisions of section 554 of Title 5, and the following pertinent sections: *Provided,* That if such employer has deviated substantially from such previously agreed to affirmative action plan, this section shall not apply: *Provided further,* That for the purposes of this section an affirmative action plan shall be deemed to have been accepted by the Government at the time the appropriate compliance agency has accepted such plan unless within forty-five days thereafter the Office of Federal Contract Compliance has disapproved such plan.

## STATUTE

**XVII.    Text of Civil Rights Act of 1991**

The following material is the text of the Civil Rights Act of 1991.*

## § 1.  SHORT TITLE.

This Act may be cited as the "Civil Rights Act of 1991."

## § 2.  FINDINGS.

The Congress finds that—

(1) additional remedies under Federal law are needed to deter unlawful harassment and intentional discrimination in the workplace;

(2) the decision of the Supreme Court in Wards Cove Packing Co. v. Atonio, 490 U.S. 642 (1989) has weakened the scope and effectiveness of Federal civil rights protections; and

(3) legislation is necessary to provide additional protections against unlawful discrimination in employment.

## § 3.  PURPOSES.

The purposes of this Act are—

(1) to provide appropriate remedies for intentional discrimination and unlawful harassment in the workplace;

(2) to codify the concepts of "business necessity" and "job related" enunciated by the Supreme Court in Griggs v. Duke Power Co., 401 U.S. 424 (1971), and in the other Supreme Court decisions prior to Wards Cove Packing Co. v. Atonio, 490 U.S. 642 (1989);

(3) to confirm statutory authority and provide statutory guidelines for the adjudication of disparate impact suits under title VII of the Civil Rights Act of 1964 (42 U.S.C. 2000e et seq.); and

(4) to respond to recent decisions of the Supreme Court by expanding the scope of relevant civil rights statutes in order to provide adequate protection to victims of discrimination.

### TITLE I—FEDERAL CIVIL RIGHTS REMEDIES

## § 101.  PROHIBITION AGAINST ALL RACIAL DISCRIMINATION IN THE MAKING AND ENFORCEMENT OF CONTRACTS.

Section 1977 of the Revised Statutes (42 U.S.C. 1981) is amended—

(1) by inserting '(a)' before 'All persons within'; and

(2) by adding at the end of the following new subsections:

"(b) For purposes of this section, the term 'make and enforce contracts' includes the making, performance, modification, and termination of contracts, and the enjoyment of all benefits, privileges, terms and conditions of the contractual relationship."

"(c) The rights protected by this section are protected against impairment by nongovernmental discrimination and impairment under color of State law."

---

*This material was obtained through the services of LEGISLATE.

## § 102.  DAMAGES IN CASES OF INTENTIONAL DISCRIMINATION.

The Revised Statutes are amended by inserting after section 1977 (42 U.S.C. 1981) the following new section:

"§ 1977A. DAMAGES IN CASES OF INTENTIONAL DISCRIMINATION IN EMPLOYMENT.

"(a) Right of Recovery.—

"(1) Civil rights.—In an action brought by a complaining party under section 706 or 717 of the Civil Rights Act of 1964 (42 U.S.C. 2000e-5) against a respondent who engaged in unlawful intentional discrimination (not an employment practice that is unlawful because of its disparate impact) prohibited under section 703, 704, or 717 of the Act (42 U.S.C. 2000e-2 or 2000e-3), and provided that the complaining party cannot recover under section 1977 of the Revised Statutes (42 U.S.C. 1981), the complaining party may recover compensatory and punitive damages as allowed in subsection (b), in addition to any relief authorized by section 706(g) of the Civil Rights Act of 1964, from the respondent.

"(2) Disability.—In an action brought by a complaining party under the powers, remedies, and procedures set forth in section 706 or 717 of the Civil Rights Act of 1964 (as provided in section 107(a) of the Americans with Disabilities Act of 1990 (42 U.S.C. 12117(a)), and section 505(a)(1) of the Rehabilitation Act of 1973 (20 U.S.C. 794a(a)(1)), respectively) against a respondent who engaged in unlawful intentional discrimination (not an employment practice that is unlawful because of its disparate impact) under section 501 of the Rehabilitation Act of 1973 (20 U.S.C. 791) and the regulations implementing section 501, or who violated the requirements of section 501 of the Act or the regulations implementing section 501 concerning the provision of a reasonable accommodation, or section 102 of the Americans with Disabilities Act of 1990 (42 U.S.C. 12112), or committed a violation of section 102(b)(5) of the Act, against an individual, the complaining party may recover compensatory and punitive damages as allowed in subsection (b), in addition to any relief authorized by section 706(g) of the Civil Rights Act of 1964, from the respondent.

"(3) Reasonable accommodation and good faith effort.—In cases where a discriminatory practice involves the provision of a reasonable accommodation pursuant to section 102(b)(5) of the Americans with Disabilities Act of 1990 or regulations implementing section 501 of the Rehabilitation Act of 1973, damages may not be awarded under this section where the covered entity demonstrates good faith efforts, in consultation with the person with the disability who has informed the covered entity that accommodation is needed, to identify and make a reasonable accommodation that would provide such individual with an equally effective opportunity and would not cause an undue hardship on the operation of the business.

"(b) Compensatory and Punitive Damages.—

"(1) Determination of punitive damages.—A complaining party may recover punitive damages under this section against a respondent (other than a government, government agency or political subdivision) if the complaining party demonstrates that the respondent engaged in a discriminatory practice or discriminatory practices with malice or with reckless indifference to the federally protected rights of an aggrieved individual.

"(2) Exclusions from compensatory damages.—Compensatory damages awarded under this section shall not include backpay, interest on backpay, or any other type of relief authorized under section 706(g) of the Civil Rights Act of 1964.

"(3) Limitations.—The sum of the amount of compensatory damages awarded under this section for future pecuniary losses, emotional pain, suffering, inconvenience, mental anguish, loss of enjoyment of life, and other nonpecuniary losses, and the amount of punitive damages awarded under this section, shall not exceed, for each complaining party—

"(A) in the case of a respondent who has more than 14 and fewer than 101 employees in each of 20 or more calendar weeks in the current or preceding calendar year, $50,000;

"(B) in the case of a respondent who has more than 100 and fewer than 201 employees in each of 20 or more calendar weeks in the current or preceding calendar year, $100,000; and

"(C) in the case of a respondent who has more than 200 and fewer than 501 employees in each of 20 or more calendar weeks in the current or preceding calendar year, $200,000; and

"(D) in the case of a respondent who has more than 500 employees in each of 20 or more calendar weeks in the current or preceding calendar year, $300,000.

"(4) Construction.—Nothing in this section shall be construed to limit the scope of, or the relief available under, section 1977 of the Revised Statutes (42 U.S.C. 1981).

"(c) Jury Trial.—If a complaining party seeks compensatory or punitive damages under this section—

"(1) any party may demand a trial by jury; and

"(2) the court shall not inform the jury of the limitations described in subsection (b)(3).

"(d) Definitions.—As used in this section:

"(1) Complaining party.—The term 'complaining party' means—

"(A) in the case of a person seeking to bring an action under subsection (a)(1), the Equal Employment Opportunity Commission, the Attorney General, or a person who may bring an action or proceeding under title VII of the Civil Rights Act of 1964 (42 U.S.C. 2000e et seq.); or

"(B) in the case of a person seeking to bring an action under subsection (a)(2), the Equal Employment Opportunity Commission, the Attorney General, a person who may bring an action or proceeding under section 505(a)(1) of the Rehabilitation Act of 1973 (29 U.S.C. 794a(a)(1)), or a person who may bring an action or proceeding under title I of the Americans with Disabilities Act of 1990 (42 U.S.C. 12101 et seq.).

"(2) Discriminatory practice.—The term 'discriminatory practice' means the discrimination described in paragraph (1), or the discrimination or the violation described in paragraph (2), of subsection (a).

## § 102.  ATTORNEY'S FEES.

The last sentence of section 722 of the Revised Statutes (42 U.S.C. 1988) is amended by inserting ", 1977A" after "1977."

## § 104.  DEFINITIONS.

Section 701 of the Civil Rights Act of 1964 (42 U.S.C. 2000e) is amended by adding at the end the following new subsections:

"(l) The term 'complaining party' means the Commission, the Attorney General, or a person who may bring an action or proceeding under this title.

"(m) The term 'demonstrates' means meets the burdens of production and persuasion.

"(n) The term 'respondent' means an employer, employment agency, labor organization, joint labor-management committee controlling apprenticeship or other training or retraining program, including an on-the-job training program, or Federal entity subject to section 717."

## § 105.   BURDEN OF PROOF IN DISPARATE IMPACT CASES.

(a) Section 703 of the Civil Rights Act of 1964 (42 U.S.C. 2000e-2) is amended by adding at the end of the following new subsection:

"(k)(l)(A) An unlawful employment practice based on disparate impact is established under this title only if—

"(i) a complaining party demonstrates that a respondent uses a particular employment practice that causes a disparate impact on the basis of race, color, religion, sex, or national origin and the respondent fails to demonstrate that the challenged practice is job related for the position in question and consistent with business necessity; or

"(ii) the complaining party makes the demonstration described in subparagraph (C) with respect to an alternative employment practice and the respondent refuses to adopt such alternative employment practice.

"(B)(i) With respect to demonstrating that a particular employment practice causes a disparate impact as described in subparagraph (A)(i), the complaining party shall demonstrate that each particular challenged employment practice causes a disparate impact, except that if the complaining party can demonstrate to the court that the elements of a respondent's decisionmaking process are not capable of separation for analysis, the decisionmaking process may be analyzed as one employment practice.

"(ii) If the respondent demonstrates that a specific employment practice does not cause the disparate impact, the respondent shall not be required to demonstrate that such practice is required by business necessity.

"(C) The demonstration referred to by subparagraph (A)(ii) shall be in accordance with the law as it existed on June 4, 1989, with respect to the concept of 'alternative employment practice'.

"(2) A demonstration that an employment practice is required by business necessity may not be used as a defense against a claim of intentional discrimination under this title.

"(3) Notwithstanding any other provision of this title, a rule barring the employment of an individual who currently and knowingly uses or possesses a controlled substance, as defined in schedules I and II of section 102(6) of the Controlled Substances Act (21 U.S.C. 802(6)), other than the use or possession of a drug taken under the supervision of a licensed health care professional, or any other use or possession authorized by the Controlled Substances Act or any other provision of Federal law, shall be considered an unlawful employment practice under this title only if such rule is adopted or applied with an intent to discriminate because of race, color, religion, sex, or national origin.".

(b) No statements other than the interpretive memorandum appearing in Vol. 137 Congressional Record S 15276 (daily ed. Oct. 25, 1991) shall be considered legislative history of, or relied upon in any way as legislative history in construing or applying, any provision of this Act that relates to Wards Cove—Business necessity/cumulation/alternative business practice.

## § 106.    PROHIBITION AGAINST DISCRIMINATORY USE OF TEST SCORES.

Section 703 of the Civil Rights Act of 1964 (42 U.S.C. 2000e-2) (as amended by section 105) is further amended by adding at the end the following new subsection:

"(l) It shall be an unlawful employment practice for a respondent, in connection with the selection or referral of applicants or candidates for employment or promotion, to adjust the scores of, use different cutoff scores for, or otherwise alter the results of, employment related tests on the basis of race, color, religion, sex, or national origin."

## § 107.    CLARIFYING PROHIBITION AGAINST IMPERMISSIBLE CONSIDERATION OF RACE, COLOR, RELIGION, SEX, OR NATIONAL ORIGIN IN EMPLOYMENT PRACTICES.

(a) In General.—Section 703 of the Civil Rights Act of 1964 (42 U.S.C. 2000e-2) (as amended by sections 105 and 106) is further amended by adding at the end the following new subsection:

"(m) Except as otherwise provided in this title, an unlawful employment practice is established when the complaining party demonstrates that race, color, religion, sex, or national origin was a motivating factor for any employment practice, even though other factors also motivated the practice."

(b) Enforcement Provisions.—Section 706(g) of such Act (42 U.S.C. 2000e-5(g)) is amended—

(1) by designating the first through third sentences as paragraph (1);

(2) by designating the fourth sentence as paragraph (2)(A) and indenting accordingly; and

(3) by adding at the end the following new subparagraph:

"(B) On a claim in which an individual proves a violation under section 703(m) and a respondent demonstrates that the respondent would have taken the same action in the absence of the impermissible motivating factor, the court—

"(i) may grant declaratory relief, injunctive relief (except as provided in clause (ii)), and attorney's fees and costs demonstrated to be directly attributable only to the pursuit of a claim under section 703(m); and

"(ii) shall not award damages or issue an order requiring any admission, reinstatement, hiring, promotion, or payment, described in subparagraph (A)."

## § 108.    FACILITATING PROMPT AND ORDERLY RESOLUTION OF CHALLENGES TO EMPLOYMENT PRACTICES IMPLEMENTING LITIGATED OR CONSENT JUDGMENTS OR ORDERS.

Section 703 of the Civil Rights Act of 1964 (42 U.S.C. 2000e-2) (as amended by sections 105, 106, and 107 of this title) is further amended by adding at the end the following new subsection:

"(n)(1)(A) Notwithstanding any other provision of law, and except as provided in paragraph (2), an employment practice that implements and is within the scope of a litigated or consent judgment or order that resolves a claim of employment discrimination under the Constitution or Federal civil rights laws may not be challenged under the circumstances described in subparagraph (B).

"(B) A practice described in subparagraph (A) may not be challenged in a claim under the Constitution or Federal civil rights laws—

"(i) by a person who, prior to the entry of the judgment or order described in subparagraph (A), had—

"(I) actual notice of the proposed judgment or order sufficient to apprise such person that such judgment or order might adversely affect the interests and legal rights of such person and that an opportunity was available to present objections to such judgment or order by a future date certain; and

"(II) a reasonable opportunity to present objections to such judgment or order; or

"(ii) by a person whose interests were adequately represented by another person who had previously challenged the judgment or order on the same legal grounds and with a similar factual situation, unless there has been an intervening change in law or fact.

"(2) Nothing in this subsection shall be construed to—

"(A) alter the standards for intervention under rule 24 of the Federal Rules of Civil Procedure or apply to the rights of parties who have successfully intervened pursuant to such rule in the proceeding in which the parties intervened;

"(B) apply to the rights of parties to the action in which a litigated or consent judgment or order was entered, or of members of a class represented or sought to be represented in such action, or of members of a group on whose behalf relief was sought in such action by the Federal Government;

"(C) prevent challenges to a litigated or consent judgment or order on the ground that such judgment or order was obtained through collusion or fraud, or is transparently invalid or was entered by a court lacking subject matter jurisdiction; or

"(D) authorize or permit the denial to any person of the due process of law required by the Constitution.

"(3) Any action not precluded under this subsection that challenges an employment consent judgment or order described in paragraph (1) shall be brought in the court, and if possible before the judge, that entered such judgment or order. Nothing in this subsection shall preclude a transfer of such action pursuant to section 1404 of title 28, United States Code."

## § 109.  PROTECTION OF EXTRATERRITORIAL EMPLOYMENT.

(a) Definition of Employee.—Section 701(f) of the Civil Rights Act of 1964 (42 U.S.C. 2000e(f)) and section 101(4) of the Americans with Disabilities Act of 1990 (42 U.S.C. 12111(4)) are each amended by adding at the end the following: "With respect to employment in a foreign country, such term includes an individual who is a citizen of the United States.".

(b) Exemption.—

(1) Civil Rights Act of 1964.—Section 702 of the Civil Rights Act of 1964 (42 U.S.C. 2000e-1) is amended—

(A) by inserting "(a)" after "§ 702."; and

(B) by adding at the end the following:

"(b) It shall not be unlawful under section 703 or 704 for an employer (or a corporation controlled by an employer), labor organization, employment agency, or joint labor-management committee controlling apprenticeship or other training or retraining (including on-the-job training programs) to take any action otherwise prohibited by such section, with respect to an employee in a workplace in a foreign country if compliance with such section would cause such employer (or such corporation), such organization, such agency, or such committee to violate the law of the foreign country in which such workplace is located.

"(c)(1) If an employer controls a corporation whose place of incorporation is a foreign country, any practice prohibited by section 703 or 704 engaged in by such corporation shall be presumed to be engaged in by such employer.

"(2) Sections 703 and 704 shall not apply with respect to the foreign operations of an employer that is a foreign person not controlled by an American employer.

"(3) For purposes of this subsection, the determination of whether an employer controls a corporation shall be based on—

"(A) the interrelation of operations;

"(B) the common management;

"(C) the centralized control of labor relations; and

"(D) the common ownership or financial control, of the employer and the corporation."

(2) Americans with disabilities act of 1990.—Section 102 of the Americans with Disabilities Act of 1990 (42 U.S.C. 12112) is amended—

(A) by redesignating subsection (c) as subsection (d); and

(B) by inserting after subsection (b) the following new subsection:

"(c) Covered Entities in Foreign Countries.—

"(1) In general.—It shall not be unlawful under this section for a covered entity to take any action that constitutes discrimination under this section with respect to an employee in a workplace in a foreign country if compliance with this section would cause such covered entity to violate the law of the foreign country in which such workplace is located.

"(2) Control of corporation.—

"(A) Presumption.—If an employer controls a corporation whose place of incorporation is a foreign country, any practice that constitutes discrimination under this section and is engaged in by such corporation shall be presumed to be engaged in by such employer.

"(B) Exception.—This section shall not apply with respect to the foreign operations of an employer that is a foreign person not controlled by an American employer.

"(C) Determination.—For purposes of this paragraph, the determination of whether an employer controls a corporation shall be based on—"(i) the interrelation of operations; "(ii) the common management; "(iii) the centralized control of labor relations; and "(iv) the common ownership or financial control, of the employer and the corporation.".

(c) Application of Amendments.—The amendments made by this section shall not apply with respect to conduct occurring before the date of the enactment of this Act.

## § 110.  TECHNICAL ASSISTANCE TRAINING INSTITUTE.

(a) Technical Assistance.—Section 705 of the Civil Rights Act of 1964 (42 U.S.C. 2000e-4) is amended by adding at the end the following new subsection:

"(j)(1) The Commission shall establish a Technical Assistance Training Institute, through which the Commission shall provide technical assistance and training regarding the laws and regulations enforced by the Commission.

"(2) An employer or other entity covered under this title shall not be excused from compliance with the requirements of this title because of any failure to receive technical assistance under this subsection.

"(3) There are authorized to be appropriated to carry out this subsection such sums as may be necessary for fiscal year 1992.".

(b) Effective Date.—The amendment made by this section shall take effect on the date of the enactment of this Act.

## § 111.  EDUCATION AND OUTREACH.

Section 705(h) of the Civil Rights Act of 1964 (42 U.S.C. 2000e-4(h)) is amended—
   (1) by inserting "(1)" after "(h)"; and
   (2) by adding at the end the following new paragraph:

"(2) In exercising its powers under this title, the Commission shall carry out educational and outreach activities (including dissemination of information in languages other than English) targeted to—

   "(A) individuals who historically have been victims of employment discrimination and have not been equitably served by the Commission; and

   "(B) individuals on whose behalf the Commission has authority to enforce any other law prohibiting employment discrimination, concerning rights and obligations under this title or such law, as the case may be."

## § 112.  EXPANSION OF RIGHT TO CHALLENGE DISCRIMINATORY SENIORITY SYSTEMS.

Section 706(e) of the Civil Rights Act of 1964 (42 U.S.C. 2000e-5(e)) is amended—
   (1) by inserting "(1)" before "A charge under this section"; and
   (2) by adding at the end the following new paragraph:

"(2) For purposes of this section, an unlawful employment practice occurs, with respect to a seniority system that has been adopted for an intentionally discriminatory purpose in violation of this title (whether or not that discriminatory purpose is apparent on the face of the seniority provision), when the seniority system is adopted, when an individual becomes subject to the seniority system, or when a person aggrieved is injured by the application of the seniority system or provision of the system."

## § 113.  AUTHORIZING AWARD OF EXPERT FEES.

(a) Revised Statutes.—Section 722 of the Revised Statutes is amended—
   (1) by designating the first and second sentences as subsections (a) and (b), respectively, and indenting accordingly; and
   (2) by adding at the end the following new subsection:

"(c) In awarding an attorney's fee under subsection (b) in any action or proceeding to enforce a provision of sections 1977 or 1977A of the Revised Statutes, the court, in its discretion, may include expert fees as part of the attorney's fee."

(b) Civil Rights Act of 1964.—Section 706(k) of the Civil Rights Act of 1964 (42 U.S.C. 2000e-5(k)) is amended by inserting "(including expert fees)" after "attorney's fee."

## § 114.  PROVIDING FOR INTEREST AND EXTENDING THE STATUTE OF LIMITATIONS IN ACTIONS AGAINST THE FEDERAL GOVERNMENT.

Section 717 of the Civil Rights Act of 1964 (42 U.S.C. 2000e-16) is amended—
   (1) in subsection (c), by striking "thirty days" and inserting "90 days"; and

(2) in subsection (d), by inserting before the period ", and the same interest to compensate for delay in payment shall be available as in cases involving nonpublic parties."

## § 115.  NOTICE OF LIMITATIONS PERIOD UNDER THE AGE DISCRIMINATION IN EMPLOYMENT ACT OF 1967.

Section 7(e) of the Age Discrimination in Employment Act of 1967 (29 U.S.C. 626(e)) is amended—

(1) by striking paragraph (2);

(2) by striking the paragraph designation in paragraph (1);

(3) by striking "Sections 6 and" and inserting "Section"; and

(4) by adding at the end the following: "If a charge filed with the Commission under this Act is dismissed or the proceedings of the Commission are otherwise terminated by the Commission, the Commission shall notify the person aggrieved. A civil action may be brought under this section by a person defined in section 11(a) against the respondent named in the charge within 90 days after the date of the receipt of such notice."

## § 116.  LAWFUL COURT-ORDERED REMEDIES, AFFIRMATIVE ACTION, AND CONCILIATION AGREEMENTS NOT AFFECTED.

Nothing in the amendments made by this title shall be construed to affect court-ordered remedies, affirmative action, or conciliation agreements, that are in accordance with the law.

## § 117.  COVERAGE OF HOUSE OF REPRESENTATIVES AND THE AGENCIES OF THE LEGISLATIVE BRANCH.

(a) Coverage of the House of Representatives.—

(1) In general.—Notwithstanding any provision of title VII of the Civil Rights Act of 1964 (42 U.S.C. 2000e et seq.) or of other law, the purposes of such title shall, subject to paragraph (2), apply in their entirety to the House of Representatives.

(2) Employment in the house.—

(A) Application.—The rights and protections under title VII of the Civil Rights Act of 1964 (42 U.S.C. 2000e et seq.) shall, subject to subparagraph (B), apply with respect to any employee in an employment position in the House of Representatives and any employing authority of the House of Representatives.

(B) Administration.—(i) In general.—In the administration of this paragraph, the remedies and procedures made applicable pursuant to the resolution described in clause (ii) shall apply exclusively.

(ii) Resolution.—The resolution referred to in clause (i) is the Fair Employment Practices Resolution (House Resolution 558 of the One Hundredth Congress, as agreed to October 4, 1988), as incorporated into the Rules of the House of Representatives of the One Hundred Second Congress as Rule LI, or any other provision that continues in effect the provisions of such resolution.

(C) Exercise of rulemaking power.—The provisions of subparagraph (B) are enacted by the House of Representatives as an exercise of the rulemaking power of the House of Representatives, with full recognition of the right of the House to change its rules, in the same manner, and to the same extent as in the case of any rule of the House.

(b) Instrumentalities of Congress.—

(1) In general.—The rights and protections under this title and title VII of the Civil Rights Act of 1964 (42 U.S.C. 2000e et seq.) shall, subject to paragraph (2), apply with respect to the conduct of each instrumentality of the Congress.

(2) Establishment of remedies and procedures by instrumentalities.—The chief official of each instrumentality of the Congress shall establish remedies and procedures to be utilized with respect to the rights and protections provided pursuant to paragraph (1). Such remedies and procedures shall apply exclusively, except for the employees who are defined as Senate employees, in section 301(c)(1).

(3) Report to Congress.—The chief official of each instrumentality of the Congress shall, after establishing remedies and procedures for purposes of paragraph (2), submit to the Congress a report describing the remedies and procedures.

(4) Definition of instrumentalities.—For purposes of this section, instrumentalities of the Congress include the following: the Architect of the Capitol, the Congressional Budget Office, the General Accounting Office, the Government Printing Office, the Office of Technology Assessment, and the United States Botanic Garden.

(5) Construction.—Nothing in this section shall alter the enforcement procedures for individuals protected under section 717 of title VII for the Civil Rights Act of 1964 (42 U.S.C. 2000e-16).

## § 118.   ALTERNATIVE MEANS OF DISPUTE RESOLUTION.

Where appropriate and to the extent authorized by law, the use of alternative means of dispute resolution, including settlement negotiations, conciliation, facilitation, mediation, factfinding, minitrials, and arbitration, is encouraged to resolve disputes arising under the Acts or provisions of Federal law amended by this title.

## TITLE II—GLASS CEILING.

## § 201.   SHORT TITLE.

This title may be cited as the "Glass Ceiling Act of 1991."

## § 202.   FINDINGS AND PURPOSE.

(a) Findings.—Congress finds that—

(1) despite a dramatically growing presence in the workplace, women and minorities remain underrepresented in management and decisionmaking positions in business;

(2) artificial barriers exist to the advancement of women and minorities in the workplace;

(3) United States corporations are increasingly relying on women and minorities to meet employment requirements and are increasingly aware of the advantages derived from a diverse work force;

(4) the "Glass Ceiling Initiative" undertaken by the Department of Labor, including the release of the report entitled "Report on the Glass Ceiling Initiative," has been instrumental in raising public awareness of—

(A) the underrepresentation of women and minorities at the management and decisionmaking levels in the United States work force;

(B) the underrepresentation of women and minorities in line functions in the United States work force;

(C) the lack of access for qualified women and minorities to credential-building developmental opportunities; and

(D) the desirability of eliminating artificial barriers to the advancement of women and minorities to such levels;

(5) the establishment of a commission to examine issues raised by the Glass Ceiling Initiative would help—

(A) focus greater attention on the importance of eliminating artificial barriers to the advancement of women and minorities to management and decisionmaking positions in business; and

(B) promote work force diversity;

(6) a comprehensive study that includes analysis of the manner in which management and decisionmaking positions are filled, the developmental and skill-enhancing practices used to foster the necessary qualifications for advancement, and the compensation programs and reward structures utilized in the corporate sector would assist in the establishment of practices and policies promoting opportunities for, and eliminating artificial barriers to, the advancement of women and minorities to management and decisionmaking positions; and

(7) a national award recognizing employers whose practices and policies promote opportunities for, and eliminate artificial barriers to, the advancement of women and minorities will foster the advancement of women and minorities into higher level positions by—

(A) helping to encourage United States companies to modify practices and policies to promote opportunities for, and eliminate artificial barriers to, and upward mobility of women and minorities; and

(B) providing specific guidance for other United States employers that wish to learn how to revise practices and policies to improve the access and employment opportunities of women and minorities.

(b) Purpose.—The purpose of this title is to establish—

(1) a Glass Ceiling Commission to study—

(A) the manner in which business fills management and decisionmaking positions;

(B) the developmental and skill-enhancing practices used to foster the necessary qualifications for advancement into such positions; and

(C) the compensation programs and reward structures currently utilized in the workplace; and (2) an annual award for excellence in promoting a more diverse skilled work force at the management and decisionmaking levels in business.

## § 203.   ESTABLISHMENT OF GLASS CEILING COMMISSION.

(a) In General.—There is established a Glass Ceiling Commission (referred to in this title as the "Commission"), to conduct a study and prepare recommendations concerning—

(1) eliminating artificial barriers to the advancement of women and minorities; and

(2) increasing the opportunities and developmental experiences of women and minorities to foster advancement of women and minorities to management and decisionmaking positions in business.

(b) Membership.—

(1) Composition.—The Commission shall be composed of 21 members, including—

(A) six individuals appointed by the President;

(B) six individuals appointed jointly by the Speaker of the House of Representatives and the Majority Leader of the Senate;

(C) one individual appointed by the Majority Leader of the House of Representatives;

(D) one individual appointed by the Minority Leader of the House of Representatives;

(E) one individual appointed by the Majority Leader of the Senate;

(F) one individual appointed by the Minority Leader of the Senate;

(G) two Members of the House of Representatives appointed jointly by the Majority Leader and the Minority Leader of the House of Representatives;

(H) two Members of the Senate appointed jointly by the Majority Leader and the Minority Leader of the Senate; and

(I) the Secretary of Labor.

(2) Considerations.—In making appointments under subparagraphs (A) and (B) of paragraph (1), the appointing authority shall consider the background of the individuals, including whether the individuals—

(A) are members of organizations representing women and minorities, and other related interest groups;

(B) hold management or decisionmaking positions in corporations or other business entities recognized as leaders on issues relating to equal employment opportunity; and

(C) possess academic expertise or other recognized ability regarding employment issues.

(3) Balance.—In making the appointments under subparagraphs (A) and (B) of paragraph (1), each appointing authority shall seek to include an appropriate balance of appointees from among the groups of appointees described in subparagraphs (A), (B), and (C) of paragraph (2).

(c) Chairperson.—The Secretary of Labor shall serve as the Chairperson of the Commission.

(d) Term of Office.—Members shall be appointed for the life of the Commission.

(e) Vacancies.—Any vacancy occurring in the membership of the Commission shall be filled in the same manner as the original appointment for the position being vacated. The vacancy shall not affect the power of the remaining members to execute the duties of the Commission.

(f) Meetings.—

(1) Meetings prior to completion of report.—The Commission shall meet not fewer than five times in connection with and pending the completion of the report described in section 204(b). The Commission shall hold additional meetings if the Chairperson or a majority of the members of the Commission request the additional meetings in writing.

(2) Meetings after completion of report.—The Commission shall meet once each year after the completion of the report described in section 204(b). The Commission shall hold additional meetings if the Chairperson or a majority of the members of the Commission request the additional meetings in writing.

(g) Quorum.—A majority of the Commission shall constitute a quorum for the transaction of business.

(h) Compensation and Expenses.—

(1) Compensation.—Each member of the Commission who is not an employee of the

Federal Government shall receive compensation at the daily equivalent of the rate specified for level V of the Executive Schedule under section 5316 of title 5, United States Code, for each day the member is engaged in the performance of duties for the Commission, including attendance at meetings and conferences of the Commission, and travel to conduct the duties of the Commission.

(2) Travel expenses.—Each member of the Commission shall receive travel expenses, including per diem in lieu of subsistence, at rates authorized for employees of agencies under subchapter I of chapter 57 of title 5, United States Code, for each day the member is engaged in the performance of duties away from the home or regular place of business of the member.

(3) Employment status.—A member of the Commission, who is not otherwise an employee of the Federal Government, shall not be deemed to be an employee of the Federal Government except for the purposes of—

(A) the tort claims provisions of chapter 171 of title 28, United States Code; and

(B) subchapter I of chapter 81 of title 5, United States Code, relating to compensation for work injuries.

### § 204.  RESEARCH ON ADVANCEMENT OF WOMEN AND MINORITIES TO MANAGEMENT AND DECISIONMAKING POSITIONS IN BUSINESS.

(a) Advancement Study.—The Commission shall conduct a study of opportunities for, and artificial barriers to, the advancement of women and minorities to management and decisionmaking positions in business. In conducting the study, the Commission shall—

(1) examine the preparedness of women and minorities to advance to management and decisionmaking positions in business;

(2) examine the opportunities for women and minorities to advance to management and decisionmaking positions in business;

(3) conduct basic research into the practices, policies, and manner in which management and decisionmaking positions in business are filled;

(4) conduct comparative research of businesses and industries in which women and minorities are promoted to management and decisionmaking positions, and businesses and industries in which women and minorities are not promoted to management and decisionmaking positions;

(5) compile a synthesis of available research on programs and practices that have successfully led to the advancement of women and minorities to management and decisionmaking positions in business, including training programs, rotational assignments, developmental programs, reward programs, employee benefit structures, and family leave policies; and

(6) examine any other issues and information relating to the advancement of women and minorities to management and decisionmaking positions in business.

(b) Report.—Not later than 15 months after the date of the enactment of this Act, the Commission shall prepare and submit to the President and the appropriate committees of Congress a written report containing—

(1) the findings and conclusions of the Commission resulting from the study conducted under subsection (a); and

(2) recommendations based on the findings and conclusions described in paragraph (1) relating to the promotion of opportunities for, and elimination of artificial barriers

to, the advancement of women and minorities to management and decisionmaking positions in business, including recommendations for—

(A) policies and practices to fill vacancies at the management and decisionmaking levels;

(B) developmental practices and procedures to ensure that women and minorities have access to opportunities to gain the exposure, skills, and expertise necessary to assume management and decisionmaking positions;

(C) compensation programs and reward structures utilized to reward and retain key employees; and

(D) the use of enforcement (including such enforcement techniques as litigation, complaint investigations, compliance reviews, conciliation, administrative regulations, policy guidance, technical assistance, training, and public education) of Federal equal employment opportunity laws by Federal agencies as a means of eliminating artificial barriers to the advancement of women and minorities in employment.

(c) Additional Study.—The Commission may conduct such additional study of the advancement of women and minorities to management and decisionmaking positions in business as a majority of the members of the Commission determines to be necessary.

## § 205.  ESTABLISHMENT OF THE NATIONAL AWARD FOR DIVERSITY AND EXCELLENCE IN AMERICAN EXECUTIVE MANAGEMENT.

(a) In General.—There is established the National Award for Diversity and Excellence in American Executive Management, which shall be evidenced by a medal bearing the inscription "Frances Perkins-Elizabeth Hanford Dole National Award for Diversity and Excellence in American Executive Management." The medal shall be of such design and materials, and bear such additional inscriptions, as the Commission may prescribe.

(b) Criteria for Qualification.—To qualify to receive an award under this section a business shall—

(1) submit a written application to the Commission, at such time, in such manner, and containing such information as the Commission may require, including at a minimum information that demonstrates that the business has made substantial effort to promote the opportunities and developmental experiences of women and minorities to foster advancement to management and decisionmaking positions within the business, including the elimination of artificial barriers to the advancement of women and minorities, and deserves special recognition as a consequence; and

(2) meet such additional requirements and specifications as the Commission determines to be appropriate.

(c) Making and Presentation of Award.—

(1) Award.—After receiving recommendations from the Commission, the President or the designated representative of the President shall annually present the award described in subsection (a) to businesses that meet the qualifications described in subsection (b).

(2) Presentation.—The President or the designated representative of the President shall present the award with such ceremonies as the President or the designated representative of the President may determine to be appropriate.

(3) Publicity.—A business that receives an award under this section may publicize the receipt of the award and use the award in its advertising, if the business agrees to

help other United States businesses improve with respect to the promotion of opportunities and developmental experiences of women and minorities to foster the advancement of women and minorities to management and decisionmaking positions.

(d) Business.—For the purposes of this section, the term "business" includes—

(1)(A) a corporation including nonprofit corporations;

(B) a partnership;

(C) a professional association;

(D) a labor organization; and

(E) a business entity similar to an entity described in subparagraphs (A) through (D);

(2) an education referral program, a training program, such as an apprenticeship or management training program or a similar program; and

(3) a joint program formed by a combination of any entities described in paragraph 1 or 2.

## § 206.   POWERS OF THE COMMISSION.

(a) In General.—The Commission is authorized to—

(1) hold such hearings and sit and act at such times;

(2) take such testimony;

(3) have such printing and binding done;

(4) enter into such contracts and other arrangements;

(5) make such expenditures; and

(6) take such other actions; as the Commission may determine to be necessary to carry out the duties of the Commission.

(b) Oaths.—Any member of the Commission may administer oaths or affirmations to witnesses appearing before the Commission.

(c) Obtaining Information from Federal Agencies.—The Commission may secure directly from any Federal agency such information as the Commission may require to carry out its duties.

(d) Voluntary Service.—Notwithstanding section 1342 of title 31, United States Code, the Chairperson of the Commission may accept for the Commission voluntary services provided by a member of the Commission.

(e) Gifts and Donations.—The Commission may accept, use, and dispose of gifts or donations of property in order to carry out the duties of the Commission.

(f) Use of Mail.—The Commission may use the United States mails in the same manner and under the same conditions as Federal agencies.

## § 207.   CONFIDENTIALITY OF INFORMATION.

(a) Individual Business Information.—

(1) In general.—Except as provided in paragraph (2), and notwithstanding section 552 of title 5, United States Code, in carrying out the duties of the Commission, including the duties described in sections 204 and 205, the Commission shall maintain the confidentiality of all information that concerns—

(A) the employment practices and procedures of individual businesses; or

(B) individual employees of the businesses.

(2) Consent.—The content of any information described in paragraph (1) may be disclosed with the prior written consent of the business or employee, as the case may be, with respect to which the information is maintained.

(b) Aggregate Information.—In carrying out the duties of the Commission, the Commission may disclose—

(1) information about the aggregate employment practices or procedures of a class or group of businesses; and

(2) information about the aggregate characteristics of employees of the businesses, and related aggregate information about the employees.

## § 208. STAFF AND CONSULTANTS.

(a) Staff.—

(1) Appointment and compensation.—The Commission may appoint and determine the compensation of such staff as the Commission determines to be necessary to carry out the duties of the Commission.

(2) Limitations.—The rate of compensation for each staff member shall not exceed the daily equivalent of the rate specified for level V of the Executive Schedule under section 5316 of title 5, United States Code for each day the staff member is engaged in the performance of duties for the Commission. The Commission may otherwise appoint and determine the compensation of staff without regard to the provisions of title 5, United States Code, that govern appointments in the competitive service, and the provisions of chapter 51 and subchapter III of chapter 53 of title 5, United States Code, that relate to classification and General Schedule pay rates.

(b) Experts and Consultants.—The Chairperson of the Commission may obtain such temporary and intermittent services of experts and consultants and compensate the experts and consultants in accordance with section 3109(b) of title 5, United States Code, as the Commission determines to be necessary to carry out the duties of the Commission.

(c) Detail of Federal Employees.—On the request of the Chairperson of the Commission, the head of any Federal agency shall detail, without reimbursement, any of the personnel of the agency to the Commission to assist the Commission in carrying out its duties. Any detail shall not interrupt or otherwise affect the civil service status or privileges of the Federal employee.

(d) Technical Assistance.—On the request of the Chairperson of the Commission, the head of a Federal agency shall provide such technical assistance to the Commission as the Commission determines to be necessary to carry out its duties.

## § 209. AUTHORIZATION OF APPROPRIATIONS.

There are authorized to be appropriated to the Commission such sums as may be necessary to carry out the provisions of this title. The sums shall remain available until expended, without fiscal year limitation.

## § 210. TERMINATION.

(a) Commission.—Notwithstanding section 15 of the Federal Advisory Committee Act (5 U.S.C. App.), the Commission shall terminate 4 years after the date of the enactment of this Act.

(b) Award.—The authority to make awards under section 205 shall terminate 4 years after the date of the enactment of this Act.

## TITLE III—GOVERNMENT EMPLOYEE RIGHTS

### § 301.   GOVERNMENT EMPLOYEE RIGHTS ACT OF 1991.

(a) Short Title.—This title may be cited as the "Government Employee Rights Act of 1991."

(b) Purpose.—The purpose of this title is to provide procedures to protect the right of Senate and other government employees, with respect to their public employment, to be free of discrimination on the basis of race, color, religion, sex, national origin, age, or disability.

(c) Definitions.—For purposes of this title:

(1) Senate employee.—The term "Senate employee" or "employee" means—

(A) any employee whose pay is disbursed by the Secretary of the Senate;

(B) any employee of the Architect of the Capitol who is assigned to the Senate Restaurants or to the Superintendent of the Senate Office Buildings;

(C) any applicant for a position that will last 90 days or more and that is to be occupied by an individual described in subparagraph (A) or (B); or

(D) any individual who was formerly an employee described in subparagraph (A) or (B) and whose claim of a violation arises out of the individual's Senate employment.

(2) Head of employing office.—The term "head of employing office" means the individual who has final authority to appoint, hire, discharge, and set the terms, conditions or privileges of the Senate employment of an employee.

(3) Violation.—The term "violation" means a practice that violates section 302 of this title.

### § 302.   DISCRIMINATORY PRACTICES PROHIBITED.

All personnel actions affecting employees of the Senate shall be made free from any discrimination based on—

(1) race, color, religion, sex, or national origin, within the meaning of section 717 of the Civil Rights Act of 1964 (42 U.S.C. 2000e-16);

(2) age, within the meaning of section 15 of the Age Discrimination in Employment Act of 1967 (29 U.S.C. 633a); or

(3) handicap or disability, within the meaning of section 501 of the Rehabilitation Act of 1973 (29 U.S.C. 791) and sections 102–104 of the Americans with Disabilities Act of 1990 (42 U.S.C. 12112–14).

### § 303.   ESTABLISHMENT OF OFFICE OF SENATE FAIR EMPLOYMENT PRACTICES.

(a) In General.—There is established, as an office of the Senate, the Office of Senate Fair Employment Practices (referred to in this title as the "Office"), which shall—

(1) administer the processes set forth in sections 305 through 307;

(2) implement programs for the Senate to heighten awareness of employee rights in order to prevent violations from occurring.

(b) Director.—

(1) In general.—The Office shall be headed by a Director (referred to in this title as the "Director") who shall be appointed by the President pro tempore, upon the recommendation of the Majority Leader in consultation with the Minority Leader. The

appointment shall be made without regard to political affiliation and solely on the basis of fitness to perform the duties of the position. The Director shall be appointed for a term of service which shall expire at the end of the Congress following the Congress during which the Director is appointed. A Director may be reappointed at the termination of any term of service. The President pro tempore, upon the joint recommendation of the Majority Leader in consultation with the Minority Leader, may remove the Director at any time.

(2) Salary.—The President pro tempore, upon the recommendation of the Majority Leader in consultation with the Minority Leader, shall establish the rate of pay for the Director. The salary of the Director may not be reduced during the employment of the Director and shall be increased at the same time and in the same manner as fixed statutory salary rates within the Senate are adjusted as a result of annual comparability increases.

(3) Annual budget.—The Director shall submit an annual budget request for the Office to the Committee on Appropriations.

(4) Appointment of Director.—The first Director shall be appointed and begin service within 90 days after the date of enactment of this Act, and thereafter the Director shall be appointed and begin service within 30 days after the beginning of the session of the Congress immediately following the termination of a Director's term of service or within 60 days after a vacancy occurs in the position.

(c) Staff of the Office.—

(1) Appointment.—The Director may appoint and fix the compensation of such additional staff, including hearing officers, as are necessary to carry out the purposes of this title.

(2) Detailees.—The Director may, with the prior consent of the Government department or agency concerned and the Committee on Rules and Administration, use on a reimbursable or nonreimbursable basis the services of any such department or agency, including the services of members or personnel of the General Accounting Office Personnel Appeals Board.

(3) Consultants.—In carrying out the functions of the Office, the Director may procure the temporary (not to exceed 1 year) or intermittent services of individual consultants, or organizations thereof, in the same manner and under the same conditions as a standing committee of the Senate may procure such services under section 202(i) of the Legislative Reorganization Act of 1946 (2 U.S.C. 72a(i)).

(d) Expenses of the Office.—In fiscal year 1992, the expenses of the Office shall be paid out of the Contingent Fund of the Senate from the appropriation account Miscellaneous Items. Beginning in fiscal year 1993, and for each fiscal year thereafter, there is authorized to be appropriated for the expenses of the Office such sums as shall be necessary to carry out its functions. In all cases, expenses shall be paid out of the Contingent Fund of the Senate upon vouchers approved by the Director, except that a voucher shall not be required for—

(1) the disbursement of salaries of employees who are paid at an annual rate;

(2) the payment of expenses for telecommunications services provided by the Telecommunications Department, Sergeant at Arms, United States Senate;

(3) the payment of expenses for stationery supplies purchased through the Keeper of the Stationery, United States Senate;

(4) the payment of expenses for postage to the Postmaster, United States Senate; and

(5) the payment of metered charges on copying equipment provided by the Sergeant

at Arms, United States Senate. The Secretary of the Senate is authorized to advance such sums as may be necessary to defray the expenses incurred in carrying out this title. Expenses of the Office shall include authorized travel for personnel of the Office.

(e) Rules of the Office.—The Director shall adopt rules governing the procedures of the Office, including the procedures of hearing boards, which rules shall be submitted to the President pro tempore for publication in the Congressional Record. The rules may be amended in the same manner. The Director may consult with the Chairman of the Administrative Conference of the United States on the adoption of rules.

(f) Representation by the Senate Legal Counsel.—For the purpose of representation by the Senate Legal Counsel, the Office shall be deemed a committee, within the meaning of title VII of the Ethics in Government Act of 1978 (2 U.S.C. 288, et seq.).

## § 304.    SENATE PROCEDURE FOR CONSIDERATION OF ALLEGED VIOLATIONS.

The Senate procedure for consideration of alleged violations consists of 4 steps as follows:

(1) Step I, counseling, as set forth in section 305.

(2) Step II, mediation, as set forth in section 306.

(3) Step III, formal complaint and hearing by a hearing board, as set forth in section 307.

(4) Step IV, review of a hearing board decision, as set forth in section 308 or 309.

## § 305.    STEP I: COUNSELING.

(a) In General.—A Senate employee alleging a violation may request counseling by the Office. The Office shall provide the employee with all relevant information with respect to the rights of the employee. A request for counseling shall be made not later than 180 days after the alleged violation forming the basis of the request for counseling occurred. No request for counseling may be made until 10 days after the first Director begins service pursuant to section 303(b)(4).

(b) Period of Counseling.—The period for counseling shall be 30 days unless the employee and the Office agree to reduce the period. The period shall begin on the date the request for counseling is received.

(c) Employees of the Architect of the Capitol and Capitol Police.—In the case of an employee of the Architect of the Capitol or an employee who is a member of the Capitol Police, the Director may refer the employee to the Architect of the Capitol or the Capitol Police Board for resolution of the employee's complaint through the internal grievance procedures of the Architect of the Capitol or the Capitol Police Board for a specific period of time, which shall not count against the time available for counseling or mediation under this title.

## § 306.    STEP II: MEDIATION.

(a) In General.—Not later than 15 days after the end of the counseling period, the employee may file a request for mediation with the Office. Mediation may include the Office, the employee, and the employing office in a process involving meetings with the parties separately or jointly for the purpose of resolving the dispute between the employee and the employing office.

(b) Mediation Period.—The mediation period shall be 30 days beginning on the date the request for mediation is received and may be extended for an additional 30 days at the discretion of the Office. The Office shall notify the employee and the head of the employing office when the mediation period has ended.

### § 307.   STEP III: FORMAL COMPLAINT AND HEARING.

(a) Formal Complaint and Request for Hearing.—Not later than 30 days after receipt by the employee of notice from the Office of the end of the mediation period, the Senate employee may file a formal complaint with the Office. No complaint may be filed unless the employee has made a timely request for counseling and has completed the procedures set forth in sections 305 and 306.

(b) Hearing Board.—A board of 3 independent hearing officers (referred to in this title as "hearing board"), who are not Senators or officers or employees of the Senate, chosen by the Director (one of whom shall be designated by the Director as the presiding hearing officer) shall be assigned to consider each complaint filed under this section. The Director shall appoint hearing officers after considering any candidates who are recommended to the Director by the Federal Mediation and Conciliation Service, the Administrative Conference of the United States, or organizations composed primarily of individuals experienced in adjudicating or arbitrating personnel matters. A hearing board shall act by majority vote.

(c) Dismissal of Frivolous Claims.—Prior to a hearing under subsection (d), a hearing board may dismiss any claim that it finds to be frivolous.

(d) Hearing.—A hearing shall be conducted—

(1) in closed session on the record by a hearing board;

(2) no later than 30 days after filing of the complaint under subsection (a), except that the Office may, for good cause, extend up to an additional 60 days the time for conducting a hearing; and

(3) except as specifically provided in this title and to the greatest extent practicable, in accordance with the principles and procedures set forth in sections 554 through 557 of title 5, United States Code.

(e) Discovery.—Reasonable prehearing discovery may be permitted at the discretion of the hearing board.

(f) Subpoena.—

(1) Authorization.—A hearing board may authorize subpoenas, which shall be issued by the presiding hearing officer on behalf of the hearing board, for the attendance of witnesses at proceedings of the hearing board and for the production of correspondence, books, papers, documents, and other records.

(2) Objections.—If a witness refuses, on the basis of relevance, privilege, or other objection, to testify in response to a question or to produce records in connection with the proceedings of a hearing board, the hearing board shall rule on the objection. At the request of the witness, the employee, or employing office, or on its own initiative, the hearing board may refer the objection to the Select Committee on Ethics for a ruling.

(3) Enforcement.—The Select Committee on Ethics may make to the Senate any recommendations by report or resolution, including recommendations for criminal or civil enforcement by or on behalf of the Office, which the Select Committee on Ethics may consider appropriate with respect to—

(A) the failure or refusal of any person to appear in proceedings under this or to produce records in obedience to a subpoena or order of the hearing board; or

(B) the failure or refusal of any person to answer questions during his or her appearance as a witness in a proceeding under this section.

For purposes of section 1365 of title 28, United States Code, the Office shall be deemed to be a committee of the Senate.

(g) Decision.—The hearing board shall issue a written decision as expeditiously as possible, but in no case more than 45 days after the conclusion of the hearing. The written decision shall be transmitted by the Office to the employee and the employing office. The decision shall state the issues raised by the complaint, describe the evidence in the record, and contain a determination as to whether a violation has occurred.

(h) Remedies.—If the hearing board determines that a violation has occurred, it shall order such remedies as would be appropriate if awarded under section 706 (g) and (k) of the Civil Rights Act of 1964 (42 U.S.C. 2000e-5 (g) and (k)), and may also order the award of such compensatory damages as would be appropriate if awarded under section 1977 and section 1977A (a) and (b)(2) of the Revised Statutes (42 U.S.C. 1981 and 1981A (a) and (b)(2)). In the case of a determination that a violation based on age has occurred, the hearing board shall order such remedies as would be appropriate if awarded under section 15(c) of the Age Discrimination in Employment Act of 1967 (29 U.S.C. 633a(c)). Any order requiring the payment of money must be approved by a Senate resolution reported by the Committee on Rules and Administration. The hearing board shall have no authority to award punitive damages.

(i) Precedent and Interpretations.—Hearing boards shall be guided by judicial decisions under statutes referred to in section 302 and subsection (h) of this section, as well as the precedents developed by the Select Committee on Ethics under section 308, and other Senate precedents.

## § 308.   REVIEW BY THE SELECT COMMITTEE ON ETHICS.

(a) In General.—An employee or the head of an employing office may request that the Select Committee on Ethics (referred to in this section as the "Committee"), or such other entity as the Senate may designate, review a decision under section 307, including any decision following a remand under subsection (c), by filing a request for review with the Office not later than 10 days after the receipt of the decision of a hearing board. The Office, at the discretion of the Director, on its own initiative and for good cause, may file a request for review by the Committee of a decision of a hearing board not later than 5 days after the time for the employee or employing office to file a request for review has expired. The Office shall transmit a copy of any request for review to the Committee and notify the interested parties of the filing of the request for review.

(b) Review.—Review under this section shall be based on the record of the hearing board. The Committee shall adopt and publish in the Congressional Record procedures for requests for review under this section.

(c) Remand.—Within the time for a decision under subsection (d), the Committee may remand a decision no more than one time to the hearing board for the purpose of supplementing the record or for further consideration.

(d) Final Decision.—

(1) Hearing board.—If no timely request for review is filed under subsection (a), the Office shall enter as a final decision, the decision of the hearing board.

(2) Select Committee on Ethics.—

(A) If the Committee does not remand under subsection (c), it shall transmit a written final decision to the Office for entry in the records of the Office. The Committee shall transmit the decision not later than 60 calendar days during which the Senate is in session after the filing of a request for review under subsection (a). The Committee may extend for 15 calendar days during which the Senate is in session the period for transmission to the Office of a final decision.

(B) The decision of the hearing board shall be deemed to be a final decision, and entered in the records of the Office as a final decision, unless a majority of the Committee votes to reverse or remand the decision of the hearing board within the time for transmission to the Office of a final decision.

(C) The decision of the hearing board shall be deemed to be a final decision, and entered in the records of the Office as a final decision, if the Committee, in its discretion, decides not to review, pursuant to a request for review under subsection (a), a decision of the hearing board, and notifies the interested parties of such decision.

(3) Entry of a final decision.—The entry of a final decision in the records of the Office shall constitute a final decision for purposes of judicial review under section 309.

(e) Statement of Reasons.—Any decision of the Committee under subsection (c) or subsection (d)(2)(A) shall contain a written statement of the reasons for the Committee's decision.

## § 309.  JUDICIAL REVIEW.

(a) In General.—Any Senate employee aggrieved by a final decision under section 308(d), or any Member of the Senate who would be required to reimburse the appropriate Federal account pursuant to the section entitled "Payments by the President or a Member of the Senate" and a final decision entered pursuant to section 308(d)(2)(B), may petition for review by the United States Court of Appeals for the Federal Circuit.

(b) Law Applicable.—Chapter 158 of title 28, United States Code, shall apply to a review under this section except that—

(1) with respect to section 2344 of title 28, United States Code, service of the petition shall be on the Senate Legal Counsel rather than on the Attorney General;

(2) the provisions of section 2348 of title 28, United States Code, on the authority of the Attorney General, shall not apply;

(3) the petition for review shall be filed not later than 90 days after the entry in the Office of a final decision under section 308(d);

(4) the Office shall be an "agency" as that term is used in chapter 158 of title 28, United States Code; and

(5) the Office shall be the respondent in any proceeding under this section.

(c) Standard of Review.—To the extent necessary to decision and when presented, the court shall decide all relevant questions of law and interpret constitutional and statutory provisions. The court shall set aside a final decision if it is determined that the decision was—

(1) arbitrary, capricious, an abuse of discretion, or otherwise not consistent with law;

(2) not made consistent with required procedures; or

(3) unsupported by substantial evidence. In making the foregoing determinations, the court shall review the whole record, or those parts of it cited by a party, and due account shall be taken of the rule of prejudicial error. The record on review shall

include the record before the hearing board, the decision of the hearing board, and the decision, if any, of the Select Committee on Ethics.

(d) Attorney's Fees.—If an employee is the prevailing party in a proceeding under this section, attorney's fees may be allowed by the court in accordance with the standards prescribed under section 706(k) of the Civil Rights Act of 1964 (42 U.S.C. 2000e-5(k)).

### § 310.  RESOLUTION OF COMPLAINT.

If, after a formal complaint is filed under section 307, the employee and the head of the employing office resolve the issues involved, the employee may dismiss the complaint or the parties may enter into a written agreement, subject to the approval of the Director.

### § 311.  COSTS OF ATTENDING HEARINGS.

Subject to the approval of the Director, an employee with respect to whom a hearing is held under this title may be reimbursed for actual and reasonable costs of attending proceedings under sections 307 and 308, consistent with Senate travel regulations. Senate Resolution 259, agreed to August 5, 1987 (100th Congress, 1st Session), shall apply to witnesses appearing in proceedings before a hearing board.

### § 312.  PROHIBITION OF INTIMIDATION.

Any intimidation of, or reprisal against, any employee by any Member, officer, or employee of the Senate, or by the Architect of the Capitol, or anyone employed by the Architect of the Capitol, as the case may be, because of the exercise of a right under this title constitutes an unlawful employment practice, which may be remedied in the same manner under this title as is a violation.

### § 313.  CONFIDENTIALITY.

(a) Counseling.—All counseling shall be strictly confidential except that the Office and the employee may agree to notify the head of the employing office of the allegations.

(b) Mediation.—All mediation shall be strictly confidential.

(c) Hearings.—Except as provided in subsection (d), the hearings, deliberations, and decisions of the hearing board and the Select Committee on Ethics shall be confidential.

(d) Final Decision of Select Committee on Ethics.—The final decision of the Select Committee on Ethics under section 308 shall be made public if the decision is in favor of the complaining Senate employee or if the decision reverses a decision of the hearing board which had been in favor of the employee. The Select Committee on Ethics may decide to release any other decision at its discretion. In the absence of a proceeding under section 308, a decision of the hearing board that is favorable to the employee shall be made public.

(e) Release of Records for Judicial Review.—The records and decisions of hearing boards, and the decisions of the Select Committee on Ethics, may be made public if required for the purpose of judicial review under section 309.

### § 314.  EXERCISE OF RULEMAKING POWER.

The provisions of this title, except for sections 309, 320, 321, and 322, are enacted by the Senate as an exercise of the rulemaking power of the Senate, with full recognition of the right of the Senate to change its rules, in the same manner, and to the same extent, as in the case of any other rule of the Senate. Notwithstanding any other provision of law, except as provided in section 309, enforcement and adjudication with respect to the discriminatory practices prohibited by section 302, and arising out of Senate employment, shall be within the exclusive jurisdiction of the United States Senate.

## § 315. TECHNICAL AND CONFORMING AMENDMENTS.

Section 509 of the Americans with Disabilities Act of 1990 (42 U.S.C. 12209) is amended—

(1) in subsection (a)—

(A) by striking paragraphs (2) through (5);

(B) by redesignating paragraphs (6) and (7) as paragraphs (2) and (3), respectively; and

(C) in paragraph (3), as redesignated by subparagraph (B) of this paragraph—(i) by striking "(2) and (6)(A)" and inserting "(2)(A)", as redesignated by subparagraph (B) of this paragraph; and (ii) by striking "(3), (4), (5), (6)(B), and (6)(C)" and inserting "(2)"; and

(2) in subsection (c)(2), by inserting ", except for the employees who are defined as Senate employees, in section 301(c)(1) of the Civil Rights Act of 1991" after "shall apply exclusively."

## § 316. POLITICAL AFFILIATION AND PLACE OF RESIDENCE.

(a) In General.—It shall not be a violation with respect to an employee described in subsection (b) to consider the—

(1) party affiliation;

(2) domicile; or

(3) political compatibility with the employing office, of such an employee with respect to employment decisions.

(b) Definition.—For purposes of this section, the term "employee" means—

(1) an employee on the staff of the Senate leadership;

(2) an employee on the staff of a committee or subcommittee;

(3) an employee on the staff of a Member of the Senate;

(4) an officer or employee of the Senate elected by the Senate or appointed by a Member, other than those described in paragraphs (1) through (3); or

(5) an applicant for a position that is to be occupied by an individual described in paragraphs (1) through (4).

## § 317. OTHER REVIEW.

No Senate employee may commence a judicial proceeding to redress discriminatory practices prohibited under section 302 of this title, except as provided in this title.

## § 318. OTHER INSTRUMENTALITIES OF THE CONGRESS.

It is the sense of the Senate that legislation should be enacted to provide the same or comparable rights and remedies as are provided under this title to employees of instrumentalities of the Congress not provided with such rights and remedies.

## § 319. RULE XLII OF THE STANDING RULES OF THE SENATE.

(a) Reaffirmation.—The Senate reaffirms its commitment to Rule XLII of the Standing Rules of the Senate, which provides as follows:

"No Member, officer, or employee of the Senate shall, with respect to employment by the Senate or any office thereof—

"(a) fail or refuse to hire an individual;

"(b) discharge an individual; or

"(c) otherwise discriminate against an individual with respect to promotion, compensation, or terms, conditions, or privileges of employment on the basis of such individual's race, color, religion, sex, national origin, age, or state of physical handicap."

(b) Authority To Discipline.—Notwithstanding any provision of this title, including any provision authorizing orders for remedies to Senate employees to redress employment discrimination, the Select Committee on Ethics shall retain full power, in accordance with its authority under Senate Resolution 338, 88th Congress, as amended, with respect to disciplinary action against a Member, officer, or employee of the Senate for a violation of Rule XLII.

## § 320.   COVERAGE OF PRESIDENTIAL APPOINTEES.

(a) In General.—

(1) Application.—The rights, protections, and remedies provided pursuant to section 302 and 307(h) of this title shall apply with respect to employment of Presidential appointees.

(2) Enforcement by administrative action.—Any Presidential appointee may file a complaint alleging a violation, not later than 180 days after the occurrence of the alleged violation, with the Equal Employment Opportunity Commission, or such other entity as is designated by the President by Executive Order, which, in accordance with the principles and procedures set forth in sections 554 through 557 of title 5, United States Code, shall determine whether a violation has occurred and shall set forth its determination in a final order. If the Equal Employment Opportunity Commission, or such other entity as is designated by the President pursuant to this section, determines that a violation has occurred, the final order shall also provide for appropriate relief.

(3) Judicial review.—

(A) In general.—Any party aggrieved by a final order under paragraph (2) may petition for review by the United States Court of Appeals for the Federal Circuit.

(B) Law applicable.—Chapter 158 of title 28, United States Code, shall apply to a review under this section except that the Equal Employment Opportunity Commission or such other entity as the President may designate under paragraph (2) shall be an "agency" as that term is used in chapter 158 of title 28, United States Code.

(C) Standard of review.—To the extent necessary to decision and when presented, the reviewing court shall decide all relevant questions of law and interpret constitutional and statutory provisions. The court shall set aside a final order under paragraph (2) if it is determined that the order was—(i) arbitrary, capricious, an abuse of discretion, or otherwise not consistent with law; (ii) not made consistent with required procedures; or (iii) unsupported by substantial evidence.

In making the foregoing determinations, the court shall review the whole record or those parts of it cited by a party, and due account shall be taken of the rule of prejudicial error.

(D) Attorney's fees.—If the presidential appointee is the prevailing party in a proceeding under this section, attorney's fees may be allowed by the court in accordance with the standards prescribed under section 706(k) of the Civil Rights Act of 1964 (42 U.S.C. 2000e-5(k)).

(b) Presidential Appointee.—For purposes of this section, the term "Presidential appointee" means any officer or employee, or an applicant seeking to become an officer or employee, in any unit of the Executive Branch, including the Executive Office of the President, whether appointed by the President or by any other appointing authority in the Executive Branch, who is not already entitled to bring an action under any of the statures referred to in section 302 but does not include any individual—

(1) whose appointment is made by and with the advice and consent of the Senate;

(2) who is appointed to an advisory committee, as defined in section 3(2) of the Federal Advisory Committee Act (5 U.S.C. App.); or

(3) who is a member of the uniformed services.

## § 321.  COVERAGE OF PREVIOUSLY EXEMPT STATE EMPLOYEES.

(a) Application.—The rights, protections, and remedies provided pursuant to section 302 and 307(h) of this title shall apply with respect to employment of any individual chosen or appointed, by a person elected to public office in any State or political subdivision of any State by the qualified voters thereof—

(1) to be a member of the elected official's personal staff;

(2) to serve the elected official on the policymaking level; or

(3) to serve the elected official as an immediate advisor with respect to the exercise of the constitutional or legal powers of the office.

(b) Enforcement by Administrative Action.—

(1) In general.—Any individual referred to in subsection (a) may file a complaint alleging a violation, not later than 180 days after the occurrence of the alleged violation, with the Equal Employment Opportunity Commission, which, in accordance with the principles and procedures set forth in sections 554 through 557 of title 5, United States Code, shall determine whether a violation has occurred and shall set forth its determination in a final order. If the Equal Employment Opportunity Commission determines that a violation has occurred, the final order shall also provide for appropriate relief.

(2) Referral to state and local authorities.—

(A) Application.—Section 706(d) of the Civil Rights Act of 1964 (42 U.S.C. 2000e-5(d)) shall apply with respect to any proceeding under this section.

(B) Definition.—For purposes of the application described in subparagraph (A), the term "any charge filed by a member of the Commission alleging an unlawful employment practice" means a complaint filed under this section.

(c) Judicial Review.—Any party aggrieved by a final order under subsection (b) may obtain a review of such order under chapter 158 of title 28, United States Code. For the purpose of this review, the Equal Employment Opportunity Commission shall be an "agency" as that term is used in chapter 158 of title 28, United States Code.

(d) Standard of Review.—To the extent necessary to decision and when presented, the reviewing court shall decide all relevant questions of law and interpret constitutional and statutory provisions. The court shall set aside a final order under subsection (b) if it is determined that the order was—

(1) arbitrary, capricious, an abuse of discretion, or otherwise not consistent with law;

(2) not made consistent with required procedures; or

(3) unsupported by substantial evidence. In making the foregoing determinations, the court shall review the whole record or those parts of it cited by a party, and due account shall be taken of the rule of prejudicial error.

(e) Attorney's Fees.—If the individual referred to in subsection (a) is the prevailing party in a proceeding under this subsection, attorney's fees may be allowed by the court in accordance with the standards prescribed under section 706(k) of the Civil Rights Act of 1964 (42 U.S.C. 2000e-5(k)).

### § 322.   SEVERABILITY.

Notwithstanding section 401 of this Act, if any provision of section 309 or 320(a)(3)is invalidated, both sections 309 and 320(a)(3) shall have no force and effect.

### § 323.   PAYMENTS BY THE PRESIDENT OR A MEMBER OF THE SENATE.

The President or a Member of the Senate shall reimburse the appropriate Federal account for any payment made on his or her behalf out of such account for a violation committed under the provisions of this title by the President or Member of the Senate not later than 60 days after the payment is made.

### § 324.   REPORTS OF SENATE COMMITTEES.

(a) Each report accompanying a bill or joint resolution of a public character reported by any committee of the Senate (except the Committee on Appropriations and the Committee on the Budget) shall contain a listing of the provisions of the bill or joint resolution that apply to Congress and an evaluation of the impact of such provisions on Congress.

(b) The provisions of this section are enacted by the Senate as an exercise of the rule-making power of the Senate, with full recognition of the right of the Senate to change its rules, in the same manner, and to the same extent, as in the case of any other rule of the Senate.

### § 325.   INTERVENTION AND EXPEDITED REVIEW OF CERTAIN APPEALS.

(a) Intervention.—Because of the constitutional issues that may be raised by section 309 and section 320, any Member of the Senate may intervene as a matter of right in any proceeding under section 309 for the sole purpose of determining the constitutionality of such section.

(b) Threshold Matter.—In any proceeding under section 309 or section 320, the United States Court of Appeals for the Federal Circuit shall determine any issue presented concerning the constitutionality of such section as a threshold matter.

(c) Appeal.—

(1) In general.—An appeal may be taken directly to the Supreme Court of the United States from any interlocutory or final judgment, decree, or order issued by the United States Court of Appeals for the Federal Circuit ruling upon the constitutionality of section 309 or 320.

(2) Jurisdiction.—The Supreme Court shall, if it has not previously ruled on the question, accept jurisdiction over the appeal referred to in paragraph (1), advance the appeal on the docket and expedite the appeal to the greatest extent possible.

### TITLE IV—GENERAL PROVISIONS

### § 401.   SEVERABILITY.

If any provision of this Act, or an amendment made by this Act, or the application of such provision to any person or circumstances is held to be invalid, the remainder of this

Act and the amendments made by this Act, and the application of such provision to other persons and circumstances, shall not be affected.

## § 402.   EFFECTIVE DATE.

(a) In General.—Except as otherwise specifically provided, this Act and the amendments made by this Act shall take effect upon enactment.

(b) Certain Disparate Impact Cases.—Notwithstanding any other provision of this Act, nothing in this Act shall apply to any disparate impact case for which a complaint was filed before March 1, 1975, and for which an initial decision was rendered after October 30, 1983.

## TITLE V—CIVIL WAR SITES ADVISORY COMMISSION

## § 501.   CIVIL WAR SITES ADVISORY COMMISSION.

Section 1205 of Public Law 101-628 is amended in subsection (a) by—
(1) striking "Three" in paragraph (4) and inserting "Four" in lieu thereof; and
(2) striking "Three" in paragraph (5) and inserting "Four" in lieu thereof.

*Appendix*

# E

# Occupational Safety and Health Act

## 29 U.S.C. §§ 651–678

### Definitions

**Sec. 3.**   (§652) For the purposes of this Act—

(1) The term "Secretary" means the Secretary of Labor.

(2) The term "Commission" means the Occupational Safety and Health Review Commission established under this Act.

(3) The term "commerce" means trade, traffic, commerce, transportation, or communication among the several States, or between a State and any place outside thereof, or within the District of Columbia, or a possession of the United States (other than the Trust Territory of the Pacific Islands), or between points in the same State but through a point outside thereof.

(4) The term "person" means one or more individuals, partnerships, associations, corporations, business trusts, legal representatives, or any organized group of persons.

(5) The term "employer" means a person engaged in a business affecting commerce who has employees, but does not include the United States or any State or political subdivision of a State.

(6) The term "employee" means an employee of an employer who is employed in a business of his employer which affects commerce.

(7) The term "State" includes a State of the United States, the District of Columbia, Puerto Rico, the Virgin Islands, American Samoa, Guam, and the Trust Territory of the Pacific Islands.

(8) The term "occupational safety and health standard" means a standard which requires conditions, or the adoption or use of one or more practices, means, methods, operations, or processes, reasonably

necessary or appropriate to provide safe or healthful employment and places of employment.

(9) The term "national consensus standard" means any occupational safety and health standard or modification thereof which (1) has been adopted and promulgated by a nationally recognized standards-producing organization under procedures whereby it can be determined by the Secretary that persons interested and affected by the scope or provisions of the standard have reached substantial agreement on its adoption, (2) was formulated in a manner which afforded an opportunity for diverse views to be considered and (3) has been designated as such a standard by the Secretary, after consultation with other appropriate Federal agencies.

(10) The term "established Federal standard" means any operative occupational safety and health standard established by any agency of the United States and presently in effect, or contained in any Act of Congress in force on the date of enactment of this Act.

## APPLICABILITY OF THIS ACT

**SEC. 4.** **(§ 653)** (a) This Act shall apply with respect to employment performed in a workplace in a State, the District of Columbia, the Commonwealth of Puerto Rico, the Virgin Islands, American Samoa, Guam, the Trust Territory of the Pacific Islands, Wake Island, Outer Continental Shelf lands defined in the Outer Continental Shelf Lands Act, Johnston Island, and the Canal Zone. The Secretary of the Interior shall, by regulation, provide for judicial enforcement of this Act by the courts established for areas in which there are no United States district courts having jurisdiction.

(b)(1) Nothing in this Act shall apply to working conditions of employees with respect to which other Federal agencies, and State agencies acting under section 274 of the Atomic Energy Act of 1954, as amended (42 U.S.C. 2021), exercise statutory authority to prescribe or enforce standards or regulations affecting occupational safety or health.

(2) The safety and health standards promulgated under the Act of June 30, 1936, commonly known as the Walsh-Healey Act (41 U.S.C. 35 et seq.), the Service Contract Act of 1965 (41 U.S.C. 351 et seq.), Public Law 91–54, Act of August 9, 1969 (40 U.S.C. 333), Public Law 85–742, Act of August 23, 1958 (33 U.S.C. 941), and the National Foundation on Arts and Humanities Act (20 U.S.C. 951 et seq.) are superseded on the effective date of corresponding standards, promulgated under this Act, which are determined by the Secretary to be more effective. Standards issued under the laws listed in this paragraph and in effect on or after the effective date of this Act shall be deemed to be occupational safety and health standards issued under this Act, as well as under such other Acts.

(3) The Secretary shall, within three years after the effective date of this Act, report to the Congress his recommendations for legislation to avoid unnecessary duplication and to achieve coordination between this Act and other Federal laws.

(4) Nothing in this Act shall be construed to supersede or in any manner affect any workmen's compensation law or to enlarge or diminish or affect in any other manner the common law or statutory rights, duties, or liabilities of employers and employees under any law with respect to injuries, diseases, or death of employees arising out of, or in the course of, employment.

## DUTIES

**SEC. 5.** **(§ 654)** (a) Each employer—

(1) shall furnish to each of his employees employment and a place of employment

which are free from recognized hazards that are causing or are likely to cause death or serious physical harm to his employees;

(2) shall comply with occupational safety and health standards promulgated under this Act.

(b) Each employee shall comply with occupational safety and health standards and all rules, regulations, and orders issued pursuant to this Act which are applicable to his own actions and conduct.

## OCCUPATIONAL SAFETY AND HEALTH STANDARDS

**SEC. 6.    (§ 655)** (a) Without regard to chapter 5 of title 5, United States Code, or to the other subsections of this section, the Secretary shall, as soon as practicable during the period beginning with the effective date of this Act and ending two years after such date, by rule promulgate as an occupational safety or health standard any national consensus standard, and any established Federal standard, unless he determines that the promulgation of such a standard would not result in improved safety or health for specifically designated employees. In the event of conflict among any such standards, the Secretary shall promulgate the standard which assures the greatest protection of the safety or health of the affected employees.

(b) The Secretary may by rule promulgate, modify, or revoke any occupational safety or health standard in the following manner:

(1) Whenever the Secretary, upon the basis of information submitted to him in writing by an interested person, a representative of any organization of employers or employees, a nationally recognized standards-producing organization, the Secretary of Health and Human Services, the National Institute for Occupational Safety and Health, or a State or political subdivision, or on the basis of information developed by the Secretary or otherwise available to him, determines that a rule should be promulgated in order to serve the objectives of this Act, the Secretary may request the recommendations of an advisory committee appointed under section 7 of this Act. The Secretary shall provide such an advisory committee with any proposals of his own or of the Secretary of Health and Human Services, together with all pertinent factual information developed by the Secretary or the Secretary of Health and Human Services, or otherwise available, including the results of research, demonstrations, and experiments. An advisory committee shall submit to the Secretary its recommendations regarding the rule to be promulgated within ninety days from the date of its appointment or within such longer or shorter period as may be prescribed by the Secretary, but in no event for a period which is longer than two hundred and seventy days.

(2) The Secretary shall publish a proposed rule promulgating, modifying, or revoking an occupational safety or health standard in the Federal Register and shall afford interested persons a period of thirty days after publication to submit written data or comments. Where an advisory committee is appointed and the Secretary determines that a rule should be issued, he shall publish the proposed rule within sixty days after the submission of the advisory committee's recommendations or the expiration of the period prescribed by the Secretary for such submission.

(3) On or before the last day of the period provided for the submission of written data or comments under paragraph (2), any interested person may file with the Secretary written objections to the proposed rule, stating the grounds therefor and requesting a public hearing on such objections. Within thirty days after the last day for filing such objections, the Secretary shall publish in the Federal Register a notice specifying the occupational

safety or health standard to which objections have been filed and a hearing requested, and specifying a time and place for such hearing.

(4) Within sixty days after the expiration of the period provided for the submission of written data or comments under paragraph (2), or within sixty days after the completion of any hearing held under paragraph (3), the Secretary shall issue a rule promulgating, modifying, or revoking an occupational safety or health standard or make a determination that a rule should not be issued. Such a rule may contain a provision delaying its effective date for such period (not in excess of ninety days) as the Secretary determines may be necessary to insure that affected employers and employees will be informed of the existence of the standard and of its terms and that employers affected are given an opportunity to familiarize themselves and their employees with the existence of the requirements of the standard.

(5) The Secretary, in promulgating standards dealing with toxic materials or harmful physical agents under this subsection, shall set the standard which most adequately assures, to the extent feasible, on the basis of the best available evidence, that no employee will suffer material impairment of health or functional capacity even if such employee has regular exposure to the hazard dealt with by such standard for the period of his working life. Development of standards under this subsection shall be based upon research, demonstrations, experiments, and such other information as may be appropriate. In addition to the attainment of the highest degree of health and safety protection for the employee, other considerations shall be the latest available scientific data in the field, the feasibility of the standards, and experience gained under this and other health and safety laws. Whenever practicable, the standard promulgated shall be expressed in terms of objective criteria and of the performance desired.

(6)(A) Any employer may apply to the Secretary for a temporary order granting a variance from a standard or any provision thereof promulgated under this section. Such temporary order shall be granted only if the employer files an application which meets the requirements of clause (B) and establishes that (i) he is unable to comply with a standard by its effective date because of unavailability of professional or technical personnel or of materials and equipment needed to come into compliance with the standard or because necessary construction or alteration of facilities cannot be completed by the effective date, (ii) he is taking all available steps to safeguard his employees against the hazards covered by the standard, and (iii) he has an effective program for coming into compliance with the standard as quickly as practicable. Any temporary order issued under this paragraph shall prescribe the practices, means, methods, operations, and processes which the employer must adopt and use while the order is in effect and state in detail his program for coming into compliance with the standard. Such a temporary order may be granted only after notice to employees and an opportunity for a hearing: Provided, That the Secretary may issue one interim order to be effective until a decision is made on the basis of the hearing. No temporary order may be in effect for longer than the period needed by the employer to achieve compliance with the standard or one year, whichever is shorter, except that such an order may be renewed not more than twice (I) so long as the requirements of this paragraph are met and (II) if an application for renewal is filed at least 90 days prior to the expiration date of the order. No interim renewal of an order may remain in effect for longer than 180 days.

(B) An application for a temporary order under this paragraph (6) shall contain:

(i) a specification of the standard or portion thereof from which the employer seeks a variance,

(ii) a representation by the employer, supported by representations from qualified persons having firsthand knowledge of the facts represented, that he is unable to comply with the standard or portion thereof and a detailed statement of the reasons therefor,

(iii) a statement of the steps he has taken and will take (with specific dates) to protect employees against the hazard covered by the standard,

(iv) a statement of when he expects to be able to comply with the standard and what steps he has taken and what steps he will take (with dates specified) to come into compliance with the standard, and

(v) a certification that he has informed his employees of the application by giving a copy thereof to their authorized representative, posting a statement giving a summary of the application and specifying where a copy may be examined at the place or places where notices to employees are normally posted, and by other appropriate means.

A description of how employees have been informed shall be contained in the certification. The information to employees shall also inform them of their right to petition the Secretary for a hearing.

(C) The Secretary is authorized to grant a variance from any standard or portion thereof whenever he determines, or the Secretary of Health, Education, and Welfare certifies, that such variance is necessary to permit an employer to participate in an experiment approved by him or the Secretary of Health and Human Services designed to demonstrate or validate new and improved techniques to safeguard the health or safety of workers.

(7) Any standard promulgated under this subsection shall prescribe the use of labels or other appropriate forms of warning as are necessary to insure that employees are apprised of all hazards to which they are exposed, relevant symptoms and appropriate emergency treatment, and proper conditions and precautions of safe use or exposure. Where appropriate, such standard shall also prescribe suitable protective equipment and control or technological procedures to be used in connection with such hazards and shall provide for monitoring or measuring employee exposure at such locations and intervals, and in such manner as may be necessary for the protection of employees. In addition, where appropriate, any such standard shall prescribe the type and frequency of medical examinations or other tests which shall be made available, by the employer or at his cost, to employees exposed to such hazards in order to most effectively determine whether the health of such employees is adversely affected by such exposure. In the event such medical examinations are in the nature of research, as determined by the Secretary of Health and Human Services, such examinations may be furnished at the expense of the Secretary of Health and Human Services. The results of such examinations or tests shall be furnished only to the Secretary or the Secretary of Health and Human Services, and, at the request of the employee, to his physician. The Secretary, in consultation with the Secretary of Health and Human Services, may by rule promulgated pursuant to section 553 of title 5, United States Code, make appropriate modifications in the foregoing requirements relating to the use of labels or other forms of warning, monitoring or measuring, and medical examinations, as may be warranted by experience, information, or medical or technological developments acquired subsequent to the promulgation of the relevant standard.

(8) Whenever a rule promulgated by the Secretary differs substantially from an existing national consensus standard, the Secretary shall, at the same time, publish in the Federal Register a statement of the reasons why the rule as adopted will better effectuate the purposes of this Act than the national consensus standard.

(c)(1) The Secretary shall provide, without regard to the requirements of chapter 5, title 5, United States Code, for an emergency temporary standard to take immediate effect upon publication in the Federal Register if he determines (A) that employees are exposed to grave danger from exposure to substances or agents determined to be toxic or physically harmful or from new hazards, and (B) that such emergency standard is necessary to protect employees from such danger.

(2) Such standard shall be effective until superseded by a standard promulgated in accordance with the procedures prescribed in paragraph (3) of this subsection.

(3) Upon publication of such standard in the Federal Register the Secretary shall commence a proceeding in accordance with section 6(b) of this Act, and the standard as published shall also serve as a proposed rule for the proceeding. The Secretary shall promulgate a standard under this paragraph no later than six months after publication of the emergency standard as provided in paragraph (2) of this subsection.

(d) Any affected employer may apply to the Secretary for a rule or order for a variance from a standard promulgated under this section. Affected employees shall be given notice of each such application and an opportunity to participate in a hearing. The Secretary shall issue such rule or order if he determines on the record, after opportunity for an inspection where appropriate and a hearing, that the proponent of the variance has demonstrated by a preponderance of the evidence that the conditions, practices, means, methods, operations, or processes used or proposed to be used by an employer will provide employment and places of employment to his employees which are as safe and healthful as those which would prevail if he complied with the standard. The rule or order so issued shall prescribe the conditions the employer must maintain, and the practices, means, methods, operations, and processes which he must adopt and utilize to the extent they differ from the standard in question. Such a rule or order may be modified or revoked upon application by an employer, employees, or by the Secretary on his own motion, in the manner prescribed for its issuance under this subsection at any time after six months from its issuance.

(e) Whenever the Secretary promulgates any standard, makes any rule, order, or decision, grants any exemption or extension of time, or compromises, mitigates, or settles any penalty assessed under this Act, he shall include a statement of the reasons for such action, which shall be published in the Federal Register.

(f) Any person who may be adversely affected by a standard issued under this section may at any time prior to the sixtieth day after such standard is promulgated file a petition challenging the validity of such standard with the United States court of appeals for the circuit wherein such person resides or has his principal place of business, for a judicial review of such standard. A copy of the petition shall be forthwith transmitted by the clerk of the court to the Secretary. The filing of such petition shall not, unless otherwise ordered by the court, operate as a stay of the standard. The determinations of the Secretary shall be conclusive if supported by substantial evidence in the record considered as a whole.

(g) In determining the priority for establishing standards under this section, the Secretary shall give due regard to the urgency of the need for mandatory safety and health standards for particular industries, trades, crafts, occupations, businesses, workplaces or work environments. The Secretary shall also give due regard to the recommendations of the Secretary of Health and Human Services regarding the need for mandatory standards in determining the priority for establishing such standards.

## INSPECTIONS, INVESTIGATIONS, AND RECORDKEEPING

**SEC. 8.**    **(§ 657)** (a) In order to carry out the purposes of this Act, the Secretary, upon presenting appropriate credentials to the owner, operator, or agent in charge, is authorized—

(1) to enter without delay and at reasonable times any factory, plant, establishment, construction site, or other area, workplace or environment where work is performed by an employee of an employer; and

(2) to inspect and investigate during regular working hours and at other reasonable times, and within reasonable limits and in a reasonable manner, any such place of employment and all pertinent conditions, structures, machines, apparatus, devices, equipment, and materials therein, and to question privately any such employer, owner, operator, agent or employee.

(b) In making his inspections and investigations under this Act the Secretary may require the attendance and testimony of witnesses and the production of evidence under oath. Witnesses shall be paid the same fees and mileage that are paid witnesses in the courts of the United States. In case of a contumacy, failure, or refusal of any person to obey such an order, any district court of the United States or the United States courts of any territory or possession, within the jurisdiction of which such person is found, or resides or transacts business, upon the application by the Secretary, shall have jurisdiction to issue to such person an order requiring such person to appear to produce evidence if, as, and when so ordered, and to give testimony relating to the matter under investigation or in question, and any failure to obey such order of the court may be punished by said court as a contempt thereof.

(c)(1) Each employer shall make, keep and preserve, and make available to the Secretary or the Secretary of Health and Human Services, such records regarding his activities relating to this Act as the Secretary, in cooperation with the Secretary of Health and Human Services, may prescribe by regulation as necessary or appropriate for the enforcement of this Act or for developing information regarding the causes and prevention of occupational accidents and illnesses. In order to carry out the provisions of this paragraph such regulations may include provisions requiring employers to conduct periodic inspections. The Secretary shall also issue regulations requiring that employers, through posting of notices or other appropriate means, keep their employees informed of their protections and obligations under this Act, including the provisions of applicable standards.

(2) The Secretary, in cooperation with the Secretary of Health and Human Services, shall prescribe regulations requiring employers to maintain accurate records of, and to make periodic reports on, work-related deaths, injuries and illnesses other than minor injuries requiring only first aid treatment and which do not involve medical treatment, loss of consciousness, restriction of work or motion, or transfer to another job.

(3) The Secretary, in cooperation with the Secretary of Health and Human Services, shall issue regulations requiring employers to maintain accurate records of employee exposures to potentially toxic materials or harmful physical agents which are required to be monitored or measured under section 6. Such regulations shall provide employees or their representatives with an opportunity to observe such monitoring or measuring, and to have access to the records thereof. Such regulations shall also make appropriate provision for each employee or former employee to have access to such records as will indicate his own exposure to toxic materials or harmful physical agents. Each employer shall promptly notify any employee who has been or is being exposed to toxic materials or harmful physical agents in concentrations or at levels which exceed those prescribed by an applicable occupational safety and health standard promulgated under section 6, and shall inform any employee who is being thus exposed of the corrective action being taken.

(d) Any information obtained by the Secretary, the Secretary of Health and Human Services, or a State agency under this Act shall be obtained with a minimum burden upon employers, especially those operating small businesses. Unnecessary duplication of efforts in obtaining information shall be reduced to the maximum extent feasible.

(e) Subject to regulations issued by the Secretary a representative of the employer and a representative authorized by his employees shall be given an opportunity to accompany the Secretary or his authorized representative during the physical inspection of any workplace under subsection (a) for the purpose of aiding such inspection. Where there is no authorized employee representative, the Secretary or his authorized representative shall consult with a reasonable number of employees concerning matters of health and safety in the workplace.

(f)(1) Any employees or representative of employees who believe that a violation of a safety or health standard exists that threatens physical harm, or that an imminent danger exists, may request an inspection by giving notice to the Secretary or his authorized representative of such violation or danger. Any such notice shall be reduced to writing, shall set forth with reasonable particularity the grounds for the notice, and shall be signed by the employees or representative of employees, and a copy shall be provided the employer or his agent no later than at the time of inspection, except that, upon the request of the person giving such notice, his name and the names of individual employees referred to therein shall not appear in such copy or on any record published, released, or made available pursuant to subsection (g) of this section. If upon receipt of such notification the Secretary determines there are reasonable grounds to believe that such violation or danger exists, he shall make a special inspection in accordance with the provisions of this section as soon as practicable, to determine if such violation or danger exists. If the Secretary determines there are no reasonable grounds to believe that a violation or danger exists he shall notify the employees or representative of the employees in writing of such determination.

(2) Prior to or during any inspection of a workplace, any employees or representative of employees employed in such workplace may notify the Secretary or any representative of the Secretary responsible for conducting the inspection, in writing, of any violation of this Act which they have reason to believe exists in such workplace. The Secretary shall, by regulation, establish procedures for information review of any refusal by a representative of the Secretary to issue a citation with respect to any such alleged violation and shall furnish the employees or representative of employees requesting such review a written statement of the reasons for the Secretary's final disposition of the case.

(g)(1) The Secretary and Secretary of Health and Human Services are authorized to compile, analyze, and publish, either in summary or detailed form, all reports or information obtained under this section.

(2) The Secretary and the Secretary of Health and Human Services shall each prescribe such rules and regulations as he may deem necessary to carry out their responsibilities under this Act, including rules and regulations dealing with the inspection of an employer's establishment.

## CITATIONS

**SEC. 9.**    **(§ 658)** (a) If, upon inspection or investigation, the Secretary or his authorized representative believes that an employer has violated a requirement of section 5 of this Act, of any standard, rule or order promulgated pursuant to section 6 of this Act, or of any regulations prescribed pursuant to this Act, he shall with reasonable promptness issue a citation to the employer. Each citation shall be in writing and shall describe with particularity the nature of the violation, including a reference to the provision of the Act, standard, rule, regulation, or order alleged to have been violated. In addition, the citation shall fix a reasonable time for the abatement of the violation. The Secretary may prescribe procedures for the issuance of a notice in lieu of a citation with respect to de minimis violations which have no direct or immediate relationship to safety or health.

(b) Each citation issued under this section, or a copy or copies thereof, shall be prominently posted, as prescribed in regulations issued by the Secretary, at or near each place a violation referred to in the citation occurred.

(c) No citation may be issued under this section after the expiration of six months following the occurrence of any violation.

PROCEDURE FOR ENFORCEMENT

**SEC. 10.**   **(§ 659)** (a) If, after an inspection or investigation, the Secretary issues a citation under section 9(a), he shall, within a reasonable time after the termination of such inspection or investigation, notify the employer by certified mail of the penalty, if any, proposed to be assessed under section 17 and that the employer has fifteen working days within which to notify the Secretary that he wishes to contest the citation or proposed assessment of penalty. If, within fifteen working days from the receipt of the notice issued by the Secretary the employer fails to notify the Secretary that he intends to contest the citation or proposed assessment of penalty, and no notice is filed by any employee or representative of employees under subsection (c) within such time, the citation and the assessment, as proposed, shall be deemed a final order of the Commission and not subject to review by any court of agency.

(b) If the Secretary has reason to believe that an employer has failed to correct a violation for which a citation has been issued within the period permitted for its correction (which period shall not begin to run until the entry of a final order by the Commission in the case of any review proceedings under this section initiated by the employer in good faith and not solely for delay or avoidance of penalties), the Secretary shall notify the employer by certified mail of such failure and of the penalty proposed to be assessed under section 17 by reason of such failure, and that the employer has fifteen working days within which to notify the Secretary that he wishes to contest the Secretary's notification or the proposed assessment of penalty. If, within fifteen working days from the receipt of notification issued by the Secretary, the employer fails to notify the Secretary that he intends to contest the notification or proposed assessment of penalty, the notification and assessment, as proposed, shall be deemed a final order of the Commission and not subject to review by any court or agency.

(c) If an employer notifies the Secretary that he intends to contest a citation issued under section 9(a) or notification issued under subsection (a) or (b) of this section, or if, within fifteen working days of the issuance of a citation under section 9(a), any employee or representative of employees files a notice with the Secretary alleging that the period of time fixed in the citation for the abatement of the violation is unreasonable, the Secretary shall immediately advise the Commission of such notification, and the Commission shall afford an opportunity for a hearing (in accordance with section 554 of title 5, United States Code, but without regard to subsection (a)(3) of such section). The Commission shall thereafter issue an order, based on findings of fact, affirming, modifying, or vacating the Secretary's citation or proposed penalty, or directing other appropriate relief, and such order shall become final thirty days after its issuance. Upon a showing by an employer of a good faith effort to comply with the abatement requirements of a citation, and that abatement has not been completed because of factors beyond his reasonable control, the Secretary, after an opportunity for a hearing as provided in this subsection, shall issue an order affirming or modifying the abatement requirements in such citation. The rules of procedure prescribed by the Commission shall provide affected employees or representatives of affected employees an opportunity to participate as parties to hearings under this subsection.

JUDICIAL REVIEW

SEC. 11.    (§ 660) (a) Any person adversely affected or aggrieved by an order of the Commission issued under subsection (c) of section 10 may obtain a review of such order in any United States court of appeals for the circuit in which the violation is alleged to have occurred or where the employer has its principal office, or in the Court of Appeals for the District of Columbia Circuit, by filing in such court within sixty days following the issuance of such order a written petition praying that the order be modified or set aside. A copy of such petition shall be forthwith transmitted by the clerk of the court to the Commission and to the other parties, and thereupon the Commission shall file in the court the record in the proceeding as provided in section 2112 of title 28, United States Code. Upon such filing, the court shall have jurisdiction of the proceeding and of the question determined therein, and shall have power to grant such temporary relief or restraining order as it deems just and proper, and to make and enter upon the pleadings, testimony, and proceedings set forth in such record a decree affirming, modifying, or setting aside in whole or in part, the order of the Commission and enforcing the same to the extent that such order is affirmed or modified. The commencement of proceedings under this subsection shall not, unless ordered by the court, operate as a stay of the order of the Commission. No objection that has not been urged before the Commission shall be considered by the court, unless the failure or neglect to urge such objection shall be excused because of extraordinary circumstances. The findings of the Commission with respect to questions of fact, if supported by substantial evidence on the record considered as a whole, shall be conclusive. If any party shall apply to the court for leave to adduce additional evidence and shall show to the satisfaction of the court that such additional evidence is material and that there were reasonable grounds for the failure to adduce such evidence in the hearing before the Commission, the court may order such additional evidence to be taken before the Commission and to be made a part of the record. The Commission may modify its findings as to the facts, or make new findings, by reason of additional evidence so taken and filed, and it shall file such modified or new findings, which findings with respect to questions of fact, if supported by substantial evidence on the record considered as a whole, shall be conclusive, and its recommendations, if any, for the modification or setting aside of its original order. Upon the filing of the record with it, the jurisdiction of the court shall be exclusive and its judgment and decree shall be final, except that the same shall be subject to review by the Supreme Court of the United States, as provided in section 1254 of title 28, United States Code. Petitions filed under this subsection shall be heard expeditiously.

(b) The Secretary may also obtain review or enforcement of any final order of the Commission by filing a petition for such relief in the United States court of appeals for the circuit in which the alleged violation occurred or in which the employer has its principal office, and the provisions of subsection (a) shall govern such proceedings to the extent applicable. If no petition for review, as provided in subsection (a), is filed within sixty days after service of the Commission's order, the Commission's findings of fact and order shall be conclusive in connection with any petition for enforcement which is filed by the Secretary after the expiration of such sixty-day period. In any such case, as well as in the case of a noncontested citation or notification by the Secretary which has become a final order of the Commission under subsection (a) or (b) of section 10, the clerk of the court, unless otherwise ordered by the court, shall forthwith enter a decree enforcing the order and shall transmit a copy of such decree to the Secretary and the employer named in the petition. In any contempt proceeding brought to enforce a decree of a court of appeals entered pursuant to this subsection or subsection (a), the court of appeals may assess the penalties provided in section 17, in addition to invoking any other available remedies.

(c)(1) No person shall discharge or in any manner discriminate against any employee because such employee has filed any complaint or instituted or caused to be instituted any proceeding under or related to this Act or has testified or is about to testify in any such proceeding or because of the exercise by such employee on behalf of himself or others of any right afforded by this Act.

(2) Any employee who believes that he has been discharged or otherwise discriminated against by any person in violation of this subsection may, within thirty days after such violation occurs, file a complaint with the Secretary alleging such discrimination. Upon receipt of such complaint, the Secretary shall cause such investigation to be made as he deems appropriate. If upon such investigation, the Secretary determines that the provisions of this subsection have been violated, he shall bring an action in any appropriate United States district court against such person. In any such action the United States district courts shall have jurisdiction, for cause shown to restrain violations of paragraph (1) of this subsection and order all appropriate relief including rehiring or reinstatement of the employee to his former position with back pay.

(3) Within 90 days of the receipt of a complaint filed under this subsection the Secretary shall notify the complainant of his determination under paragraph 2 of this subsection.

## THE OCCUPATIONAL SAFETY AND HEALTH REVIEW COMMISSION

**SEC. 12.**   **(§ 661)** (a) The Occupational Safety and Health Review Commission is hereby established. The Commission shall be composed of three members who shall be appointed by the President, by and with the advice and consent of the Senate, from among persons who by reason of training, education, or experience are qualified to carry out the functions of the Commission under this Act. The President shall designate one of the members of the Commission to serve as Chairman.

(b) The terms of members of the Commission shall be six years except that (1) the members of the Commission first taking office shall serve, as designated by the President at the time of appointment, one for a term of two years, one for a term of four years, and one for a term of six years, and (2) a vacancy caused by the death, resignation, or removal of a member prior to the expiration of the term for which he was appointed shall be filled only for the remainder of such unexpired term. A member of the Commission may be removed by the President for inefficiency, neglect of duty, or malfeasance in office.

(j) An administrative law judge appointed by the Commission shall hear, and make a determination upon, any proceeding instituted before the Commission and any motion in connection therewith, assigned to such administrative law judge by the Chairman of the Commission, and shall make a report of any such determination which constitutes his final disposition of the proceedings. The report of the administrative law judge shall become the final order of the Commission within thirty days after such report by the administrative law judge unless within such period any Commission member has directed that such report shall be reviewed by the Commission.

## PROCEDURES TO COUNTERACT IMMINENT DANGERS

**SEC. 13.**   **(§ 662)** (a) The United States district courts shall have jurisdiction, upon petition of the Secretary, to restrain any conditions or practices in any place of employment which are such that a danger exists which could reasonably be expected to cause death or serious physical harm immediately or before the imminence of such danger can be eliminated through the enforcement procedures otherwise provided by this Act. Any order issued under this section may require such steps to be taken as may be necessary to avoid, correct, or remove such imminent danger and prohibit the employment or presence of any individual in locations or under conditions where such imminent danger exists, except individuals whose presence is necessary to avoid, correct, or remove such imminent danger or to maintain the capacity of a continuous process operation to resume normal operations without a complete cessation of operations, or where a cessation of operations is necessary, to permit such to be accomplished in a safe and orderly manner.

(b) Upon the filing of any such petition the district court shall have jurisdiction to grant such injunctive relief or temporary restraining order pending the outcome of an enforcement proceeding pursuant to this Act. The proceeding shall be as provided by Rule 65 of the Federal Rules, Civil Procedure, except that no temporary restraining order issued without notice shall be effective for a period longer than five days.

(c) Whenever and as soon as an inspector concludes that conditions or practices described in subsection (a) exist in any place of employment, he shall inform the affected employees and employers of the danger and that he is recommending to the Secretary that relief be sought.

(d) If the Secretary arbitrarily or capriciously fails to seek relief under this section, any employee who may be injured by reason of such failure, or the representative of such employees, might bring an action against the Secretary in the United States district court for the district in which the imminent danger is alleged to exist or the employer has its principal office, or for the District of Columbia, for a writ of mandamus to compel the Secretary to seek such an order and for such further relief as may be appropriate.

## REPRESENTATION IN CIVIL LITIGATION

**SEC. 14.**    **(§ 663)** Except as provided in section 518(a) of title 28, United States Code, relating to litigation before the Supreme Court, the Solicitor of Labor may appear for and represent the Secretary in any civil litigation brought under this Act but all such litigation shall be subject to the direction and control of the Attorney General.

## CONFIDENTIALITY OF TRADE SECRETS

**SEC. 15.**    **(§ 664)** All information reported to or otherwise obtained by the Secretary or his representative in connection with any inspection or proceeding under this Act which contains or which might reveal a trade secret referred to in section 1905 of title 18 of the United States Code shall be considered confidential for the purpose of that section, except that such information may be disclosed to other officers or employees concerned with carrying out this Act or when relevant in any proceeding under this Act. In any such proceeding the Secretary, the Commission, or the court shall issue such orders as may be appropriate to protect the confidentiality of trade secrets.

## VARIATIONS, TOLERANCES, AND EXEMPTIONS

**SEC. 16.**    **(§ 665)** The Secretary, on the record, after notice and opportunity for a hearing may provide such reasonable limitations and may make such rules and regulations allowing reasonable variations, tolerances, and exemptions to and from any or all provisions of this Act as he may find necessary and proper to avoid serious impairment of the national defense. Such action shall not be in effect for more than six months without notification to affected employees and an opportunity being afforded for a hearing.

## PENALTIES

**SEC. 17.**    **(§ 666)** (a) Any employer who willfully or repeatedly violates the requirements of section 5 of this Act, any standard, rule, or order promulgated pursuant to section 6 of this Act, or regulations prescribed pursuant to this Act, may be assessed a civil penalty of not more than $70,000 for each violation, but not less than $5,000 for each willful violation.

(b) Any employer who has received a citation for a serious violation of the requirements of section 5 of this Act, of any standard, rule, or order promulgated pursuant to section 6 of this Act, or of any regulations prescribed pursuant to this Act, shall be assessed a civil penalty of up to $7,000 for each such violation.

(c) Any employer who has received a citation for a violation of the requirements of section 5 of this Act, of any standard, rule, or order promulgated pursuant to section 6 of this Act, or of regulations prescribed pursuant to this Act, and such violation is specifically determined not to be of a serious nature, may be assessed a civil penalty of up to $7,000 for each such violation.

(d) Any employer who fails to correct a violation for which a citation has been issued under section 9(a) within the period permitted for its correction (which period shall not begin to run until the date of the final order of the Commission in the case of any review proceeding under section 10 initiated by the employer in good faith and not solely for delay or avoidance of penalties), may be assessed a civil penalty of not more than $7,000 for each day during which such failure or violation continues.

(e) Any employer who willfully violates any standard, rule, or order promulgated pursuant to section 6 of this Act, or of any regulations prescribed pursuant to this Act, and that violation caused death to any employee, shall, upon conviction, be punished by a fine of not more than $10,000 or by imprisonment for not more than six months, or by both: except that if the conviction is for a violation committed after a first conviction of such person, punishment shall be by a fine of not more than $20,000 or by imprisonment for not more than one year, or by both.

(f) Any person who gives advance notice of any inspection to be conducted under this Act, without authority from the Secretary or his designees, shall, upon conviction, be punished by a fine of not more than $1,000 or by imprisonment for not more than six months, or by both.

(g) Whoever knowingly makes any false statement, representation, or certification in any application, record, report, plan, or other document filed or required to be maintained pursuant to this Act shall, upon conviction, be punished by a fine of not more than $10,000, or by imprisonment for not more than six months, or by both.

(i) Any employer who violates any of the posting requirements, as prescribed under the provisions of this Act, shall be assessed a civil penalty of up to $7,000 for each violation.

(j) The Commission shall have authority to assess all civil penalties provided in this section, giving due consideration to the appropriateness of the penalty with respect to the size of the business of the employer being charged, the gravity of the violation, the good faith of the employer, and the history of previous violations.

(k) For purposes of this section, a serious violation shall be deemed to exist in a place of employment if there is a substantial probability that death or serious physical harm could result from a condition which exists, or from one or more practices, means, methods, operations, or processes which have been adopted or are in use, in such place of employment unless the employer did not, and could not with the exercise of reasonable diligence, know of the presence of the violation.

(*l*) Civil penalties owed under this Act shall be paid to the Secretary for deposit into the Treasury of the United States and shall accrue to the United States and may be recovered in a civil action in the name of the United States brought in the United States district court for the district where the violation is alleged to have occurred or where the employer has its principal office.

## STATE JURISDICTION AND STATE PLANS

**SEC. 18.**    **(§ 667)** (a) Nothing in this Act shall prevent any State agency or court from asserting jurisdiction under State law over any occupational safety or health issue with respect to which no standard is in effect under section 6.

(b) Any State which, at any time, desires to assume responsibility for development and enforcement therein of occupational safety and health standards relating to any occupational safety or health issue with respect to which a Federal standard has been promulgated under section 6 shall submit a State plan for the development of such standards and their enforcement.

(c) The Secretary shall approve the plan submitted by a State under subsection (b), or any modification thereof, if such plan in his judgment—

(1) designates a State agency or agencies as the agency or agencies responsible for administering the plan throughout the State,

(2) provides for the development and enforcement of safety and health standards relating to one or more safety or health issues, which standards (and the enforcement of which standards) are or will be at least as effective in providing safe and healthful employment and places of employment as the standards promulgated under section 6 which relate to the same issues, and which standards, when applicable to products which are distributed or used in interstate commerce, are required by compelling local conditions and do not unduly burden interstate commerce,

(3) provides for a right of entry and inspection of all workplaces subject to the Act which is at least as effective as that provided in section 8, and includes a prohibition on advance notice of inspections,

(4) contains satisfactory assurances that such agency or agencies have or will have the legal authority and qualified personnel necessary for the enforcement of such standards,

(5) gives satisfactory assurances that such State will devote adequate funds to the administration and enforcement of such standards,

(6) contains satisfactory assurances that such State will, to the extent permitted by its law, establish and maintain an effective and comprehensive occupational safety and health program applicable to all employees of public agencies of the State and its political subdivisions, which program in as effective as the standards contained in an approved plan,

(7) requires employers in the State to make reports to the Secretary in the same manner and to the same extent as if the plan were not in effect, and

(8) provides that the State agency will make such reports to the Secretary in such form and containing such information, as the Secretary shall from time to time require.

(d) If the Secretary rejects a plan submitted under subsection (b), he shall afford the State submitting the plan due notice and opportunity for a hearing before so doing.

(e) After the Secretary approves a State plan submitted under subsection (b), he may, but shall not be required to, exercise his authority under sections 8, 9, 10, 13, and 17 with respect to comparable standards promulgated under section 6, for the period specified in the next sentence. The Secretary may exercise the authority referred to above until he determines, on the basis of actual operations under the State plan, that the criteria set forth in subsection (c) are being applied, but he shall not make such determination for at least three years after the plan's approval under subsection (c). Upon making the determination referred to in the preceding sentence, the provisions of sections 5(a) (2), 8 (except for the purpose of carrying out subsection (f) of this section), 9, 10, 13, and 17, and standards promulgated under section 6 of this Act, shall not apply with respect to any occupational safety or health issues covered under the plan, but the Secretary may retain jurisdiction under the above provisions in any proceeding commenced under section 9 or 10 before the date of determination.

(f) The Secretary shall, on the basis of reports submitted by the State agency and his own inspections make a continuing evaluation of the manner in which each State having a plan approved under this section is carrying out such plan. Whenever the Secretary finds, after affording due notice and opportunity for a hearing, that in the administration of the State plan there is a failure to comply substantially with any provision of the State plan (or any assurance contained therein), he shall notify the State agency of his withdrawal of approval of such plan and upon receipt of such notice such plan shall cease to be in effect, but the State may retain jurisdiction in any case commenced before the withdrawal of the plan in order to enforce standards under the plan whenever the issues involved do not relate to the reasons for the withdrawal of the plan.

(g) The State may obtain a review of a decision of the Secretary withdrawing approval of or rejecting its plan by the United States court of appeals for the circuit in which the State is located by filing in such court within thirty days following receipt of notice of such decision a petition to modify or set aside in whole or in part the action of the Secretary. A copy of such petition shall forthwith be served upon the Secretary, and thereupon the Secretary shall certify and file in the court the record upon which the decision complained of was issued as provided in section 2112 of title 28, United States Code. Unless the court finds that the Secretary's decision in rejecting a proposed State plan or withdrawing his approval of such a plan is not supported by substantial evidence the court shall affirm the Secretary's decision. The judgment of the court shall be subject to review by the Supreme Court of the United States upon certiorari or certification as provided in section 1254 of title 28, United States Code.

(h) The Secretary may enter into an agreement with a State under which the State will be permitted to continue to enforce one or more occupational health and safety standards in effect in such State until final action is taken by the Secretary with respect to a plan submitted by a State under subsection (b) of this section, or two years from the date of enactment of this Act, whichever is earlier.

*Appendix*

# F

# Age Discrimination in Employment Act

## 29 U.S.C. §§ 621–634

STATEMENT OF FINDINGS AND PURPOSE

**SEC. 2   (§ 621)** (a) The Congress hereby finds and declares that—

(1) in the face of rising productivity and affluence, older workers find themselves disadvantaged in their efforts to retain employment, and especially to regain employment when displaced from jobs;

(2) the setting of arbitrary age limits regardless of potential for job performance has become a common practice, and certain otherwise desirable practices may work to the disadvantage of older persons;

(3) the incidence of unemployment, especially long-term unemployment with resultant deterioration of skill, morale, and employer acceptability is, relative to the younger ages, high among older workers; their numbers are great and growing; and their employment problems grave;

(4) the existence in industries affecting commerce of arbitrary discrimination in employment because of age burdens commerce and the free flow of goods in commerce.

(b) It is therefore the purpose of this Act to promote employment of older persons based on their ability rather than age; to prohibit arbitrary age discrimination in employment; to help employers and workers find ways of meeting problems arising from the impact of age on employment.

**SEC. 4   (§ 623)   Prohibition of age discrimination**

(a) It shall be unlawful for an employer—

(1) to fail or refuse to hire or to discharge any individual or otherwise discriminate against any indi-

vidual with respect to his compensation, terms, conditions, or privileges of employment, because of such individual's age;

(2) to limit, segregate, or classify his employees in any way which would deprive or tend to deprive any individual of employment opportunities or otherwise adversely affect his status as an employee, because of such individual's age; or

(3) to reduce the wage rate of any employee in order to comply with this chapter.

(b) It shall be unlawful for an employment agency to fail or refuse to refer for employment, or otherwise to discriminate against, any individual because of such individual's age, or to classify or refer for employment any individual on the basis of such individual's age.

(c) It shall be unlawful for a labor organization—

(1) to exclude or to expel from its membership, or otherwise to discriminate against, any individual because of his age;

(2) to limit, segregate, or classify its membership, or to classify or fail or refuse to refer for employment any individual, in any way which would deprive or tend to deprive any individual of employment opportunities, or would limit such employment opportunities or otherwise adversely affect his status as an employee or as an applicant for employment, because of such individual's age;

(3) to cause or attempt to cause an employer to discriminate against an individual in violation of this section.

(d) It shall be unlawful for an employer to discriminate against any of his employees or applicants for employment, for an employment agency to discriminate against any individual, or for a labor organization to discriminate against any member thereof or applicant for membership, because such individual, member or applicant for membership has opposed any practice made unlawful by this section, or because such individual, member or applicant for membership has made a charge, testified, assisted, or participated in any manner in an investigation, proceeding, or litigation under this chapter.

(e) It shall be unlawful for an employer, labor organization, or employment agency to print or publish, or cause to be printed or published, any notice or advertisement relating to employment by such an employer or membership in or any classification or referral for employment by such a labor organization, or relating to any classification or referral for employment by such an employment agency, indicating any preference, limitation, specification, or discrimination, based on age.

(f) It shall not be unlawful for an employer, employment agency, or labor organization—

(1) to take any action otherwise prohibited under subsections (a), (b), (c), or (e) of this section where age is a bona fide occupational qualification reasonably necessary to the normal operation of the particular business, or where the differentiation is based on reasonable factors other than age, or where such practices involve an employee in a workplace in a foreign country, and compliance with such subsections would cause such employer, or a corporation controlled by such employer, to violate the laws of the country in which such workplace is located;

(2) to observe the terms of a bona fide seniority system or any bona fide employee benefit plan such as a retirement, pension, or insurance plan, which is not a subterfuge to evade the purposes of this chapter, except that no such employee benefit plan shall excuse the failure to hire any individual, and no such seniority system or employee benefit plan shall require or permit the involuntary retirement of any individual specified by section 631(a) of this title because of the age of such individual; or

(3) to discharge or otherwise discipline an individual for good cause.

(g)(1) For purposes of this section, any employer must provide that any employee aged 65 through 69, and any employee's spouse aged 65 through 69, shall be entitled to coverage under any group health plan offered to such employees under the same conditions as any employee, and the spouse of such employee, under age 65.

(2) For purposes of paragraph (1), the term "group health plan" has the meaning given to such term in section 162(i)(2) of Title 26.

(h)(1) If an employer controls a corporation whose place of incorporation is in a foreign country, any practice by such corporation prohibited under this section shall be presumed to be such practice by such employer.

(2) The prohibitions of this section shall not apply where the employer is a foreign person not controlled by an American employer.

(3) For the purpose of this subsection the determination of whether an employer controls a corporation shall be based upon the—

    (A) interrelation of operations,

    (B) common management,

    (C) centralized control of labor relations, and

    (D) common ownership or financial control, of the employer and the corporation.

(i)(1) Except as otherwise provided in this subsection, it shall be unlawful for an employer, an employment agency, a labor organization, or any combination thereof to establish or maintain an employee pension benefit plan which requires or permits—

    (A) in the case of a defined benefit plan, the cessation of an employee's benefit accrual, or the reduction of the rate of an employee's benefit accrual, because of age, or

    (B) in the case of a defined contribution plan, the cessation of allocations to an employee's account, or the reduction of the rate at which amounts are allocated to an employee's account, because of age.

(2) Nothing in this section shall be construed to prohibit an employer, employment agency, or labor organization from observing any provision of an employee pension benefit plan to the extent that such provision imposes (without regard to age) a limitation on the amount of benefits that the plan provides or a limitation on the number of years of service or years of participation which are taken into account for purposes of determining benefit accrual under the plan.

[Editor's note: Both the previous section and the following section were designated as Section 4(i) by the language of the public laws that added them.]

(i) It shall not be unlawful for an employer which is a State, a political subdivision of a State, an agency or instrumentality of a State or a political subdivision of a State, or an interstate agency to fail refuse to hire or to discharge any individual because of such individual's age if such action is taken—

(1) with respect to the employment of an individual as a firefighter or as a law enforcement officer and the individual has attained the age of hiring or retirement in effect under applicable State or local law on March 3, 1983, and

(2) pursuant to a bona fide hiring or retirement plan that is not a subterfuge to evade the purposes of this Act. [The preceding section was added by Public Law 99–592, effective January 1, 1987, through December 31, 1993. It does not apply to any causes of action arising under ADEA as in effect before January 1, 1987. Section 5 of Public Law 99–592 directed EEOC and the Labor Department to conduct a study and make recommendations on the use of physical and mental fitness tests to measure the ability and competence of police officers and firefighters. In addition, by November, 1991, EEOC must propose guidelines for the administration and use of such tests.]

ADMINISTRATION

**SEC. 6  (§ 625)**  The Secretary shall have the power—

(a) to make delegations, to appoint such agents and employees, and to pay for technical assistance on a fee-for-service basis, as he deems necessary to assist him in the performance of his functions under this Act;

(b) to cooperate with regional, State, local, and other agencies, and to cooperate with and furnish technical assistance to employers, labor organizations, and employment agencies to aid in effectuating the purposes of this Act.

RECORDKEEPING, INVESTIGATION, AND ENFORCEMENT

**SEC. 7  (§ 626)**  (a) The Equal Employment Opportunity Commission shall have the power to make investigations and require the keeping of records necessary or appropriate for the administration of this Act in accordance with the powers and procedures provided in sections 9 and 11 of the Fair Labor Standards Act of 1938, as amended (29 U.S.C. 209 and 211).

(b) The provisions of this Act shall be enforced in accordance with the powers, remedies, and procedures provided in sections 11(b), 16 (except for subsection (a) thereof), and 17 of the Fair Labor Standards Act of 1938, as amended (29 U.S.C. 211(b), 216, 217) and subsection (c) of this section. Any act prohibited under section 4 of this Act shall be deemed to be a prohibited act under section 15 of the Fair Labor Standards Act of 1938, as amended (29 U.S.C. 215). Amounts owing to an individual as a result of a violation of this Act shall be deemed to be unpaid minimum wages or unpaid overtime compensation for purposes of sections 16 and 17 of the Fair Labor Standards Act of 1938, as amended (29 U.S.C. 216, 217): Provided, that liquidated damages shall be payable only in cases of willful violations of this Act. In any action brought to enforce this Act the court shall have jurisdiction to grant such legal or equitable relief as may be appropriate to effectuate the purposes of this Act, including without limitation judgments compelling employment, reinstatement or promotion, or enforcing the liability for amounts deemed to be unpaid minimum wages or unpaid overtime compensation under this section. Before instituting any action under this section, the Equal Employment Opportunity Commission shall attempt to eliminate the discriminatory practice or practices alleged, and to effect voluntary compliance with the requirements of this Act through informal methods of conciliation, conference, and persuasion.

(c)(1) Any person aggrieved may bring a civil action in any court of competent jurisdiction for such legal or equitable relief as will effectuate the purposes of this Act: Provided, that the right of any person to bring such action shall terminate upon the commencement of an action by the Equal Employment Opportunity Commission to enforce the right of such person under this Act.

(2) In an action brought under paragraph (1), a person shall be entitled to a trial by jury of any issue of fact in any such action for recovery of amounts owing as a result of a violation of this Act, regardless of whether equitable relief is sought by any party in such action.

(d) No civil action may be commenced by an individual under this section until 60 days after a charge alleging unlawful discrimination has been filed with the Equal Employment Opportunity Commission. Such a charge shall be filed—

(1) within 180 days after the alleged unlawful practice occurred; or

(2) in a case to which section 14(b) applies, within 300 days after the alleged unlawful

practice occurred, or within 30 days after receipt by the individual of notice of termination of proceedings under State law, whichever is earlier.

Upon receiving such a charge, the Commission shall promptly notify all persons named in such charge as prospective defendants in the action and shall promptly seek to eliminate any alleged unlawful practice by informal methods of conciliation, conference, and persuasion.

(e)(1) Sections 6 and 10 of the Portal-to-Portal Act of 1947 shall apply to actions under this Act.

(2) For the period during which the Equal Employment Opportunity Commission is attempting to effect voluntary compliance with requirements of this Act through informal methods of conciliation, conference, and persuasion pursuant to subsection (b), the statute of limitations as provided in section 6 of the Portal-to-Portal Act of 1947 shall be tolled, but in no event for a period in excess of one year.

Notwithstanding section 7(e), a civil action may be brought under section 7 by the Commission or an aggrieved person, during the 540–day period beginning on the date of enactment of this Act [April 7, 1988] if—

(1) with respect to the alleged unlawful practice on which the claim in such civil action is based, a charge was timely filed under such Act with the Commission after December 31, 1983,

(2) the Commission did not, within the applicable period set forth in section 7(e) either—

(A) eliminate such alleged unlawful practice by informal methods of conciliation, conference, and persuasion, or

(B) notify such person, in writing, of the disposition of such charge and of the right of such person to bring a civil action on such claim,

(3) the statute of limitations applicable under such section 7(e) to such claim ran before the date of enactment of this Act, and

(4) a civil action on such claim was not brought by the Commission or such person before the running of the statute of limitations.

## NOTICES TO BE POSTED

**SEC. 8    (§ 627)**   Every employer, employment agency, and labor organization shall post and keep posted in conspicuous places upon its premises a notice to be prepared or approved by the Equal Employment Opportunity Commission setting forth information as the Commission deems appropriate to effectuate the purposes of this Act.

## CRIMINAL PENALTIES

**SEC. 10    (§ 629)**   Whoever shall forcibly resist, oppose, impede, intimidate, or interfere with a duly authorized representative of the Equal Employment Opportunity Commission while it is engaged in the performance of duties under this Act shall be punished by a fine of not more than $500 or by imprisonment for not more than one year, or by both: Provided, however, that no person shall be imprisoned under this section except when there has been a prior conviction hereunder.

## DEFINITIONS

**SEC. 11    (§ 630)**   For the purposes of this Act—

(a) The term "person" means one or more individuals, partnerships, associations, labor organizations, corporations, business trusts, legal representatives, or any organized groups of persons.

(b) The term "employer" means a person engaged in an industry affecting commerce who has twenty or more employees for each working day in each of twenty or more calendar weeks in the current or preceding calendar year: Provided, that prior to June 30, 1968, employers having fewer than fifty employees shall not be considered employers. The term also means (1) any agent of such a person, and (2) a State or political subdivision of a State and any agency or instrumentality of a State or a political subdivision of a State, and any interstate agency but such term does not include the United States, or a corporation wholly owned by the Government of the United States.

(c) The term "employment agency" means any person regularly undertaking with or without compensation to procure employees for an employer and includes an agent of such a person; but shall not include an agency of the United States.

(d) The term "labor organization" means a labor organization engaged in an industry affecting commerce, and any agent of such an organization, and includes any organization of any kind, any agency, or employee representation committee, group, association, or plan so engaged in which employees participate and which exists for the purpose, in whole or in part, of dealing with employers concerning grievances, labor disputes, wages, rates of pay, hours, or other terms or conditions of employment, and any conference, general committee, joint or system board, or joint council so engaged which is subordinate to a national or international labor organization.

(e) A labor organization shall be deemed to be engaged in an industry affecting commerce if (1) it maintains or operates a hiring hall or hiring office which procures employees for an employer or procures for employees opportunities to work for an employer, or (2) the number of its members (or, where it is a labor organization composed of other labor organizations or their representatives, if the aggregate number of the members of such other labor organization) is fifty or more prior to July 1, 1968, or twenty-five or more on or after July 1, 1968, and such labor organization—

(1) is the certified representative of employees under the provisions of the National Labor Relations Act, as amended, or the Railway Labor Act, as amended; or

(2) although not certified, is a national or international labor organization or a local labor organization recognized or acting as the representative of employees of an employer or employers engaged in an industry affecting commerce; or

(3) has chartered a local labor organization or subsidiary body which is representing or actively seeking to represent employees or employers within the meaning of paragraph (1) or (2); or

(4) has been chartered by a labor organization representing or actively seeking to represent employees within the meaning of paragraph (1) or (2) as the local or subordinate body through which such employees may enjoy membership of become affiliated with such labor organization; or

(5) is a conference, general committee, joint or system board or joint council subordinate to a national or international labor organization, which includes a labor organization engaged in an industry affecting commerce within the meaning of any of the preceding paragraphs of this subsection.

(f) The term "employee" means an individual employed by an employer except that the term "employee" shall not include any person elected to public office in any State or political subdivision of any State by the qualified voters thereof, or any person chosen by such officer to be on such officer's personal staff, or an appointee on the policy-making level or an immediate adviser with respect to the exercise of the constitutional or legal powers of the office. The exemption set forth in the preceding sentence shall not include employees subject to the civil service laws of a State government, governmental agency, or political subdivision. The term "employee" includes any individual who is a citizen of the United States employed by an employer in a workplace in a foreign country.

(g) The term "commerce" means trade, traffic, commerce, transportation, transmission, or communication among the several States, or between a State and any place outside thereof; or within the District of Columbia, or a possession of the United States, or between points in the same State but through a point outside thereof.

(h) The term "industry affecting commerce" means any activity, business, or industry in commerce or in which a labor dispute would hinder or obstruct commerce or the free flow of commerce and includes any activity or industry "affecting commerce" within the meaning of the Labor-Management Reporting and Disclosure Act of 1959.

(i) The term "State" includes a State of the United States, the District of Columbia, Puerto Rico, the Virgin Islands, American Samoa, Guam, Wake Island, the Canal Zone, and Outer Continental Shelf Lands defined in the Outer Continental Shelf Lands Act.

(j) The term "firefighter" means an employee, the duties of whose position are primarily to perform work directly connected with the control and extinguishment of fires or the maintenance and use of firefighting apparatus and equipment, including an employee engaged in this activity who is transferred to a supervisory or administrative position.

(k) The term "law enforcement officer" means an employee, the duties of whose position are primarily the investigation, apprehension, or detention of individuals suspected or convicted of offenses against the criminal laws of a State, including an employee engaged in this activity who is transferred to a supervisory or administration position. For the purpose of this subsection, "detention" includes the duties of employees assigned to guard individuals incarcerated in any penal institution.

LIMITATION

SEC. 12    (§ 631) (a) The prohibitions in this chapter shall be limited to individuals who are at least 40 years of age.

(b) In the case of any personnel action affecting employees or applicants for employment which is subject to the provisions of section 15 of this Act, the prohibitions established in section 15 of this Act shall be limited to individuals who are at least 40 years of age.

(c)(1) Nothing in this chapter shall be construed to prohibit compulsory retirement of any employee who has attained 65 years of age, and who, for the two-year period immediately before retirement, is employed in a bona fide executive or a high policymaking position, if such employee is entitled to an immediate nonforfeitable annual retirement benefit from a pension, profitsharing, savings, or deferred compensation plan, or any combination of such plans, of the employer of such employee, which equals, in aggregate, at least $44,000.

(2) In applying the retirement benefit test of paragraph (1) of this subsection, if any such retirement benefit is in a form other than a straight life annuity (with no ancillary benefits), or if employees contribute to any such plan or make rollover contributions, such benefit shall be adjusted in accordance with regulations prescribed by the Equal Employment Opportunity Commission, after consultation with the Secretary of the Treasury, so that the benefit is the equivalent of a straight life annuity (with no ancillary benefits) under a plan

to which employees do not contribute and under which no rollover contributions are made.

(d) Nothing in this Act shall be construed to prohibit compulsory retirement of any employee who has attained 70 years of age, and who is serving under a contract of unlimited tenure (or similar arrangement providing for unlimited tenure) at an institution of higher education (as defined by section 1141(a) of Title 20).

## FEDERAL–STATE RELATIONSHIP

**SEC. 14  (§ 633)** (a) Nothing in this Act shall affect the jurisdiction of any agency of any State performing like functions with regard to discriminatory employment practices on account of age except that upon commencement of an action under this Act such action shall supersede any State action.

(b) In the case of an alleged unlawful practice occurring in a State which has a law prohibiting discrimination in employment because of age and establishing or authorizing a State authority to grant or seek relief from such discriminatory practice, no suit may be brought under section 7 of this Act before the expiration of sixty days after proceedings have been commenced under the State law, unless such proceedings have been earlier terminated, provided that such sixty-day period shall be extended to one hundred and twenty days during the first year after the effective date of such State law. If any requirement for the commencement of such proceedings is imposed by a State authority other than a requirement of the filing of a written and signed statement of the facts upon which the proceeding is based, the proceeding shall be deemed to have been commenced for the purposes of this subsection at the time such statement is sent by registered mail to the appropriate State authority.

[Section 15 of the ADEA, 29 U.S.C. § 633a, prohibits discrimination on account of age in federal government employment, on bases similar to the rest of the Act.]

[In addition to the more widely known ADEA of 1967, the Age Discrimination Act of 1975, 42 U.S.C. § 6101 et seq., prohibits discrimination on the basis of age by recipients of federal financial assistance. This act does not in any way affect enforcement of the ADEA of 1967. The act does not specify age limits for coverage.]

# National Labor Relations Act

## 29 U.S.C. §§ 151–169

### RIGHTS OF EMPLOYEES

**SEC. 7.**   **(§ 157)** Employees shall have the right to self-organization, to form, join, or assist labor organizations, to bargain collectively through representatives of their own choosing, and to engage in other concerted activities for the purpose of collective bargaining or other mutual aid or protection, and shall also have the right to refrain from any or all of such activities except to the extent that such right may be affected by an agreement requiring membership in a labor organization as a condition of employment as authorized in section 8(a)(3).

### UNFAIR LABOR PRACTICES

**SEC. 8.**   **(§ 158)** (a) It shall be an unfair labor practice for an employer—

(1) to interfere with, restrain, or coerce employees in the exercise of the rights guaranteed in section 7;

(2) to dominate or interfere with the formation or administration of any labor organization or contribute financial or other support to it: *Provided,* That subject to rules and regulations made and published by the Board pursuant to section 6, an employer shall not be prohibited from permitting employees to confer with him during working hours without loss of time or pay;

(3) by discrimination in regard to hire or tenure of employment or any term or condition of

employment to encourage or discourage membership in any labor organization: *Provided,* That nothing in this Act, or in any other statute of the United States, shall preclude an employer from making an agreement with a labor organization (not established, maintained, or assisted by any action defined in this Act as an unfair labor practice) to require as a condition of employment membership therein on or after the thirtieth day following the beginning of such employment or the effective date of such agreement, whichever is the later, (i) if such labor organization is the representative of the employees as provided in section 9(a), in the appropriate collective-bargaining unit covered by such agreement when made, and (ii) unless following an election held as provided in section 9(e) within one year preceding the effective date of such agreement, the Board shall have certified that at least a majority of the employees eligible to vote in such election have voted to rescind the authority of such labor organization to make such an agreement: *Provided further,* That no employer shall justify any discrimination against an employee for nonmembership in a labor organization (A) if he has reasonable grounds for believing that such membership was not available to the employee on the same terms and conditions generally applicable to other members, or (B) if he has reasonable grounds for believing that membership was denied or terminated for reasons other than the failure of the employee to tender the periodic dues and the initiation fees uniformly required as a condition of acquiring or retaining membership;

(4) to discharge or otherwise discriminate against an employee because he has filed charges or given testimony under this Act;

(5) to refuse to bargain collectively with the representatives of his employees, subject to the provisions of section 9(a).

(b) It shall be an unfair labor practice for a labor organization or its agents—

(1) to restrain or coerce (A) employees in the exercise of the rights guaranteed in section 7: *Provided,* That this paragraph shall not impair the right of a labor organization to prescribe its own rules with respect to the acquisition or retention of membership therein; or (B) an employer in the selection of his representatives for the purposes of collective bargaining or the adjustment of grievances;

(2) to cause or attempt to cause an employer to discriminate against an employee in violation of subsection (a)(3) or to discriminate against an employee with respect to whom membership in such organization has been denied or terminated on some ground other than his failure to tender the periodic dues and the initiation fees uniformly required as a condition of acquiring or retaining membership;

(3) to refuse to bargain collectively with an employer, provided it is the representative of his employees subject to the provisions of section 9(a);

(4)(i) to engage in, or to induce or encourage any individual employed by any person engaged in commerce or in an industry affecting commerce to engage in, a strike or a refusal in the course of his employment to use, manufacture, process, transport, or otherwise handle or work on any goods, articles, materials, or commodities or to perform any services; or (ii) to threaten, coerce, or restrain any person engaged in commerce or in an industry affecting commerce, where in either case an object thereof is:

(A) forcing or requiring any employer or self-employed person to join any labor or employer organization or to enter into any agreement which is prohibited by section 8(e);

(B) forcing or requiring any person to cease using, selling, handling, transporting, or otherwise dealing in the products of any other producer, processor, or manufacturer, or to cease doing business with any other person, or forcing or requiring any other employer to recognize or bargain with a labor organization as the representative of his employees unless such labor organization has been certified as the representative of such employees under the provisions of section 9: *Provided,* That nothing contained in this clause (B) shall be construed to make unlawful, where not otherwise unlawful, any primary strike or primary picketing;

(C) forcing or requiring any employer to recognize or bargain with a particular labor organization as the representative of his employees if another labor organization has been certified as the representative of such employees under the provisions of section 9;

(D) forcing or requiring any employer to assign particular work to employees in a particular labor organization or in a particular trade, craft, or class rather than to employees in another labor organization or in another trade, craft, or class, unless such employer is failing to conform to an order or certification of the Board determining the bargaining representative for employees performing such work:

*Provided,* That nothing contained in this subsection [8](b) shall be construed to make unlawful a refusal by any person to enter upon the premises of any employer (other than his own employer), if the employees of such employer are engaged in a strike ratified or approved by a representative of such employees whom such employer is required to recognize under this Act: *Provided further,* That for the purposes of this paragraph (4) only, nothing contained in such paragraph shall be construed to prohibit publicity, other than picketing, for the purpose of truthfully advising the public, including consumers and members of a labor organization, that a product or products are produced by an employer with whom the labor organization has a primary dispute and are distributed by another employer, as long as such publicity does not have an effect of inducing any individual employed by any person other than the primary employer in the course of his employment to refuse to pick up, deliver, or transport any goods, or not to perform any services, at the establishment of the employer engaged in such distribution;

(5) to require of employees covered by an agreement authorized under subsection (a)(3) the payment, as a condition precedent to becoming a member of such organization, of a fee in an amount which the Board finds excessive or discriminatory under all the circumstances. In making such a finding, the Board shall consider, among other relevant factors, the practices and customs of labor organizations in the particular industry, and the wages currently paid to the employees affected;

(6) to cause or attempt to cause an employer to pay or deliver or agree to pay or deliver any money or other thing of value, in the nature of an exaction, for services which are not performed or not to be performed; and

(7) to picket or cause to be picketed, or threaten to picket or cause to be picketed, any employer where an object thereof is forcing or requiring an employer to recognize or bargain with a labor organization as the representative of his employees, or forcing or requiring the employees of an employer to accept or select such labor organization as their collective bargaining representative, unless such labor organization is currently certified as the representative of such employees:

(A) where the employer has lawfully recognized in accordance with this Act any other labor organization and a question concerning representation may not appropriately be raised under section 9(c) of this Act,

(B) where within the preceding twelve months a valid election under section 9(c) of this Act has been conducted, or

(C) where such picketing has been conducted without a petition under section 9(c) being filed within a reasonable period of time not to exceed thirty days from the commencement of such picketing: *Provided,* That when such a petition has been filed the Board shall forthwith, without regard to the provisions of section 9(c)(1)or the absence of a showing of a substantial interest on the part of the labor organization, direct an election in such unit as the Board finds to be appropriate and shall certify the results thereof: *Provided further,* That nothing in this subparagraph (C) shall be construed to prohibit any picketing or other publicity for the purpose of truthfully advising the public (including consumers) that an employer does not employ members of, or have a contract with, a labor organization, unless an effect of such picketing is to induce any individual employed by any other person in the course of his employment, not to pick up, deliver or transport any goods or not to perform any services.

Nothing in this paragraph (7) shall be construed to permit any act which would otherwise be an unfair labor practice under this section 8(b).

(c) The expressing of any views, argument, or opinion, or the dissemination thereof, whether in written, printed, graphic, or visual form, shall not constitute or be evidence of an unfair labor practice under any of the provisions of this Act, if such expression contains no threat of reprisal or force or promise of benefit.

(d) For the purposes of this section, to bargain collectively is the performance of the mutual obligation of the employer and the representative of the employees to meet at reasonable times and confer in good faith with respect to wages, hours, and other terms and conditions of employment, or the negotiation of an agreement, or any question arising thereunder, and the execution of a written contract incorporating any agreement reached if requested by either party, but such obligation does not compel either party to agree to a proposal or require the making of a concession: *Provided,* That where there is in effect a collective bargaining contract covering employees in an industry affecting commerce, the duty to bargain collectively shall also mean that no party to such contract shall terminate or modify such contract, unless the party desiring such termination or modification—

(1) serves a written notice upon the other party to the contract of the proposed termination or modification sixty days prior to the expiration date thereof, or in the event such contract contains no expiration date, sixty days prior to the time it is proposed to make such termination or modification;

(2) offers to meet and confer with the other party for the purpose of negotiating a new contract or a contract containing the proposed modifications;

(3) notifies the Federal Mediation and Conciliation Service within thirty days after such notice of the existence of a dispute, and simultaneously therewith notifies any State or Territorial agency established to mediate and conciliate disputes within the State or Territory where the dispute occurred, provided no agreement has been reached by that time; and

(4) continues in full force and effect, without resorting to strike or lockout, all the terms and conditions of the existing contract for a period of sixty days after such notice is given or until the expiration date of such contract, whichever occurs later:

The duties imposed upon employers, employees, and labor organizations by paragraphs (2)–(4) of this subsection shall become inapplicable upon an intervening certification of the Board, under which the labor organization or individual, which is a party to the contract, has been superseded as or ceased to be the representative of the employees subject to the provisions of section 9(a) of this Act, and the duties so imposed shall not be construed as requiring either party to discuss or agree to any modification of the terms and conditions contained in a contract for a fixed period, if such modification is to become effective before such terms and conditions can be reopened under the provisions of the contract. Any employee who engages in a strike within any notice period specified in this subsection, or who engages in any strike within the appropriate period specified in subsection (g) of this section, shall lose his status as an employee of the employer engaged in the particular labor dispute, for the purposes of sections 8 to 10 of this Act, but such loss of status for such employee shall terminate if and when he is reemployed by such employer. Whenever the collective bargaining involves employees of a health care institution, the provisions of this subsection shall be modified as follows:

(A) The notice of paragraph (1) of this subsection shall be ninety days; the notice of paragraph (3) of this subsection shall be sixty days; and the contract period of paragraph (4) of this subsection shall be ninety days.

(B) Where the bargaining is for an initial agreement following certification or recognition, at least thirty days' notice of the existence of a dispute shall be given by the labor organization to the agencies set forth in paragraph (3) of this subsection.

(C) After notice is given to the Federal Mediation and Conciliation Service under either clause (A) or (B) of this sentence, the Service shall promptly communicate with the parties and use its best efforts, by mediation and conciliation, to bring them to agreement. The parties shall participate fully and promptly in such meetings as may be undertaken by the Service for the purpose of aiding in a settlement of the dispute.

(e) It shall be an unfair labor practice for any labor organization and any employer to enter into any contract or agreement, express or implied, whereby such employer ceases or refrains or agrees to cease or refrain from handling, using, selling, transporting or otherwise dealing in any of the products of any other employer, or to cease doing business with any other person, and any contract or agreement entered into heretofore or hereafter containing such an agreement shall be to such extent unenforceable and void: *Provided,* That nothing in this subsection (e) shall apply to an agreement between a labor organization and an employer in the construction industry relating to the contracting or subcontracting of work to be done at the site of the construction, alteration, painting, or repair of a building, structure, or other work: *Provided further,* That for the purposes of this subsection (e) and section 8(b)(4)(B) the terms "any employer", "any person engaged in commerce or in industry affecting commerce", and "any person" when used in relation to the terms "any other producer, processor, or manufacturer", "any other employer", or "any other person" shall not include persons in the relation of a jobber, manufacturer, contractor, or subcontractor working on the goods or premises of the jobber or manufacturer or performing parts of an integrated process of production in the apparel and clothing industry: *Provided further,* That nothing in this Act shall prohibit the enforcement of any agreement which is within the foregoing exception.

(f) It shall not be an unfair labor practice under subsections (a) and (b) of this section for an employer engaged primarily in the building and construction industry to make an agreement covering employees engaged (or who, upon their employment, will be engaged) in the building and construction industry with a labor organization of which building and construction employees are members (not established, maintained, or assisted by any action defined in section 8(a) of this Act as an unfair labor practice) because (1) the majority status of such labor organization has not been established under the provisions of section 9 of this Act prior to the making of such agreement, or (2) such agreement requires as a condition of employment, membership in such labor organization after the seventh day following the beginning of such employment or the effective date of the agreement, whichever is later, or (3) such agreement requires the employer to notify such labor organization of opportunities for employment with such employer, or gives such labor organization an opportunity to refer qualified applicants for such employment, or (4) such agreement specifies minimum training or experience qualifications for employment or provides for priority in opportunities for employment based upon length of service with such employer, in the industry or in the particular geographical area: *Provided,* That nothing in this subsection shall set aside the final proviso to section 8(a)(3) of this Act: *Provided further,* That any agreement which would be invalid, but for clause (1) of this subsection, shall not be a bar to a petition filed pursuant to section 9(c) or 9(e).

(g) A labor organization before engaging in any strike, picketing, or other concerted refusal to work at any health care institution shall, not less than ten days prior to such action, notify the institution in writing and the Federal Mediation and Conciliation Service of that intention, except that in the case of bargaining for an initial agreement following certification or recognition the notice required by this subsection shall not be given until the expiration of the period specified in clause (B) of the last sentence of subsection (d) of this section. The notice shall state the date and time that such action will commence. The notice, once given, may be extended by the written agreement of both parties.

REPRESENTATIVES AND ELECTIONS

**SEC. 9.    (§ 159)** (a) Representatives designated or selected for the purposes of collective bargaining by the majority of the employees in a unit appropriate for such purposes, shall be the exclusive representatives of all the employees in such unit for the purposes of collective bargaining in respect to rates of pay, wages, hours of employment, or other conditions of employment: *Provided,* That any individual employee or a group of employees shall have the right at any time to present grievances to their employer and to have such grievances adjusted, without the intervention of the bargaining representative, as long as the adjustment is not inconsistent with the terms of a collective-bargaining contract or agreement then in effect: *Provided further,* That the bargaining representative has been given opportunity to be present at such adjustment.

(b) The Board shall decide in each case whether, in order to assure to employees the fullest freedom in exercising the rights guaranteed by this Act, the unit appropriate for the purposes of collective bargaining shall be the employer unit, craft unit, plant unit, or subdivision thereof: *Provided,* That the Board shall not (1) decide that any unit is appropriate for such purposes if such unit includes both professional employees and employees who are not professional employees unless a majority of such professional employees vote for inclusion in such unit; or (2) decide that any craft unit is inappropriate for such purposes on the ground that a different unit has been established by a prior Board determination, unless a majority of the employees in the proposed craft unit vote against separate representation or (3) decide that any unit is appropriate for such purposes if it includes, together with other employees, any individual employed as a guard to enforce against employees and other persons rules to protect property of the employer or to protect the safety or persons on the employer's premises; but not labor organization shall be certified as the representative of employees in a bargaining unit of guards if such organization admits to membership, or is affiliated directly or indirectly with an organization which admits to membership, employees other than guards.

(c)(1) Wherever a petition shall have been filed, in accordance with such regulations as may be prescribed by the Board—

(A) by an employee or group of employees or any individual or labor organization acting in their behalf alleging that a substantial number of employees (i) wish to be represented for collective bargaining and that their employer declines to recognize their representative as the representative defined in section 9(a), or (ii) assert that the individual or labor organization, which has been certified or is being currently recognized by their employer as the bargaining representative, is no longer a representative as defined in section 9(a);

(B) by an employer, alleging that one or more individuals or labor organizations have presented to him a claim to be recognized as the representative defined in section 9(a);

the Board shall investigate such petition and if it has reasonable cause to believe that a question of representation affecting commerce exists shall provide for an appropriate hearing upon due notice. Such hearing may be conducted by an officer or employee of the regional office, who shall not make any recommendations with respect thereto. If the Board finds upon the record of such hearing that such a question of representation exists, it shall direct an election by secret ballot and shall certify the results thereof.

(2) In determining whether or not a question of representation affecting commerce exists, the same regulations and rules of decision shall apply irrespective of the identity of the persons filing the petition or the kind of relief sought and in no case shall the Board deny a labor organization a place on the ballot by reason of an order with respect to such labor organization or its predecessor not issued in conformity with section 10(c).

(3) No election shall be directed in any bargaining unit or any subdivision within which, in the preceding twelve-month period, a valid election shall have been held. Employees engaged in an economic strike who are not entitled to reinstatement shall be eligible to vote under such regulations as the Board shall find are consistent with the purposes and provisions of this Act in any election conducted within twelve months after the commencement of the strike. In any election where none of the choices on the ballot receives a majority, a run-off shall be conducted, the ballot providing for a selection between the two choices receiving the largest and second largest number of valid votes cast in the election.

(4) Nothing in this section shall be construed to prohibit the waiving of hearings by stipulation for the purpose of a consent election in conformity with regulations and rules of decision of the Board.

(5) In determining whether a unit is appropriate for the purposes specified in subsection (b) the extent to which the employees have organized shall not be controlling.

(d) Whenever an order of the Board made pursuant to section 10(c) is based in whole or in part upon facts certified following an investigation pursuant to subsection (c) of this section and there is a petition for the enforcement or review of such order, such certification and the record of such investigation shall be included in the transcript of the entire record required to be filed under section 10(e) or 10(f), and thereupon the decree of the court enforcing, modifying, or setting aside in whole or in part the order of the Board shall be made and entered upon the pleadings, testimony, and proceedings set forth in such transcript.

(e)(1) Upon the filing with the Board, by 30 per centum or more of the employees in a bargaining unit covered by an agreement between their employer and a labor organization made pursuant to section 8(a)(3), of a petition alleging they desire that such authority be rescinded, the Board shall take a secret ballot of the employees in such unit and certify the results thereof to such labor organization and to the employer.

(2) No election shall be conducted pursuant to this subsection in any bargaining unit or any subdivision within which, in the preceding twelve-month period, a valid election shall have been held.

LIMITATIONS

**SEC. 13.**   **(§ 163)** Nothing in this Act, except as specifically provided for herein, shall be construed so as either to interfere with or impede or diminish in any way the right to strike, or to affect the limitations or qualifications on that right.

**SEC. 14.**   **(§ 164)** (a) Nothing herein shall prohibit any individual employed as a supervisor from becoming or remaining a member of a labor organization, but no employer subject to this Act shall be compelled to deem individuals defined herein as supervisors as employees for the purpose of any law, either national or local, relating to collective bargaining.

(b) Nothing in this Act shall be construed as authorizing the execution or application of agreements requiring membership in a labor organization as a condition of employment in any State or Territory in which such execution or application is prohibited by State or Territorial law.

(c) (1) The Board, in its discretion, may, by rule of decision or by published rules adopted pursuant to the Administrative Procedure Act, decline to assert jurisdiction over any labor dispute involving any class or category of employees, where, in the opinion of the Board, the effect of such labor dispute on commerce is not sufficiently substantial to warrant the exercise of its jurisdiction: *Provided,* That the Board shall not decline to assert jurisdiction over any labor dispute over which it would assert jurisdiction under the standards prevailing upon August 1, 1959.

(2) Nothing in this Act shall be deemed to prevent or bar any agency or the courts of any State or Territory (including the Commonwealth of Puerto Rico, Guam, and the Virgin Islands), from assuming and asserting jurisdiction over labor disputes over which the Board declines, pursuant to paragraph (1) of this subsection, to assert jurisdiction.

**SEC. 19.**   **(§ 169)** Any employee who is a member of and adheres to established and traditional tenets or teachings of a bona fide religion, body, or sect which has historically held conscientious objections to joining or financially supporting labor organizations shall not be required to join or financially support any labor organization as a condition of employment; except that such employee may be required in a contract between such employees' employer and a labor organization in lieu of periodic dues and initiation fees, to pay sums equal to such dues and initiation fees to a nonreligious, nonlabor organization charitable fund exempt from taxation under section 591(c)(3) of Title 26, chosen by such employee from a list of at least three such funds, designated in such contract or if the contract fails to designate such funds, then to any such fund chosen by the employee. If such employee who holds conscientious objections pursuant to this section requests the labor organization to use grievance-arbitration procedure on the employee's behalf, the labor organization is authorized to charge the employee for the reasonable cost of using such procedure.

# Glossary

**Affirm**  To confirm positively.

**Agency Shop**  A place of employment where union membership is not required, but payment of union fees is mandatory.

**Americans with Disabilities Act**  The federal statute that protects disabled workers in many environments.

**ANSI**  American National Standards Institute.

**Appellate**  Pertaining to or having cognizance of appeals and other proceedings for the judicial review of adjudications. The term has a general meaning, and it has a specific meaning indicating the distinction between original jurisdiction and appellate jurisdiction.

**Assault**  Any willful attempt or threat to inflict injury upon the person of another, when coupled with an apparent present ability so to do, and any intentional display of force such as would give the victim reason to fear or expect immediate bodily harm constitutes an assault. An assault may be committed without actually touching, or striking, or doing bodily harm, to the person of another.

**Assumption of the Risk Doctrine**  A concept whereby a person may not redeem for an injury resulting from voluntarily entering into a situation known to be dangerous.

**Attractive Nuisance**  An attractive object or feature—created by someone either on his or her own premises, a public place, or another's premises—that may be reasonably observed as being a source of danger thereby requiring that precautions be taken.

**Battery**  Intentional and wrongful physical contact with a person without his or her consent that entails some injury or offensive touching.

**Breach**  The breaking or violating of a law, right, obligation, engagement, or duty, either by commission or omission. Exists where one party to contract fails to carry out term, promise, or condition of the contract.

**Burden of Proof**  The necessity or duty to prove the facts positively in a dispute on the issue raised between the parties in a cause.

**Caveat Emptor**  Let the buyer beware.

**CFR**  Code of Federal Regulations.

**Case Law**  The aggregate of reported cases as forming a body of jurisprudence, or the law of a particular subject as evidenced or formed by the adjudged cases, in distinction to statutes and other sources of law. It includes the aggregate of reported cases that interpret statutes, regulations, and constitutional provisions. (*See* Common Law.)

**Circuit Court**  Courts whose jurisdiction extends over several counties or districts, and of which terms are held in the various counties or districts to which their jurisdiction extends.

**Citation**  A writ issued by the court requiring the person named to appear on a specific day or to show just cause as to why he or she should not.

**Civil Law**  The body of law that every particular nation, commonwealth, or city has established peculiarly for itself, more properly called *municipal* law, to distinguish it from the law of nature, and from international law. Laws concerned with civil or private rights and remedies as contrasted with criminal laws.

**Civil Rights Act of 1964**  Legislation barring discrimination based on race, color, religion, national origin, or sex in virtually all settings.

**Closed Shop**  A place of employment where prospective employees are required to become union members prior to being hired as a result of an agreement between the union and employer.

**Code**  A system of principles or rules usually required by law.

**Collective Bargaining Agreement** A contract that regulates the terms and conditions between an employer and the labor union.

**Common Law**  As distinguished from statutory law created by the enactment of legislatures, the common law comprises the body of those principles and rules of action relating to the government and security of persons and property that derive their authority solely from usages and customs of immemorial antiquity, or from the judgments and decrees of the courts recognizing, affirming, and enforcing such usages and customs; and, in this sense, particularly the ancient unwritten law of England. In general, it is a body of law that develops and derives through judicial decisions, as distinguished from legislative enactments. The common law is all the statutory and case law background of England and the American colonies before the American revolution. It consists of those principles, usage and rules of action applicable to government and security of persons and property that do not rest for their authority upon any express and positive declaration of the will of the legislature.

**Compensatory Damages**  Payment awarded to an injured party for the restoration of his or her position prior to injury.

**Contract**  An agreement between two or more persons that creates an obligation to do or not to do a particular thing, as defined in Restatement. Second, Contracts § 3: "A contract is a promise or a set of promises for the breach of which the law gives a remedy, or the performance of which the law in some way recognizes as a duty." A legal relationship consisting of the rights and duties of the contracting parties, a promise or set of promises constituting an agreement between the parties that gives each a legal duty to the other and also the right to seek a remedy for the breach of those duties. Its essentials are competent parties, subject

matter, a legal consideration, mutuality of agreement, and mutuality of obligation.

**Criminal Law**  The substantive criminal law is that law that, for the purpose of preventing harm to society, (a) declares what conduct is criminal, and (b) prescribes the punishment to be imposed for such conduct. It includes the definition of specific offenses and general principles of liability. Substantive criminal laws are commonly codified into criminal or penal codes; e.g., U.S.C.A. Title 18, California Penal Code, Model Penal Code.

**Deep Pocket**  A person or corporation of substantial wealth and resources from which a claim or judgment may be made. Under the deep pocket theory in antitrust law, parent corporation's substantial assets will have an impact on competition in which subsidiary is engaged.

**Defamation**  An intentional false communication, either published or publicly spoken, that injures another's reputation or good name. Holding up of a person to ridicule, scorn, or contempt in a respectable and considerable part of the community; may be criminal as well as civil. Includes both libel and slander.

**Defendant**  The person defending or denying; the party against whom relief or recovery is sought in an action or suit or the accused in a criminal case.

**Dicta**  Opinions of a judge that do not embody the resolution or determination of the specific case before the court. Expressions in court's opinion that go beyond the facts before court and therefore are individual views of author of opinion and not binding in subsequent cases as legal precedent.

**District Court**  Each state is comprised of one or more federal judicial districts, and in each district there is a district court. 28 U.S.C.A. § 81 et seq. The United States district courts are the trial courts with general federal jurisdiction over cases involving federal laws or offenses and actions between citizens or different states. Each state has at least one district court, although many have several judicial districts (e.g., northern, southern, middle districts) or divisions. There is also a U.S. District Court in the District of Columbia. In addition, the Commonwealth of Puerto Rico has a U.S. District Court with jurisdiction corresponding to that of district courts in the various states. Only one judge is usually required to hear and decide a case in a district court, but in some kinds of cases it is required that three judges be called together to comprise the court (28 U.S.C.A. § 2284). In districts with more than one judge, the judge senior in commission who has not reached his seventieth birthday acts as the chief judge.

**EEOC**   Equal Employment Opportunity Commission.

**Employment at Will**   A common law rule that is the basis on which most employment situations are analyzed.

**Fair Labor Standards Act (FLSA)**   A 1938 federal act that established a standard minimum wage, and that regulates the hours and type of work performed in industries involved in interstate commerce.

**Family and Medical Leave Act of 1993**   A law protecting employees who work for qualified employers that gives them the right to request a specified period of time away from the job for a specific circumstance.

**Immunity**   Exemption, as from serving in an office or performing duties that the law generally requires other citizens to perform, e.g., exemption from paying taxes. Freedom or exemption from penalty, burden, or duty. Special privilege.

**Inchoate**   Imperfect; partial; unfinished; begun, but not completed; as a contract not executed by all the parties.

**Indemnification**   In corporate law, the practice by which corporations pay expenses of officers or directors who are named as defendants in litigation relating to corporate affairs. In some instances, corporations may indemnify officers and directors for fines, judgments, or amounts paid in settlement, as well as expenses.

**Interrogatory**   A set or series of written questions drawn up for the purpose of being propounded to a party, witness, or other person having information of interest in the case.

**Joint Tortefeasors**   Two or more persons jointly or severally liable in tort for the same injury to person or property.

**Jurisdiction**   The perimeters of a particular court's power or range to hear a case.

**Law of Tort**   Violation of duty in a private or civil wrong.

**Legislative Law**   General rules of conduct consciously, formally, and solemnly written by an authorized branch of government; may be made on the national level by Congress, on the state level by a state legislature, or on a local level by a county commission or city council.

**LEXIS**   A computer-assisted legal research service provided by Mead Data Central. LEXIS provides on-line access to database of legal information including federal and state caselaw, statutes, and administrative, regulatory, and secondary materials.

**Litigation**   A lawsuit or legal action.

**Malicious**   Characterized by, or involving, malice; having, or done with, wicked, evil, or mischievous intentions or motives; wrongful and done intentionally without just cause or excuse or as a result of ill will.

**Motion**   A document submitted to a court or judge for the purpose of acquiring a ruling or order directing some act to be done in favor of the person filing the motion.

**National Labor Relations Board (NLRB)**   The governing body established to administer and enforce the laws in the labor relations area.

**NFPA**   National Fire Protection Association.

**Negligence**   The omission to do something that a reasonable man, guided by those ordinary considerations that ordinarily regulate human affairs, would do, or the doing of something that a reasonable and prudent person would not do. The failure to use such care as a reasonably prudent and careful person would use under similar circumstances; it is the doing of some act that a person of ordinary prudence would not have done under similar circumstances or failure to do what a person of ordinary prudence would have done under similar circumstances. Conduct that falls below the standard established by law for the protection of others against unreasonable risk of harm, it is a departure from the conduct expectable of a reasonably prudent person under like circumstances.

**NIOSH**   National Institute of Occupation Health and Safety.

**Occupational Safety and Health Act**   A federal law dispensed by the Occupational Safety and Health Administration to decrease the incidents of injuries, illnesses, and death among workers as a result of the work environment.

**OSHA**   Occupational Safety and Health Administration.

**OSHRC**   Occupational Safety and Health Review Commission.

**Omission**   The neglect to perform what the law requires. The intentional or unintentional failure to act which may or may not impose criminal liability depending upon the existence, vel non, of a duty to act under the circumstances.

**Permanent Total Rating**   Compensation to an employee for a permanently incurred injury or illness.

**Plaintiff**   A person who brings an action; the party who complains or sues in a civil action and is so named on the record. A person who seeks remedial relief for an injury to rights; it designates a complainant. The prosecution (i.e., state or federal) in a criminal case.

**Portal-to-Portal Act**   Federal statute that regulates the pay for an employee's required nonproductive time.

**Proprietary**   Belonging to ownership; owned by a particular person; belonging or pertaining to a proprietor; relating to a certain owner or proprietor.

**Punitive Damages**   Payment awarded to an injured party over and above the compensatory damages awarded.

**Quasi**   As if; almost as it were; analogous to. This term is used in legal phraseology to indicate that one subject resembles another, with which it is compared, in certain characteristics, but that there are intrinsic and material differences between them. A term used to mark a resemblance, and supposes a difference between two objects. It is exclusively a term of classification. It implies that conception to which it serves as index is connected with conception with which comparison is instituted by strong superficial analogy or resemblance. Moreover it negates idea of identity, but points out that the conceptions are sufficiently similar for one to be classed as the equal of the other. It is often prefixed to English words, implying mere appearance or want of reality or having some resemblance to given thing.

**Regulations**   Rules, orders, and the like, issued by various governmental departments to carry out the intent of the law. Agencies issue regulations to guide the activity of those regulated by the agency and of their own employees and to ensure uniform application of the law. Regulations are not the work of the legislature and do not have the effect of law in theory. In practice however, because of the intricacies of judicial review of administrative action, regulations can have an important effect in determining the outcome of cases involving regulatory activity.

**Remand**   To order back to the lower court for new or further action.

**Reporter** A person who reports the decisions of a court of record; also, published volumes of decisions by a court or group of courts. The court reporter is the person who records court proceedings in court and later transcribes such.

**Rescue Doctrine**   An invitation to rescue, with danger present, providing for liability in the case of injuries sustained by the rescuer

***Respondeat Superior***   Concept by which the master is responsible for the actions of the servant; the employer for the employees.

**Reverse**   To repeal a judgment or sentence of a lower court by an appellate court.

**Serious Health Condition** An illness, injury, impairment, or physical or mental condition involving either inpatient care or continuing treatment by a health care provider.

**Serious Violation** An infraction in which there is substantial probability that death or serious harm could result.

**Standard** Usually, minimum requirements established, often through a consensus of opinion, that may or may not be required by law.

**Statutes** A formal written enactment of a legislative body, whether federal, state, city, or county. An act of the legislature declaring, commanding, or prohibiting something; a particular law enacted and established by the will of the legislative department of government; the written will of the legislature, solemnly expressed according to the forms necessary to constitute it the law of the state. Such may be public or private, declaratory, mandatory, directory, or enabling, in nature.

**Tort** A private or civil wrong or injury, including action for bad faith breach of contract, for which the court will provide a remedy in the form of an action for damages. A violation of a duty imposed by general law or otherwise upon all persons occupying the relation to each other that is involved in a given transaction. There must always be a violation of some duty owing to plaintiff, and generally such duty must arise by operation of law and not by mere agreement of the parties.

**Tort-Feasor** A wrong-doer; an individual or business that commits or is guilty of a tort.

**Undue Hardship** An action requiring significant difficulty or expense.

**Union Shop** A place of employment where employees are required to become union members within a specified time period following hiring as a result of an agreement between the union and employer.

**Vacate** To void.

**Westlaw** A computer-assisted legal research service provided by West Publishing Company. Westlaw provides on-line access to a database of legal information including federal and state caselaw, statutes, and administrative, regulatory, and secondary materials.

**Workers' Compensation** Federal and state statutes that provide compensation to employees in the event of an employment-related injury or illness.

**Working Time** Time for which an employee is entitled to compensation, including all time an employee is required to be on duty either on the employer's premises or at a prescribed work place.

# Index